Studies in Systems, Decision and Control

Volume 603

The series "Studies in Systems, Decision and Control" (SSDC) covers both new developments and advances, as well as the state of the art, in the various areas of broadly perceived systems, decision making and control–quickly, up to date and with a high quality. The intent is to cover the theory, applications, and perspectives on the state of the art and future developments relevant to systems, decision making, control, complex processes and related areas, as embedded in the fields of engineering, computer science, physics, economics, social and life sciences, as well as the paradigms and methodologies behind them. The series contains monographs, textbooks, lecture notes and edited volumes in systems, decision making and control spanning the areas of Cyber-Physical Systems, Autonomous Systems, Sensor Networks, Control Systems, Energy Systems, Automotive Systems, Biological Systems, Vehicular Networking and Connected Vehicles, Aerospace Systems, Automation, Manufacturing, Smart Grids, Nonlinear Systems, Power Systems, Robotics, Social Systems, Economic Systems and other. Of particular value to both the contributors and the readership are the short publication timeframe and the world-wide distribution and exposure which enable both a wide and rapid dissemination of research output.

Indexed by SCOPUS, DBLP, WTI Frankfurt eG, zbMATH, SCImago.

All books published in the series are submitted for consideration in Web of Science.

Zhiru Cao · Yugang Niu

Finite-Time Sliding Mode Control for Markov Jump Systems

Transient Performance Analysis and Design

Springer

Zhiru Cao
School of Mechatronic Engineering
and Automation
Shanghai University
Shanghai, China

Yugang Niu
School of Information Science
and Engineering
East China University of Science
and Technology
Shanghai, China

ISSN 2198-4182 ISSN 2198-4190 (electronic)
Studies in Systems, Decision and Control
ISBN 978-981-96-7650-7 ISBN 978-981-96-7651-4 (eBook)
https://doi.org/10.1007/978-981-96-7651-4

This Springer imprint is published by the registered company Springer Nature Singapore Pte Ltd.
The registered company address is: 152 Beach Road, #21-01/04 Gateway East, Singapore 189721, Singapore

To my esteemed supervisor, Professor Niu, whose profound wisdom and unwavering support have not only illuminated my path into the captivating world of sliding mode control research but have also been the driving force behind my groundbreaking advancements in sliding mode control of Markov jump systems. Your visionary guidance has been the cornerstone of my academic journey, inspiring me to push the boundaries of knowledge and achieve new horizons in control theory.

—Zhiru Cao

Preface

The pursuit of robust control methodologies for Markov jump systems (MJSs) has long been intertwined with the challenges posed by their inherent stochasticity and abrupt structural transitions. While asymptotic stability and steady-state performance have dominated classical control theory, modern engineering demands require rigorous guarantees over finite-time horizons, particularly in safety-critical domains such as aerospace, power systems, and networked robotics. This monograph emerges from a decade of research aimed at reconciling these transient performance requirements with the unique complexities of MJSs, culminating in a systematic framework for finite-time sliding mode control (SMC) that balances theoretical depth with practical applicability.

The impetus for this work arose from a critical gap in existing literature, namely, the lack of methodologies addressing both stochastic mode transitions and transient behavior within finite intervals. Traditional Lyapunov-based approaches, though indispensable for asymptotic analysis, often neglect the short-term dynamics that dictate system resilience during critical operational phases. For MJSs, where mode-dependent uncertainties and external disturbances compound these challenges, the need for a dedicated finite-time control paradigm becomes undeniable. This realization spurred the development of novel SMC strategies tailored to MJSs, integrating stochastic stability theory with finite-time performance metrics to address real-world constraints such as actuator faults, cyber-physical attacks, and resource limitations.

At its core, this book establishes a cohesive framework for finite-time SMC design, anchored by two foundational innovations. First, the "partition–integration" technique decouples the analysis of reaching and sliding phases, enabling explicit characterization of transient bounds under stochastic mode transitions. Second, the co-design of transition rate (TR) matrices and sliding mode dynamics transcends conventional assumptions of prior TR knowledge, offering a versatile approach for systems with partially or fully unknown switching patterns. These theoretical advancements are complemented by pragmatic solutions to physical and cyber-physical constraints, including component-wise controllers for actuator nonlinearities, neural network-aided attack mitigation, and adaptive fault-tolerant designs.

Structured to guide readers from principles to applications, the monograph begins with a rigorous exposition of MJS fundamentals, finite-time stability theory, and SMC mechanics in Chap. 1, and provides basic design frameworks for stochastic finite-time boundedness (SFTB) and input-output stochastic finite-time stabilization (IO-SFTS) in Chap. 2. Subsequent chapters delve into constraint-specific control synthesis, and actuator degradation, actuator nonlinearities and controller gain perturbations are addressed through fault-tolerant, component-wise and non-fragile SMC methods in Chaps. 3–5; while cyber-physical threats such as false data injection and deception attacks are countered via probability-dependent and neural network-enhanced strategies in Chaps. 6–8. A pivotal shift occurs in Chap. 9, where TR matrix synthesis is integrated with sliding mode dynamics, emancipating controller design from reliance on pre-specified transition probabilities. The concluding chapter reflects on unresolved challenges, such as adaptive event-triggered mechanisms, chattering suppression, and multi-agent extensions, while advocating for cross-disciplinary collaborations to harness emerging technologies like AI-driven attack detection.

This work is intended for researchers and practitioners grappling with the dual demands of stochastic stability and transient precision in MJSs. Graduate students in control theory will find its blend of rigorous proofs and numerical case studies instructive, while engineers in aerospace, energy, and robotics may adapt its methodologies to real-world systems constrained by finite operational windows or cyber-physical vulnerabilities.

As control systems evolve amidst escalating cyber threats and resource constraints, the quest for adaptive, finite-time solutions will only intensify. It is our hope that this monograph not only equips readers with actionable tools but also ignites curiosity toward uncharted territories, where stochastic control, transient optimization, and intelligent resilience converge to redefine the boundaries of modern engineering.

Shanghai, China
March 2025

Zhiru Cao
Yugang Niu

Acknowledgements There are numerous individuals without whose constructive comments, useful suggestions, and wealth of ideas, this monograph could not have been completed. Special thanks go to Prof. Zidong Wang, Brunel University London; Prof. James Lam, The University of Hong Kong; Prof. Hak-Keung Lam, King's College London, for their valuable suggestions, constructive comments, and support. Also, the authors would like to express their sincere gratitude to the editors of the book for their time and kind help. Next, our acknowledgments go to East China University of Science & Technology and Shanghai University for providing the conducive environment to carry out the research reported in the monograph. Finally, the authors would like to thank the students of Prof. Niu's Research Group and many colleagues who have offered support and encouragement throughout this research effort.

The monograph was supported in part by the National Natural Science Foundation of China (62203288, 62333006, 624702287, 62473153), in part by Shanghai Pujiang Programme (23PJD033), and in part by Shanghai Natural Science Foundation under Grant 24ZR1454700.

Shanghai, China
March 2025

Zhiru Cao
Yugang Niu

Competing Interests The authors have no competing interests to declare that are relevant to the content of this manuscript.

Contents

Acronyms

Abbreviations

MJS	Markov jump system
SMC	Sliding mode control
SFTS	Stochastic finite-time stability/stochastic finite-time stable
SFTB	Stochastic finite-time boundedness/stochastically finite-time bounded
IO-SFTS	Input-output stochastic finite-time stabilization/input-output stochastically finite-time stable
TP	Transition probability
LMI	Linear matrix inequality

Symbols

$X > 0$	The symmetric positive-definite matrix
$X < 0$	The symmetric negative-definite matrix
$\mathbb{R}$	The set of real numbers
$\mathbb{R}^+$	The set of nonnegative real numbers
$\mathbb{R}^n$	The n dimensional Euclidean space
$\mathbb{R}^{n\times m}$	The set of $n \times m$ matrices with elements in $\mathbb{R}$
$\mathbb{N}$	The set of non-negative integers, i.e., $\mathbb{N} \triangleq \{0, 1, 2, \ldots\}$
X^{T}	The transpose of matrix X
$\mathrm{diag}\{\cdots\}$	The block-diagonal matrix
$\mathrm{tr}(X)$	The trace of matrix X
$\mathrm{rank}(X)$	The rank of matrix X
$\mathrm{He}(X)$	The matrix operation $X + X^{\mathrm{T}}$
$\lambda_{\max}(X)(\lambda_{\min}(X))$	The maximum (minimum) eigenvalue of the matrix X
$\mathbb{E}\{x\}$	The mathematical expectation of random variable x
$\mathbb{E}\{x \mid y\}$	The conditional expectation of random variable x on y
$\Pr\{x\}$	The probability of random variable x

$\Pr\{x \mid y\}$	The conditional probability of random variable x on y
$(\Omega, \mathscr{F}, \Pr)$	Probability space with Ω the sample space $\mathscr{F}$ the σ-algebra of subsets of Ω
$\lvert x \rvert$	The 1-norm of the vector $x \triangleq [x_1\, x_2 \cdots x_n]^{\mathrm{T}} \in \mathbb{R}^n$ defined as $\lvert x\rvert \triangleq \lvert x_1\rvert + \lvert x_2\rvert + \cdots + \lvert x_n\rvert$
$\lVert x \rVert$	The 2-norm of the vector $x = [x_1\, x_2 \cdots x_n]^{\mathrm{T}} \in \mathbb{R}^n$ defined as $\lVert x\rVert^2 \triangleq \sum_{i=1}^{n} x_i^2$
$\lVert X \rVert$	The 2-norm of the matrix X defined as $\lVert X\rVert^2 \triangleq \mathrm{tr}(X^{\mathrm{T}}X)$
$\mathrm{sgn}(s)$	The sign function of vector $s \triangleq [s_1\, s_2 \cdots s_m]^{\mathrm{T}}$ defined as $\mathrm{sgn}(s) \triangleq [\mathrm{sgn}(s_1)\ \mathrm{sgn}(s_2)\ \ldots\ \mathrm{sgn}(s_m)\ ]^{\mathrm{T}}$ with $\mathrm{sgn}(s_i) \triangleq \begin{cases} 1, & s_i > 0 \\ 0, & s_i = 0\ i = 1, 2, \ldots, m \\ -1, & s_i < 0 \end{cases}$

List of Figures

List of Tables

Chapter 1
Introduction

1.1 Background

In the wave of advancements within the control theory field, a significant progress has been made in the precise models of complex systems. This development is crucial as it determines whether the theoretical methods under investigation can be effectively applied to real-world systems. Over the past few decades, practical systems have often been modeled as *simple* linear systems based on numerous assumptions that overlook the degradation of system components and the influence of environmental factors. *Simple* linear system models are becoming increasingly weak to deal with increasingly complex and changeable systems, especially, while inevitably facing sudden environmental disturbance, component maintenance and/or subsystem connection failure issues, all of which may lead to abrupt changes in system dynamics or structure, severely impacting the system's dynamic behaviors. Markov stochastic processes are capable of characterizing the types of systems that experience dynamic or structural changes. The system governed by a Markov process is so-called Markov jump system (MJS), which can provide a clear framework for understanding and analyzing the dynamic behaviors of these systems, thereby enhancing the feasibility and effectiveness of control strategies. MJSs have long been a research hotspot in the control field and find widespread applications in robotic systems [1], wind power generation systems [2], economic systems [3], and aerospace systems [4].

The control issues associated with MJSs have witnessed substantial advancements, encompassing a variety of methodologies such as state feedback control [5, 6], output feedback control [7–9], model predictive control [10–12], active disturbance rejection control [13–15], and sliding mode control (SMC) [16–18]. Among these approaches, SMC method has garnered extensive attention within both the realms of control theory and engineering applications largely due to its remarkable properties, including invariance to matched disturbances, robust performance against unmatched disturbances, and rapid convergence characteristics. SMC is a specific type of variable structure control and is characterized by the sliding surface that intersects with

Z. Cao and Y. Niu, *Finite-Time Sliding Mode Control for Markov Jump Systems*,
Studies in Systems, Decision and Control 603,
https://doi.org/10.1007/978-981-96-7651-4_1

the system's equilibrium point. The sliding mode controller is meticulously designed to drive the system trajectory toward this predefined sliding surface and subsequently guide it to the equilibrium point. This method is particularly advantageous in scenarios where system dynamics are subject to uncertainties or external disturbances. As an effective robust control strategy, SMC has not only advanced the development of robust control techniques but also has opened avenues for deeper exploration into systems characterized by unmodeled dynamics and parameter uncertainties.

Most existing investigations of SMC methods for MJSs focus on asymptotic stability that is an *infinite-time* performance. Specifically, for a given system, focusing solely on its asymptotic stability implies that one can disregard the intermediate variations in the system's state, provided that it ultimately converges to the desired target. While asymptotic stability plays a pivotal role in system analysis and engineering applications, the common oversight of transient state conditions and the lack of constraints on settling time often render this approach unsuitable in some critical domains such as aerospace engineering. In aerospace applications, where safety and performance are paramount, it is essential to consider how the system state/output behaves during the transient phase. The settling time, defined as the duration required for the system's response to enter and remain within a specified error band around its final value, is particularly significant. This parameter is crucial for evaluating system performance, especially, in scenarios involving rapid changes or disturbances.

Moreover, transient responses can exhibit complex behaviors influenced by various factors, including system dynamics, damping ratios, and external disturbances. For systems that frequently operate under stringent conditions, understanding these transient characteristics is essential for ensuring reliability and effectiveness. Consequently, finite-time stability (FTS), as a form of transient performance, has stirred significant interest and been development frontiers in recent years. Given the two-phase nature of SMC and the stochastic jumping characteristics of MJSs, ensuring the FTS of MJSs through the design of appropriate sliding mode controllers represents a series of interesting issues that this book aims to address. In light of these considerations, this chapter provides a comprehensive overview of the concepts, features, methodologies, and current research status concerning SMC for MJSs.

Section 1.2 presents the modeling of MJSs, introduces the concepts of finite-time performance, and reviews the research status on FTS for MJSs. Section 1.3 discusses the design principles of SMC methods and summarizes the current advancements in SMC for MJSs. Section 1.4 offers relevant definitions and fundamental lemmas that will be utilized in subsequent chapters. Finally, Sect. 1.5 outlines the structure of this book, providing a roadmap for readers to navigate through the topics covered.

1.2 Markov Jump Systems and Related Research Issues

1.2.1 Markov Jump Systems

In 1961, scholars N. N. Krasovskii and E. A. Lidskii first utilized Markov stochastic processes to describe structural mutations in practical systems. Since then, MJSs have attracted considerable attention from researchers both domestically and internationally [19]. MJSs can be regarded as a specialized type of switching system, comprising several subsystems. The fundamental principle of switching systems dictates that the system state adheres to specific switching rules, allowing transitions to the corresponding subsystem. In contrast, the switching of MJSs among subsystems occurs randomly according to certain statistical characteristics defined by the Markov process. In the context of MJSs, these transitions between subsystems are commonly referred to "mode jumping".

For instance, consider a system with five distinct working modes as shown in Fig. 1.1, i.e., "best", "good", "ordinary", "poor", and "extremely poor". If this system is in the "best" mode at time k, there is a 40% probability that the system will remain in this working mode at the next instant; a 30% probability of jumping to the "good" working mode; a 20% probability of moving to the "ordinary" working mode; a 7% probability of shifting to the "poor" working mode; and a 3% probability of descending to the "extremely poor" working mode.

This probabilistic framework provides a comprehensive understanding of system behavior under varying conditions and highlights the inherent uncertainties in practical applications. By employing Markov stochastic processes, researchers can effectively model and analyze the dynamic transitions that characterize MJSs, thereby

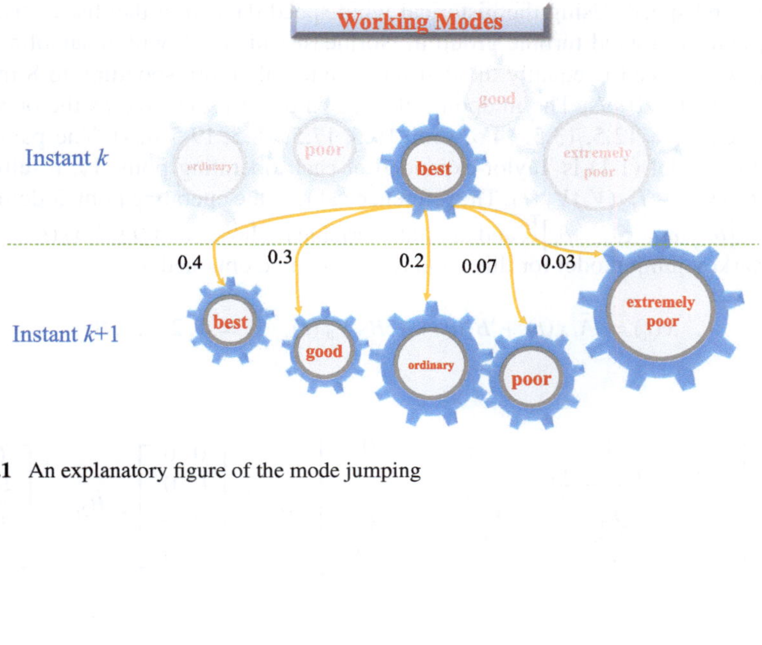

Fig. 1.1 An explanatory figure of the mode jumping

enhancing their applicability across various fields such as engineering, finance, and biological systems.

Taking the wind power generation system in [2] as an example, we introduce the modeling process of an MJS. The wind turbine with the parameters in Table 1.1 can be modeled as:

$$\dot{x}_h(t) = f\left(x_h(t), t\right) + g\left(x_h(t)\right) u(t), \tag{1.1}$$

where the state variable $x_h(t) \triangleq [\theta, \omega_r, \omega_g, \beta]^{\mathrm{T}}$, the control variable $u(t) \triangleq [\beta_r, \omega_z]^{\mathrm{T}}$,

$$f\left(x_h(t), t\right) \triangleq \begin{bmatrix} \omega_r - \frac{1}{N_g}\omega_g \\ \frac{1}{J_r}T_r - \frac{K_d}{J_r}\omega_r + \frac{K_d}{J_r N_g}\omega_g - \frac{B_s}{J_r}\theta \\ \frac{K_d}{J_r N_g}\omega_r - \left(\frac{K_d}{J_g N_g^2} + \frac{B_g}{J_g}\right)\omega_g + \frac{B_s}{J_g N_g}\theta \\ -\frac{1}{\tau}\beta \end{bmatrix},$$

$$g(x_h(t)) \triangleq \begin{bmatrix} 0 & 0 & 0 & \frac{1}{\tau} \\ 0 & 0 & \frac{B_g}{J_g} & 0 \end{bmatrix}^{\mathrm{T}}, \quad T_r \triangleq \frac{1}{2}C_p(\lambda, \beta)\rho\pi R^2 \frac{V^3}{\omega_r},$$

$$C_p(\lambda, \beta) \triangleq c_1\left(\frac{c_2}{\lambda^*} - c_3\beta - c_4\right)e^{-\frac{c_5}{\lambda^*}} + c_6\lambda,$$

$$\lambda^* \triangleq \left(\frac{1}{\lambda + c_7\beta} - \frac{c_8}{\beta^3 + 1}\right)^{-1}, \quad \lambda \triangleq \frac{\omega_r R}{V},$$

and the parameters in the wind energy utilization coefficient $C_p(\lambda, \beta)$ are $c_1 = 0.5176$, $c_2 = 116$, $c_3 = 0.4$, $c_4 = 5$, $c_5 = 21$, $c_6 = 0.0068$, $c_7 = 0.08$, and $c_8 = 0.035$.

The actual wind speed is divided into mean wind speed and perturbation wind speed, i.e., $V = V_s + V_\omega(t)$ with V_s the mean wind speed and $V_\omega(t)$ the perturbation wind speed. Using the historical wind speed data from the distributed control system of a wind turbine group in Northeast China [2], with a sampling time 1s, the wind speed is equally divided into 8 intervals (corresponding to 8 modes) from 12 m/s to 20 m/s. The midpoint of each partition is selected as the operating point (i.e., $V_s = 12.5, 13.5, 14.5, 15.5, 16.5, 17.5, 18.5, 19.5$ m/s). The parameter T_r in the model (1.1) is Taylor-expanded at each operating point V_s, resulting in $T_r = T_{r1}(V_s) + T_{r2}(V_s)V_w(t)$. The reference value at the operating point is defined as $\bar{x}_i(t) \triangleq \begin{bmatrix} \bar{\theta}_i & \bar{\omega}_{ri} & \bar{\omega}_{gi} & \bar{\beta}_i \end{bmatrix}^{\mathrm{T}}$, and the state variable is chosen as $x(t) \triangleq x_h(t) - \bar{x}_i(t)$. The Markov jump model for this wind turbine can be obtained as

$$\dot{x}(t) = \bar{A}_i x(t) + \bar{B}_1 u(t) + \bar{B}_{2i} V_w(t), \quad (i = 1, 2, \ldots, 8) \tag{1.2}$$

where

$$\bar{A}_i \triangleq \begin{bmatrix} 0 & 1 & -\frac{1}{N_g} & 0 \\ -\frac{B_s}{J_r} & \frac{1}{J_r}\frac{\partial T_{r1}}{\partial \omega_r} - \frac{K_d}{J_r} & \frac{K_d}{J_r N_g} & \frac{1}{J_r}\frac{\partial T_{r1}}{\partial \beta} \\ \frac{B_s}{I_g N_g} & \frac{K_d}{J_g N_g} & -\frac{K_d}{J_g N_g^2} - \frac{B_g}{J_g} & 0 \\ 0 & 0 & 0 & -\frac{1}{\tau} \end{bmatrix}, \quad \bar{B}_1 \triangleq \begin{bmatrix} 0 & 0 \\ 0 & 0 \\ 0 & \frac{B_g}{J_g} \\ \frac{1}{\tau} & 0 \end{bmatrix}, \quad \bar{B}_{2i} \triangleq \begin{bmatrix} 0 \\ \frac{T_{r2}}{J_r} \\ 0 \\ 0 \end{bmatrix}.$$

Table 1.1 The parameters in the system (1.1)

Parameter	Physical meaning	Value
R	Rotor radius	40 m
ρ	Air density	1.25 kg/m^3
τ	Time constant	50 ms
J_r	Equivalent rotational inertia of the wind turbine rotor	4.95×10^6 kg/m^2
J_g	Equivalent rotational inertia of the generator rotor	90 kg/m^2
N_g	Generator rotor coefficient	83.531
B_s	Equivalent torsional stiffness of the shaft	1.14×10^8 N/m^2
K_d	Equivalent damping coefficient of the shaft	7.55658×10^5 Nm/(rad/s)
B_g	Slope of the torque-speed curve of the induction generator	400 Nm/(rad/s)
T_r	Wind turbine rotor torque	
V	Wind speed	
ω_r	Wind turbine rotor speed	
ω_g	Generator rotor speed	
ω_z	Control output	
λ	Tip speed ratio	
θ	Equivalent torsional angle of the shaft	
β	Spiral angle	
β_r	Reference spiral angle	

Furthermore, for each operating point $\bar{x}_i(t)$, using the Euler discretization method with a sampling time $T_s = 1$s, the discrete-time model of MJS (1.2) can be obtained as

$$x(k+1) = A_i x(k) + B_1 u(k) + B_{2i} V_w(k), \quad (i = 1, 2, \ldots, 8) \tag{1.3}$$

with

$$A_i \triangleq \begin{bmatrix} 1 & T_s & -\frac{T_s}{N_g} & 0 \\ -\frac{B_s T_s}{J_r} & 1 + \frac{T_s}{J_r}\frac{\partial T_{r1}}{\partial \omega_r} - \frac{K_d T_s}{J_r} & \frac{K_d T_s}{J_r N_g} & \frac{T_s}{J_r}\frac{\partial T_{r1}}{\partial \beta} \\ \frac{B_s T_s}{J_g N_g} & \frac{K_d T_s}{J_g N_g} & 1 - \frac{K_d T_s}{J_g N_g^2} - \frac{B_g T_s}{J_g} & 0 \\ 0 & 0 & 0 & 1 - \frac{T_s}{\tau} \end{bmatrix},$$

$$B_1 \triangleq \begin{bmatrix} 0 & 0 \\ 0 & 0 \\ 0 & \frac{B_g T_s}{J_g} \\ \frac{1}{\tau} & 0 \end{bmatrix}, \quad B_{2i} \triangleq \begin{bmatrix} 0 \\ \frac{T_{r2} T_s}{J_r} \\ 0 \\ 0 \end{bmatrix}.$$

The random characteristics of the mean wind speed can be described by a Markov process. According to the historical wind speed data measured by the distributed

Table 1.2 The transition among the average wind speeds

Wind speed partition (m/s)	[12, 13)	[13, 14)	[14, 15)	[15, 16)	[16, 17)	[17, 18)	[18, 19)	[19, 20]
[12, 13)	2672	182	0	0	0	0	0	0
[13, 14)	183	1993	148	0	0	0	0	0
[14, 15)	0	148	1610	118	0	0	0	0
[15, 16)	0	0	118	1038	72	0	0	0
[16, 17)	0	0	0	72	699	60	0	0
[17, 18)	0	0	0	0	60	349	35	0
[18, 19)	0	0	0	0	0	35	282	21
[19, 20]	0	0	0	0	0	0	21	136

control system of a wind turbine group in Northeast China [2], the transition of the mean wind speed between adjacent partitions is shown in Table 1.2.

From Table 1.2, it can be seen that when the wind speed in the previous instant is within the interval [12, 13) m/s, the number of times that the wind speed remains within [12, 13) m/s in the current instant is 2672, and the number of times that the wind speed changes to the interval [13, 14) m/s is 182. The number of times that the wind speed changes to other intervals is 0, and the rest of the data in the table has a similar meaning. Next, the transition probabilities are calculated. Taking the first row of Table 1.2 as an example, the probability that the wind speed remains within [12, 13) m/s is

$$\pi_{11} = \frac{2672}{2672 + 182} = 0.9362.$$

The probability that the wind speed jumps from the interval [12, 13) m/s to the interval [13, 14) m/s is

$$\pi_{12} = \frac{182}{2672 + 182} = 0.0638.$$

The probability that the wind speed jumps from the interval [12, 13) m/s to other intervals is $\pi_{1j} = 0\ (j = 3, \ldots, 8)$. Using the same method to calculate all transition probabilities, the transition probability matrix for the discrete-time MJS (1.3) is obtained as

$$\Pi = \begin{bmatrix} 0.9362 & 0.0638 & 0 & 0 & 0 & 0 & 0 & 0 \\ 0.0787 & 0.8576 & 0.0637 & 0 & 0 & 0 & 0 & 0 \\ 0 & 0.0789 & 0.8582 & 0.0629 & 0 & 0 & 0 & 0 \\ 0 & 0 & 0.0961 & 0.8453 & 0.0586 & 0 & 0 & 0 \\ 0 & 0 & 0 & 0.0866 & 0.8412 & 0.0722 & 0 & 0 \\ 0 & 0 & 0 & 0 & 0.1351 & 0.7860 & 0.0788 & 0 \\ 0 & 0 & 0 & 0 & 0 & 0.1036 & 0.8343 & 0.0621 \\ 0 & 0 & 0 & 0 & 0 & 0 & 0.1338 & 0.8662 \end{bmatrix}.$$

According to the modeling process of the above MJS (1.3), we can summarize the general state-space representations of a linear MJS with the external disturbance as follows.

► **Continuous-time MJS defined on a complete probability space ($\Omega, \mathscr{F}, \Pr$):**

$$\begin{cases} \dot{x}(t) = A(r_t)x(t) + B(r_t)u(t) + D(r_t)w(t) \\ y(t) = C(r_t)x(t) \end{cases} \tag{1.4}$$

where $x(t) \in \mathbb{R}^n$, $u(t) \in \mathbb{R}^m$, and $y(t) \in \mathbb{R}^p$ represent the state vector, control input, and output of the system, respectively; $w(t) \in \mathbb{R}^q$ means the unmatched external disturbance; and $A(r_t)$, $B(r_t)$, $C(r_t)$, $D(r_t)$ are system matrices with appropriate dimensions. $\{r_t\}_{t\geq 0}$ represents a homogeneous Markov process defined in $t \in [0, \infty)$ and takes values in a finite set $\mathbb{S} \triangleq \{1, 2, \ldots, N\}$. r_t describes the modes of system (1.4) and the ith subsystem is for the mode $r_t = i$. The transition probability (TP) between different modes is

$$\Pr\{r_{t+\Delta} = j \mid r_t = i\} = \begin{cases} \lambda_{ij}\Delta + o(\Delta), & i \neq j \\ 1 + \lambda_{ij}\Delta + o(\Delta), & i = j \end{cases}$$

where $\Delta > 0$ is the dwell time, $o(\Delta)$ means infinitesimal value on Δ satisfying $\lim_{\Delta\to 0^+} \frac{o(\Delta)}{\Delta} = 0$, and λ_{ij} is the transition rate (TR) of the system state from mode i to mode j satisfying $\lambda_{ij} \geq 0$ $(i \neq j)$ and $\lambda_{ii} = -\sum_{i\neq j}\lambda_{ij}$. Then, TR matrix is defined as $\Lambda \triangleq [\lambda_{ij}]_{N\times N}$ $(i, j \in \mathbb{S})$.

► **Discrete-time MJS defined on a complete probability space** ($\Omega, \mathscr{F}, \Pr$):

$$\begin{cases} x(k+1) = A(r_k)x(k) + B(r_k)u(k) + D(r_k)w(k) \\ y(k) = C(r_k)x(k) \end{cases} \tag{1.5}$$

where $x(k) \in \mathbb{R}^n$, $u(k) \in \mathbb{R}^m$, and $y(k) \in \mathbb{R}^p$ represent the state vector, control input, and output of the system, respectively; $w(k) \in \mathbb{R}^q$ means the unmatched external disturbance; and $A(r_k)$, $B(r_k)$, $C(r_k)$, $D(r_k)$ are system matrices with appropriate dimensions. $\{r_k\}_{k\in\mathbb{N}}$ is a homogeneous Markov process defined in instant series $\mathbb{N}$ and takes values in a finite set $\mathbb{S} \triangleq \{1, 2, \ldots, N\}$ and its TP is given as

$$\Pr\{r_{k+1} = j \mid r_k = i\} = \pi_{ij}$$

where $0 \leq \pi_{ij} \leq 1$, $\sum_{j=1}^{N} \pi_{ij} = 1$, and the TP matrix is defined as $\Pi \triangleq [\pi_{ij}]_{N\times N}$ $(i, j \in \mathbb{S})$.

From the aforementioned continuous- and discrete-time MJSs, it is evident that an MJS consists of multiple subsystems, with transitions between each mode governed by a Markov process. Specifically, the probability distribution of the current system mode being i and the subsequent mode being j follows this Markov process. This implies that the mode of the system at the next time step is solely dependent on its current mode and is independent of its past modes. Furthermore, the above models of MJSs indicate that the performance of an MJS is influenced not only by the intrinsic performance characteristics of each subsystem but also by the TR/TP that govern mode switching.

1.2.2 Stability Analysis for Markov Jump Systems

1.2.2.1 Transition Rate/Probability Issues

MJSs have garnered significant attention for their ability to provide effective analytical methods for systems experiencing sudden structural variations [4]. Existing control schemes can be categorized into two types based on whether the controller is dependent on the system mode, i.e., mode-independent and mode-dependent control schemes. In mode-independent control strategies, a common controller is designed for all system modes. This approach is simpler and easier to implement; however, it results in conservative analysis outcomes due to ignoring all mode information. In contrast, mode-dependent controllers effectively incorporate mode-specific details into the controller design, thereby enhancing performance and responsiveness to varying operational conditions. By recognizing the distinct characteristics of each mode, mode-dependent control strategies can optimize system behavior more effectively than their mode-independent counterparts. This adaptability is particularly crucial in applications where system dynamics vary significantly between modes, allowing for a more nuanced approach that considers the unique challenges posed by each subsystem.

As specific parameters of MJS, TRs/TPs depict the stochastic uncertainties associated with switching or jumping among subsystems, and this information plays a critical role in the control design and analysis of MJSs. Under the assumption that TRs/TPs are fully known, significant research progress has been made in addressing control problems in MJSs [7–12]. In practice, however, the TPs of MJSs are often obtained through physical experiments or numerical simulations, such as Monte Carlo methods [20, 21]. Due to constraints on measurement costs or environmental conditions, these TPs may not be precisely determined; in some cases, only partial information on TPs is available. Consequently, research on the control of MJSs under partially known TPs holds practical significance. Issues related to uncertain TPs can generally be divided into two categories: partly unknown case and complete unavailable case. In [22, 23], control problems have been addressed for both continuous- and discrete-time MJS with norm-bounded uncertain TRs and convex polytope-based uncertain TPs. In [24], a continuous-time MJS has been studied under the scenario where parts of the TR structure and bounds are entirely unknown, in which the stochastic stability of the closed-loop system has been analyzed utilizing the property that the rows of the TR matrix sum to one. This body of work underscores the importance of developing robust control strategies that can effectively handle uncertainties in TPs, thereby enhancing the reliability and performance of MJSs in practical applications.

Regardless of whether the TRs/TPs are completely known or partially unknown, both scenarios operate under the same premise that all/parts of TR/TP information is provided as prior information. Within this framework, controllers are designed to ensure the desired performance of MJSs under the specified TR/TP rules. Inspired by the design idea of the switching rule for switched systems [25], recent years have reported researchers optimizing system performance through the design of TRs/TPs. Notably, an early study in [26] addressed a co-design problem involving TRs and static output feedback controllers for continuous-time MJSs, establishing necessary and sufficient conditions for TR design. Building on this co-design approach, various control problems have been tackled in [27–29]—including stochastic stability, H_∞ control, and guaranteed cost control, by co-designing the TR matrix alongside state feedback controller. Subsequently, the exponential stability of discrete-time MJS has been addressed in [30] through the co-design of TPs and output feedback controller gains. It is worth mentioning that these studies typically assume the absence of parameter uncertainties or external disturbances within the controlled system. When such uncertainties or disturbances are present, the feasibility of existing methods requires further investigation.

1.2.2.2 Resource Constraint Issues

The above parts have examined the characteristics of several main control methods for MJSs, including mode-independent and mode-dependent control methods, as well as the co-design of TRs/TPs with controllers. Those aforementioned works are primarily based on traditional "point-to-point" system structures with "healthy" system components (e.g., sensors, controllers, actuators). When the system components become "unhealthy" due to limited physical resources or the communication among various components via "network", not "point-to-point", control design and analysis of MJSs is faced with more complex solutions.

Physical constraints: Under this case, the practical components may not always be "perfect". Namely, factors such as aging from prolonged use, low device accuracy, and limited energy efficiency can lead to some adverse phenomena in actuator operations, controller calculations and sensor measurements. These will significantly affect the performance and reliability of controlled systems. Specifically, the common physical component problems involve actuator/sensor failures, actuator/sensor nonlinearities (including dead zones and saturations), controller gain perturbations.

Actuator/sensor failures commonly arise from wear and tear, environmental factors, or design flaws. Such failures may result in the inability of actuators to respond appropriately to control signals or the sensor's incapacity to provide accurate measurements. An effective method is to detect and diagnose faults for promptly identifying them, based on which the detected faults can be timely removed and intervened to maintain system performance [31–33]. Another method is to design robust controllers for tolerating faults instead of intervening them. It should be noted that the actuator/sensor failures would become more complex when occurring in MJSs. For an MJS subject to the mode-dependent actuator failures, a fault-tolerant controller has been designed in [34] by using the known upper bounds of actuator faults, which could reduce the negative effects of faults to some extent. Relaxing the assumption of known bounds for actuator failure, the fault-tolerant controller has been improved via proposing an adaptive rule to estimate the fault bounds in [35], which increasing robustness to tolerate faults.

Actuator/sensor nonlinearity refers to a specific nonlinear behavior commonly observed in actuators and sensors, characterized by the presence of dead zones, hysteresis, or saturations effects within certain input ranges. This phenomenon arises due to imperfections in mechanical components, manufacturing tolerances, and the inherent characteristics of the materials used in these devices. Wherein, dead zones refer to input ranges where no output is produced, while hysteresis describes the dependency of the output on the input history. These nonlinearities can lead to significant discrepancies between expected and actual system behavior. Recent studies have focused on developing robust control strategies that account for these nonlinearities, ensuring stable performance even in their presence [36, 37]. Saturation occurs when an actuator reaches its physical limits, preventing it from producing the desired output even when commanded to do so. This phenomenon can degrade system performance even result in instability [38]. Various strategies have been proposed to manage actuator saturation, including anti-windup techniques in control design that aim to mitigate the effects of saturation on system dynamics [39].

Controller gain perturbations refer to variations in the controller's parameters due to environmental changes or component aging. Such perturbations can adversely affect system stability and dynamic performance. In the synthesis and analysis of control systems, the kind of gain perturbation is usually modeled as parameter uncertainty in the designed controller, correspondingly, the kind of controller with gain perturbation is often called non-fragile controller [40]. When facing the gain perturbations in mode-dependent controllers for MJSs, the gain perturbations are usually considered to be mode-dependent in [41].

Communication constraints: The data lacks adequate security protections when transmitting via network due to computer technical limitations. The interconnection of large-scale networked components complicates the defense and mitigation of inherent physical vulnerabilities, making the data transmitted over networks vulnerable to cyber attacks [42]. Network attacks can be highly disruptive and may even result in complete shutdowns of physical systems. As such, it is critical to ensuring the security, robustness and resilience of controlled systems [43, 44].

Cyber attacks can generally be classified into three types, i.e., Denial-of-Service (DoS) attacks, replay attacks, and deception attacks. In the case of DoS attacks, an attacker floods the network with large volumes of invalid data to obstruct transmission, which maliciously consumes limited communication resources and prevents the recipient from receiving valid data, thereby disrupting system operations. Depending on the nature of DoS attacks, it can be modeled as either a deterministic or a stochastic DoS attack. For instance, a deterministic DoS attack has been considered for MJSs in [45], under which the closed-loop MJS has been transformed into a time-delay MJS.

The intermittent DoS attacks have been addressed in [46] by analyzing the relationship between attack dwell time and stability of MJS. Additionally, a Bernoulli process has been employed to describe randomly occurring DoS attacks and the performance of the closed-loop MJS under Bernoulli-type DoS attacks has been examined in [47], akin to handling packet loss. Furthermore, a kind of DoS attack has been modeled for MJS in [48], where the coupling Markov processes from the controlled system and DoS attack have been dealt with by using the mapping technique.

Essentially, DoS attacks are highly disruptive yet relatively easy to be detected. In contrast, replay attacks and deception attacks exhibit greater stealth, making them more challenging to identify and mitigate. Replay attacks involve intercepting data within a specific time interval and repeatedly injecting it into the system. Due to the unknown upper bounds and variation rates of replay attacks, handling such attacks is challenging [43].

Deception attacks, also known as false data injection (FDI) attacks, involve injecting false information into the transmitted system data during inter-component communication, compromising data integrity and thereby destabilizing the system. Common types of deception attacks include deterministic and stochastic deception attacks [49]. In existing literature on MJSs subject to the deterministic deception attack, when the attack occurs in the Controller-to-Actuator (C/A) channel, it is typically treated as a matched disturbance [50]; when the attack occurs in the Sensor-to-Controller (S/C) channel, it is regarded as bounded measurement noise [51]. For stochastic deception attacks, additive-type deception attack models have been established in [52, 53] for MJSs, where the attacker injects an additional signal into the original data, so the received data is the sum of the original and injected data. Differently, in [54, 55], alternate-type deception attack models have been built for MJSs, where the original data is intercepted and only the injected signal is sent, with the received data consisting solely of the injected data by attackers. Though described differently, both models characterize an attacker injecting false information into the communication channel with a certain probability.

From the above discussions, it is evident that substantial progresses have been made in the analysis and synthesis of MJSs across various domains, including TR synthesis, physical structure constraints, and network resource limitations. However, it is important to note that the objective of stability results mentioned primarily focus on the ultima system state and do not account for state variations during the operation process. This limitation renders them unsuitable for many practical systems that have stringent requirements for transient performance.

1.2.3 Finite-Time Boundedness Theory

In modern control theory, Lyapunov stability holds a very important position, focusing on the construction of appropriate Lyapunov functions or functionals. As mentioned in above sections, Lyapunov stability concerns the steady-state performance of systems, which is only applicable to dynamic systems that operate for a sufficiently long time (even infinitely long), and thus cannot be applied to the performance analysis of some dynamic systems with a limited, or even very short, operating time. Based on engineering practice, in the 1960s, G. Kamenkov and A. Lebedev first noticed the above fact and proposed the concept of finite short-time stability in Russian literature [56, 57], which is also called finite-time stability (FTS) or short-time stability. The so-called FTS refers to the system state trajectory not departing from a predetermined constraint range within a specified finite-time interval, starting from an initial domain. Obviously, FTS has an essential difference from Lyapunov asymptotic stability. Namely, FTS focuses on the transient performance of dynamic systems within a finite-time interval, while Lyapunov asymptotic stability focuses on the steady-state performance of dynamic systems in sufficiently long time.

In 1961, P. Dorato first introduced the concept of FTS into English literature [58], providing the short-time stability conditions for linear time-varying systems, which has greatly promoted the research and development of FTS worldwide. Subsequently, in [59, 60], L. Weiss and E. F. Infante have further introduced the concept of FTS to nonlinear systems with external disturbances. A. Michel has not only first proposed the concept of FTS for discrete-time systems [61], but also introduced the concept of practical stability for discontinuous systems for the first time [62]. Based on the linear matrix inequalities (LMIs) technique, P. Dorato and colleagues have addressed the finite-time control problem of continuous-time linear time-varying systems in [63]. Similar to the aforementioned idea of FTS, S. P. Bhat and D. S. Bernstein have proposed the concept of finite-time convergence and solved the finite-time control problem for continuous-time autonomous systems [64].

Taking a continuous-time linear system as an example, we give the general definition of the FTS as follows:

Definition 1.1 (*Finite-Time Stability, FTS*) The continuous-time linear system

$$\dot{x}(t) = Ax(t)$$

is said to be finite-time stable (FTS) with respect to the given quaternity $(c_1, c_2, [0, T], R)$ with $0 < c_1 \leq c_2$ and $R > 0$, if one has

$$x^{\mathrm{T}}(0)Rx(0) \leq c_1 \Rightarrow x^{\mathrm{T}}(t)Rx(t) < c_2,\ \forall\, 0 \leq t \leq T.$$

Since the early twenty-first century, F. Amato has made many pioneering contributions to the field of FTS and control synthesis. Considering the presence of external disturbances in the system, F. Amato has first extended the concept of FTS to the concept of finite-time boundedness (FTB) in [65], and solved the FTB conditions for a class of parameter-varying systems by proposing an LMI-based controller design method.

The definition of the FTB for the continuous-time system with external disturbance is given as follows:

Definition 1.2 (*Finite-Time Boundedness, FTB*) The continuous-time linear system

$$\dot{x}(t) = Ax(t) + Dw(t)$$

subject to an external disturbance $w(t)$ satisfying

$$\mathbb{W}_{[0,T],\delta} \triangleq \left\{ w^{\mathrm{T}}(t)w(t) \leq \delta^2 \right\},$$

is said to be finite-time bounded (FTB) with respect to the given quintuplet (c_1, c_2, $[0, T]$, $R, \mathbb{W}_{[0,T],\delta}$) with $0 < c_1 \leq c_2$ and $R > 0$, if one has

$$x^{\mathrm{T}}(0)Rx(0) \leq c_1 \text{ with } w(t) \in \mathbb{W}_{[0,T],\delta} \Rightarrow x^{\mathrm{T}}(t)Rx(t) < c_2, \ \forall\, 0 \leq t \leq T.$$

Subsequently, in [66], F. Amato et al. have addressed the dynamic output feedback FTB problem for continuous-time linear time-invariant systems. In [67], F. Amato and M. Ariola have extended the concept of FTB to discrete-time systems for the first time, solving the FTB problem for discrete-time linear time-invariant systems.

Based on the idea of FTS, F. Amato has first proposed the concept and definition of input-output FTS (IO-FTS) in [68, 69], and solved the IO-FTS problem for a class of time-varying systems.

The definition of the IO-FTS for the continuous-time linear system is given as follows:

Definition 1.3 (*Input-Output Finite-Time Stability, IO-FTS*) The continuous-time linear system

$$\begin{cases} \dot{x}(t) = Ax(t) + Dw(t) \\ y(t) = Cx(t) \end{cases}$$

with zero initial condition $x(0) \equiv 0$ and the external disturbance $w(t)$ satisfying

$$\mathbb{W}_{[0,T],\delta} \triangleq \left\{ w^{\mathrm{T}}(t)w(t) \leq \delta^2 \right\},$$

is said to be input-output FTS (IO-FTS) with respect to the given triplet ($[0, T]$, R, $\mathbb{W}_{[0,T],\delta}$) and $R > 0$, if one has

$$w(t) \in \mathbb{W}_{[0,T],\delta} \Rightarrow y^{\mathrm{T}}(t)Ry(t) < 1, \ \forall\, 0 \leq t \leq T.$$

1.2.4 Transient Performance Analysis for Markov Jump Systems

In recent years, the issues of FTS and control synthesis have been extended to an increasing number of complex dynamic systems [70, 71], and thereby the transient performances depicted by FTS/FTB have been paid attention for MJSs. The concept of FTB has been introduced into MJSs for the first time in [72], in which the control problems have been solved under both continuous and discrete conditions. Building on this work, the FTB control/filtering issues of nonlinear, Itô stochastic, time-varying, and 2-D MJSs have been further studied in [73–76], respectively.

The application of finite-time control synthesis for MJSs in practical engineering has also attracted the attention [85–88]. The FTB problem of the DC-DC buck converter with metal-oxide-semiconductor field effect transistor has been investigated in [85], in which the transient behaviors of resistance voltage and the inductor current have been examined and noticed. In [86, 87], a single-link robotic arm has been modeled as a T-S fuzzy MJS, whose angular position and angular velocity have been ensured to be FTB governed by the designed fuzzy controller. For a wind turbine with a Markov stochastic process in [88], the referenced pitch angle and input have been designed by using the measured outputs to achieve the satisfactory transient performance of rotor speed and torsional angle during a short-time interval.

1.3 Sliding Mode Control and Its Application in Markov Jump Systems

1.3.1 Design Idea of Sliding Mode Control

SMC is a kind of variable structure control method proposed in the mid-1950s, which changes the system dynamics through switching control [89]. Due to the nature of switching control, SMC is a non-continuous control with high-frequency switching terms, whose fundamental concept is to drive the state trajectory of the system onto a specified smooth manifold passing through the zero point of the state space and remain it on the smooth manifold until reaching the zero point. This specified smooth manifold is usually called the sliding surface. The motion of the system state trajectory on the sliding surface is called the sliding motion and the corresponding dynamics are called sliding mode dynamics. The most essential difference between the SMC and other control methods lies in the sliding surface. On the one hand, the sliding motion is completely insensitive to matched disturbances and parameter uncertainties when the system trajectory slides on the sliding surface, that is, it can completely eliminate their effects. On the other hand, the sliding surface limits the state trajectory, enabling it has a faster convergence speed. As such, the selection of the sliding surface is crucial [90]. For classical SMC, there are usually two types of sliding surface: the linear sliding surface and integral sliding surface. The linear sliding surface is simple to design, and the integral sliding surface can make the system trajectory be on the sliding surface from the initial moment, thereby eliminating non-robust effects before reaching the sliding surface.

As shown in Fig. 1.2, the state trajectory under SMC can be divided into two phases, i.e., the reaching phase and the sliding motion phase. During the reaching phase, the emphasis is on the reachability of the sliding surface, whereas in the sliding phase, the focus shifts to the stability of the sliding mode dynamics. Accordingly, the design and implementation of SMC also unfold from these two objectives, that is, specify a sliding surface to ensure the stability of the sliding mode dynamics and design a sliding mode controller to achieve the reachability of the specified sliding surface. Noted that it is an ideal situation for the sliding mode controller to drive the system state to reach the sliding surface and slide along the sliding surface, as the left figure in Fig. 1.3.

In practice, the ideal sliding motion might not be attained, especially, data transmission in network environments. Specifically, when the measured output/state signals are transmitted via network and suffer from some network-induced problems or attacks, the signals received by the sliding mode controller are fairly different from original ones. This phenomenon will further result in the differences between the constructed sliding function and the one used by the controller, which directly influences the switching behaviors. As such, the resultant closed-loop system cannot achieve precise sliding motion. The restricted switching behaviors cause the system state trajectory to move around the sliding manifold instead of moving on the sliding surface, thus making the reachability and sliding motion require to be redefined. D. Milosavljevic used quasi-sliding motion (QSM) to describe the oscillation of the sliding variable in the small neighborhood of the sliding manifold [91]. Inspired by the QSM, B. Bandyopadhyay proposed the concept of practical sliding motion (PSM) to represent a general sliding motion [92], namely, sliding trajectories remain bounded in the vicinity of sliding manifold with the bound as practical sliding mode band (PSMB). Precisely due to this generality of PSM, SMC problems during network environments have important research significance and have currently been applied to various complex systems such as switched systems, fuzzy systems, multi-agent systems, and MJSs.

1.3.2 Sliding Mode Control for Markov Jump Systems

For the MJSs under the traditional "point-to-point" transmission, a mode-dependent sliding function has been constructed in [93] and the corresponding mode-dependent sliding mode controller has been designed to achieve stochastic stability for the

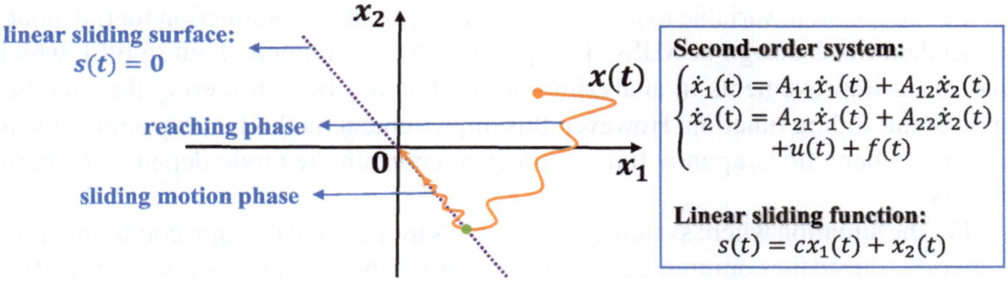

Fig. 1.2 An example of state trajectory for second-order system under SMC

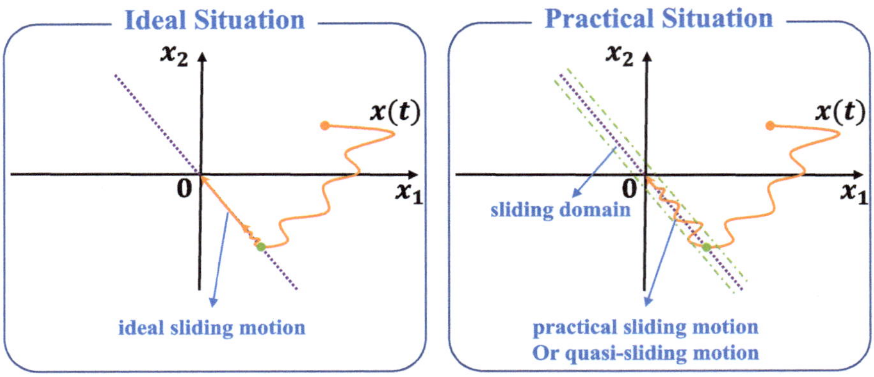

Fig. 1.3 An explanatory figure on ideal and practical sliding motions

considered MJSs. Considering that the system mode may not be measurable, a mode-independent sliding function has been constructed in [94], an asynchronous sliding mode controller has been designed by utilizing the detected mode information. For MJSs, the mode-dependent sliding surface can fully utilize the information of each mode and are more flexible in controller design, under which a rare but realistic situation might be encountered that the sliding motion is hard to be remained due to frequent switching. On the contrary, a mode-independent sliding surface (also called common sliding surface) can avoid frequent switching problem with simple operation, but ignore the internal connection among the modes and also increase the conservatism of the control design.

The above studies all have been carried out under the assumption that all TRs of MJSs are completely known, which obviously is a desired situation. When considering the case where the TR is partially unknown, the reachability of the sliding surface and the stability of the sliding mode dynamic have been analyzed in [95] according to whether the diagonal elements of TR matrix are known or not, by which a piece-wise mode-dependent sliding mode controller has designed to ensure the stability of MJSs. Noting that when using the Lyapunov method to find the infinitesimal operator to analyze the reachability, most of the existing literature ignores the fact that the sliding function are related to system mode via system states and treats it as a mode-independent variable to process, which is a possible misdirection for the mode-dependent SMC design of MJSs. This problem has been pointed out in [96], which has made improvements in analyzing the reachability by considering the randomness of the sliding function. However, this improvement method is only applicable to the case where the Lyapunov function does not contain the mode-dependent weight matrix.

For the situation where system information is transmitted through communication networks, due to the communication constraints and the special structures of the MJSs and SMC, the existing SMC methods for MJSs under the "point-to-point" transmission are no longer applicable. For this reason, researches on SMC of MJSs under communication constraints have faced new issues and difficulties, such as packet loss [97, 98], quantization [99, 100], and channel fading [101]. When considering that data cannot avoid being affected by cyber attacks via network transmission, random DoS attacks have been considered in [102] for MJSs with different state components transmitted via different S/C channels, where mode-dependent SMC schemes have been designed by means of signal compensation technique to robustly ensure system security. The intermittent DoS attack issue has been handled in [103] for MJSs by designing a sliding mode controller with switching feature according to the attack occurrence situation. The FDI attacks in the C/A channel have been effectively addressed in [104, 105] by designing a sliding mode controller in virtue of known attack upper-bound information.

Through the discussions of the research status on MJSs, it becomes evident that while substantial advancements have been reported regarding SMC methods under physical/communication resource constraints, significant research gaps and challenges remain to be addressed:

(1) Most studies have focused on the asymptotic stability performance of MJSs over long operational durations. Consequently, the controller design approaches are primarily based on the asymptotic stability analysis of sliding motion dynamics. However, when investigating transient performance under finite-time constraints in practice, an exclusive focus on sliding motion dynamics is insufficient. There is a pressing need to ensure reachability within a predetermined time interval and design controller that guarantee real-time transient performance of system states/outputs, despite existing mode transitions.
(2) As previously noted in this subsection, existing frameworks for analyzing the reachability of mode-dependent sliding surfaces do not account for the randomness inherent in sliding variables. This oversight could lead to controller failures in critical scenarios. Enhancing the analytical framework to incorporate these stochastic influences will be crucial for enhancing the theory of mode-dependent SMC for MJSs.
(3) When implementing SMC to achieve finite-time transient performance in MJSs, the consideration of physical/communication resource constraints significantly influences the design of control strategies. Addressing various comprehensive constraints and developing corresponding finite-time SMC solutions warrants further investigation.

These points highlight the need for ongoing research to overcome existing limitations and enhance the robustness and applicability of SMC methods for MJSs.

1.4 Relevant Assumptions, Definitions and Lemmas

In this section, we will present the required assumptions, key definitions and essential lemmas that will be utilized throughout this book, setting the foundation for the subsequent presentation of the main research findings.

Consider the continuous-time MJS defined on a complete probability space $(\Omega, \mathscr{F}, \Pr)$ as follows:

$$\begin{cases} \dot{x}(t) = \big(A(r_t) + \Delta A(t, r_t)\big)x(t) + B(r_t)\big(u(t) + f(x(t), r_t)\big) + D(r_t)w(t) \\ y(t) = C(r_t)x(t) \end{cases} \tag{1.6}$$

where $x(t) \in \mathbb{R}^n$, $u(t) \in \mathbb{R}^m$, and $y(t) \in \mathbb{R}^p$ represent the state vector, control input, and output of the system, respectively; $\Delta A(t, r_t) \in \mathbb{R}^{n\times n}$ denotes the parameter uncertainty; $f(x(t), r_t) \in \mathbb{R}^m$ stands for the matched nonlinear perturbation; $w(t) \in \mathbb{R}^q$ means the unmatched external disturbance; and $A(r_t)$, $B(r_t)$, $C(r_t)$, $D(r_t)$ are system matrices with appropriate dimensions. $\{r_t\}_{t\geq 0}$ represents a homogeneous Markov process defined in $t \in [0, \infty)$ and takes values in a finite set $\mathbb{S} \triangleq \{1, 2, \ldots, N\}$. The TP of modes is

$$\Pr\{r_{t+\Delta} = j \mid r_t = i\} = \begin{cases} \lambda_{ij}\Delta + o(\Delta), & i \neq j \\ 1 + \lambda_{ij}\Delta + o(\Delta), & i = j \end{cases} \tag{1.7}$$

where $\Delta > 0$ is the dwell time; $o(\Delta)$ means infinitesimality satisfying $\lim_{\Delta\to 0^+} \frac{o(\Delta)}{\Delta} = 0$; and λ_{ij} is the TR of the system state from mode i to mode j meeting $\lambda_{ij} \geq 0$ $(i \neq j)$ and $\lambda_{ii} = -\sum_{i\neq j}\lambda_{ij}$. Then TR matrix is defined as $\Lambda \triangleq [\lambda_{ij}]_{N\times N}$ $(i, j \in \mathbb{S})$. For each $i \triangleq r_t \in \mathbb{S}$, $A_i \triangleq A(r_t)$, $B_i \triangleq B(r_t)$, $C_i \triangleq C(r_t)$, $D_i \triangleq D(r_t)$, $\Delta A_i(t) \triangleq \Delta A(t, r_t)$, and $f_i(x(t)) \triangleq f(x(t), r_t)$.

Assumption 1.1 *The matrix B_i $(i \in \mathbb{S})$ is full-ranked, that is,* $\operatorname{rank}(B_i) = m$.

Assumption 1.2 *The parameter uncertainty $\Delta A_i(t)$ is norm-bounded and can be expressed as*

$$\Delta A_i(t) = M_i F_i(t) N_i,$$

where M_i, N_i are known real constant matrices, and $F_i(t)$ is an unknown matrix function satisfying $F_i^{\mathrm{T}}(t)F_i(t) \leq I$.

Assumption 1.3 *The nonlinear perturbation $f_i(x(t))$ is norm-bounded and satisfies*

$$\|f_i(x(t))\| \leq \epsilon_i \|x(t)\|,$$

where $\epsilon_i \geq 0$ is a known constant.

Assumption 1.4 *The external disturbance $w(t)$ is time-varying and satisfies*

$$\mathbb{W}_{[0,T],\delta} \triangleq \left\{w^{\mathrm{T}}(t)w(t) \leq \delta^2\right\},$$

where $\delta \geq 0$ in $[0, T]$ is a known constant.

Definition 1.4 (*Infinitesimal Operator*) [106] Let $C_2(\mathbb{R}^n \times \mathbb{S}; \mathbb{R}^+)$ denote the family of all nonnegative functions $V(x(t), r_t)$ on $\mathbb{R}^n \times \mathbb{S}$ that are continuously twice differentiable in $x(t)$. For $V(x(t), r_t) \in C_2(\mathbb{R}^n \times \mathbb{S}; \mathbb{R}^+)$ and $r_t = i$, define an infinitesimal operator $\mathscr{L}V(x(t), i)$ as

$$\mathscr{L}V(x(t), i) = \lim_{\Delta \to 0^+} \frac{1}{\Delta} \left[\mathbb{E}\{V(x(t+\Delta), r_{t+\Delta})\} - V(x(t), i)\right].$$

When investigating the finite-time transient performance of MJSs, the random characteristic of the considered MJSs cannot be ignored. Thus, we define the following stochastic finite-time boundedness (SFTB) and input-output stochastic finite-time stability (IO-SFTS) for the MJS (1.6).

Definition 1.5 (*Stochastic Finite-Time Boundedness, SFTB*) Give a time interval $[0, T]$, two positive scalars c_1, c_2 $(0 < c_1 < c_2)$, and a weighted matrix $R > 0$. The considered MJS (1.6) with $u(t) = 0$ and $w(t) \in \mathbb{W}_{[0,T],\delta}$ is said to be stochastically finite-time bounded (SFTB) with respect to the quintuplet $\aleph_{[0,T]}$, if one has for $\forall t \in [0, T]$

$$\mathbb{E}\{x^{\mathrm{T}}(0)Rx(0)\} \leq c_1 \Rightarrow \mathbb{E}\{x^{\mathrm{T}}(t)Rx(t)\} < c_2,$$

with the quintuplet as $\aleph_{[0,T]} \triangleq (c_1, c_2, [0, T], R, \mathbb{W}_{[0,T],\delta})$.

Definition 1.6 (*Input-Output Stochastic Finite-Time Stability, IO-SFTS*) Give a time interval $[0, T]$, explicit output constraint scalars β_1, β_2 $(0 < \beta_1 < \beta_2)$, and a weighted matrix $R > 0$. The considered MJS (1.6) with $u(t) = 0$ and $w(t) \in \mathbb{W}_{[0,T],\delta}$ is said to be input-output stochastically finite-time stable (IO-SFTS) with respect to the quintuplet $\Re_{[0,T]}$ if one has for $\forall t \in [0, T]$

$$\mathbb{E}\{y^{\mathrm{T}}(0)Ry(0)\} < \beta_1 \Rightarrow \mathbb{E}\{y^{\mathrm{T}}(t)Ry(t)\} < \beta_2, \tag{1.8}$$

with the quintuplet as $\Re_{[0,T]} \triangleq (\beta_1, \beta_2, [0, T], R, \mathbb{W}_{[0,T],\delta})$.

Remark 1.1 It should be mentioned that, the definition of IO-FTS given in [107] assumes that the controlled systems has a zero initial condition $x(0) = 0$, which is not required in Definition 1.6. The main difference between Definition 1.6 and the one in existing works is that the extended definition on IO-SFTS depends on the initial constraint in mean square, i.e., $\mathbb{E}\{y^{\mathrm{T}}(0)Ry(0))\} < \beta_1$, which may avoid the limitation and conservativeness arising from zero initial condition.

Remark 1.2 The scalars c_1, c_2, matrix R in Definition 1.5 and the scalars β_1, β_2, matrix R in Definition 1.6 are known. In practice, the scalar c_2 or β_2 and the matrix R are determined directly according to the transient constraints of physical systems, and the scalar c_1 or β_1 are chosen according to the initial conditions (see [108, 109]).

During the theoretical analysis process, the following lemmas on matrix inequality will be used.

Lemma 1.1 *For any real scalar* $\alpha > 0$, *one has*

$$MF(t)N + N^{\mathrm{T}}F^{\mathrm{T}}(t)M^{\mathrm{T}} \leq \alpha^{-1}MM^{\mathrm{T}} + \alpha N^{\mathrm{T}}N,$$

where M, N *are known constant matrices with appropriate dimensions, and* $F(t)$ *is an unknown matrix function satisfying* $F^{\mathrm{T}}(t)F(t) \leq I$.

Lemma 1.2 ([110]) *For any real scalar* $\alpha > 0$ *and matrix* $X > 0$ *satisfying* $\alpha I - M^{\mathrm{T}}XM > 0$, *one has*

$$\begin{aligned}&(A + MF(t)N)^{\mathrm{T}}X(A + MF(t)N)\\ &\leq A^{\mathrm{T}}XA + A^{\mathrm{T}}XM(\alpha I - M^{\mathrm{T}}XM)^{-1}M^{\mathrm{T}}XA + \alpha N^{\mathrm{T}}N,\end{aligned}$$

where A, M, N *are known constant matrices with appropriate dimensions, and* $F(t)$ *is an unknown matrix function satisfying* $F^{\mathrm{T}}(t)F(t) \leq I$.

Lemma 1.3 (Schur Complement) *For symmetric matrix* $S = \begin{bmatrix} S_{11} & S_{12} \\ S_{12}^{\mathrm{T}} & S_{22} \end{bmatrix}$ *with* $S_{11} \in \mathbb{R}^{r\times r}$, *one has*

$$S < 0 \Leftrightarrow S_{22} < 0 \text{ and } S_{11} - S_{12}S_{22}^{-1}S_{12}^{\mathrm{T}} < 0$$

It is known that there are two phases for state trajectories of an SMC system: the reaching phase and the sliding motion phase. When considering the SFTB via SMC method during the whole given interval $[0, T]$, it is supposed that $T^* \in (0, T)$ represents the reaching time onto the sliding surface. Then, due to the transient performance requirements, reaching phase during $[0, T^*]$ and the sliding motion phase during $[T^*, T]$ should be analyzed, simultaneously. Inspired by the work in [108], we propose a phase-based partition strategy adapted to the framework of MJSs as follows.

Before providing the partition strategy, we first denote a quintuplet as $\aleph_{[0,T^*]} \triangleq (c_1, c^*, [0, T^*], R, \mathbb{W}_{[0,T^*],\delta})$, $\aleph_{[T^*,T]} \triangleq (c^*, c_2, [T^*, T], R, \mathbb{W}_{[T^*,T],\delta})$, $\Re_{[0,T^*]} \triangleq (\beta_1, \beta^*, [0, T^*], R, \mathbb{W}_{[0,T^*],\delta})$, and $\Re_{[T^*,T]} \triangleq (\beta^*, \beta_2, [T^*, T], R, \mathbb{W}_{[T^*,T],\delta})$.

Lemma 1.4 (Partition Strategy) *For the system* (1.6) *with the assigned parameters* $0 < c_1 < c_2$, $T > 0$, $R > 0$, $\delta \leq 0$, *the closed-loop system under the designed controller* $u(t)$ *is SFTB with respect to* $\aleph_{[0,T]}$, *if only if there exist auxiliary scalars* c_i^* $(i \in \mathbb{S})$ *satisfying* $c_1 < c_i^* < c_2$ *such that the closed-loop system is SFTB with respect to* $\aleph_{[0,T^*]}$ *during reaching phase and the sliding mode dynamics are SFTB with respect to* $\aleph_{[T^*,T]}$ *during sliding motion phase, where* $c^* \triangleq \max_{i\in\mathbb{S}}\{c_i^*\}$.

Proof The following proof will be made from sufficiency and necessity aspects. It is observed from Fig. 1.4 that during $[0, T]$, the motion trajectory is divided into two parts. For ith subsystem $(i \in \mathbb{S})$, the intermediate state constraint is denoted as c_i^*, and the maximum constraint for all modes is defined as $c^* \triangleq \max_{i\in\mathbb{S}}\{c_i^*\}$.

► **Sufficiency**. During reaching phase for $\forall t \in [0, T^*]$, one has

$$\mathbb{E}\{x^{\mathrm{T}}(0)Rx(0)\} \leq c_1 \Rightarrow \mathbb{E}\{x^{\mathrm{T}}(t)Rx(t)\} < c_i^* \Rightarrow \mathbb{E}\{x^{\mathrm{T}}(t)Rx(t)\} < c^*, \quad (1.9)$$

Also, during sliding motion phase for $\forall t \in [T^*, T]$, we have

$$\mathbb{E}\{x^{\mathrm{T}}(t)Rx(t)\} < c^* \Rightarrow \mathbb{E}\{x^{\mathrm{T}}(t)Rx(t)\} < c_2. \quad (1.10)$$

Combining (1.9) and (1.10), we obtain

$$\mathbb{E}\{x^{\mathrm{T}}(0)Rx(0)\} \leq c_1 \Rightarrow \mathbb{E}\{x^{\mathrm{T}}(t)Rx(t)\} < c_2, \quad \forall t \in [0, T].$$

Hence, according to Definition 1.5, the closed-loop system is SFTB with respect to $\aleph_{[0,T]}$.

► **Necessity**. Suppose the closed-loop system is SFTB with respect to $\aleph_{[0,T]}$. Then, by continuity arguments, there must exist some sufficiently small constants d_i $(i \in \mathbb{S})$, such that during reaching phase for $\forall t \in [0, T^*]$, we have

$$\mathbb{E}\{x^{\mathrm{T}}(0)Rx(0)\} \leq c_1 \Rightarrow \mathbb{E}\{x^{\mathrm{T}}(t)Rx(t)\} < c_2 - d_i. \quad (1.11)$$

It is obtained from (1.11) that $\mathbb{E}\{x^{\mathrm{T}}(T^*)Rx(T^*)\} < c_i^*$, with $c_i^* \triangleq c_2 - d_i$, which is just the initial condition during sliding motion phase. On the other hand, because the closed-loop system is SFTB with respect to $\aleph_{[0,T]}$, we have $\mathbb{E}\{x^{\mathrm{T}}(t)Rx(t)\} < c_2$, $\forall t \in [0, T]$, which yields $\mathbb{E}\{x^{\mathrm{T}}(t)Rx(t)\} < c_2$, $\forall t \in [T^*, T]$. Hence during sliding motion phase for $\forall t \in [T^*, T]$, one has

$$\mathbb{E}\{x^{\mathrm{T}}(T^*)Rx(T^*)\} < c^* \Rightarrow \mathbb{E}\{x^{\mathrm{T}}(t)Rx(t)\} < c_2, \quad (1.12)$$

where $c^* \triangleq \max_{i \in \mathbb{S}}\{c_i^*\}$, $\forall i \in \mathbb{S}$. This completes the proof.

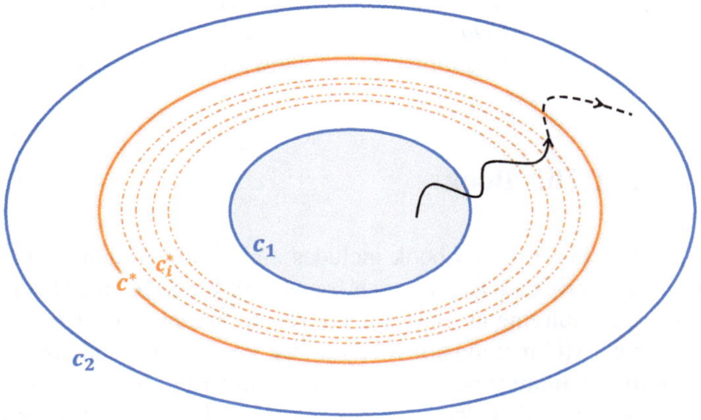

Fig. 1.4 Schematic diagram of the separation strategy

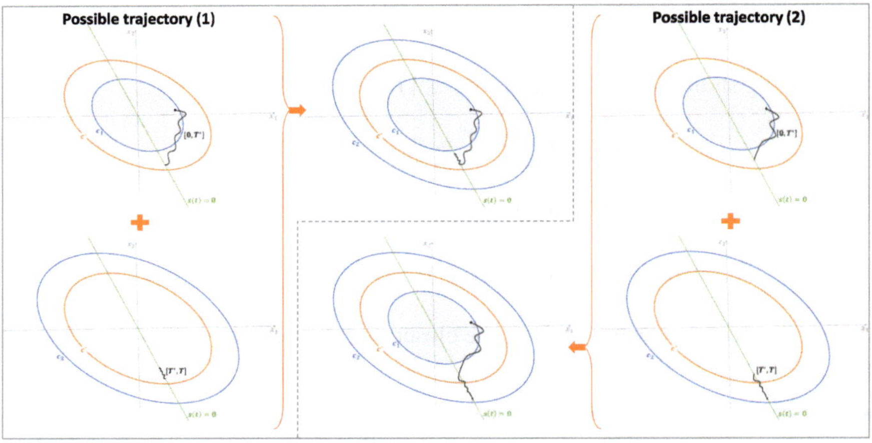

Fig. 1.5 Schematic diagram of possible state trajectories under finite-time SMC

It is shown from Lemma 1.4 that the reaching phase within $[0, T^*]$ and the sliding motion phase within $[T^*, T]$ can be connected by parameters c^* and T^*. Therefore, when SFTB during the reaching phase and the sliding motion phase can be ensured simultaneously, it can be guaranteed over the whole finite-time interval $[0, T]$. The partition-strategy-based design idea can be shown in Fig. 1.5, which can be summarized as "partition–integration" technique.

As the similar line of Lemma 1.4, the partition strategy for IO-SFTS of MJSs can be given as follows:

Lemma 1.5 (Partition Strategy) *For the system* (1.6) *with the assigned parameters* $0 < \beta_1 < \beta_2$, $T > 0$, $R > 0$, $\delta \leq 0$, *the closed-loop system under the designed controller* $u(t)$ *is IO-SFTS with respect to* $\Re_{[0,T]}$, *if only if there exist auxiliary scalars* β_i^* $(i \in \mathbb{S})$ *satisfying* $\beta_1 < \beta_i^* < \beta_2$ *such that the closed-loop system is IO-SFTS with respect to* $\Re_{[0,T^*]}$ *during reaching phase and the sliding mode dynamics are SFTB with respect to* $\Re_{[T^*,T]}$ *during sliding motion phase, where* $\beta^* \triangleq \max_{i\in\mathbb{S}}\{\beta_i^*\}$.

1.5 Outline of This Book

As exhibited in Fig. 1.6, this book includes eight main contents. For the given TR matrix, Chap. 2 gives a basic research framework for SMC of MJSs under transient performance requirements. Based on this research framework, Chaps. 3–5 focus on the finite-time SMC methods of MJSs under various physical constraints, and Chaps. 6–8 study the finite-time SMC schemes countering different types of cyber attacks. To broaden the SMC technologies for MJSs, Chap. 9 further proposes the TR stabilization methods and TR-fuzed hybrid SMC approaches for the SFTS.

Specifically, the outline of this book is listed as follows:

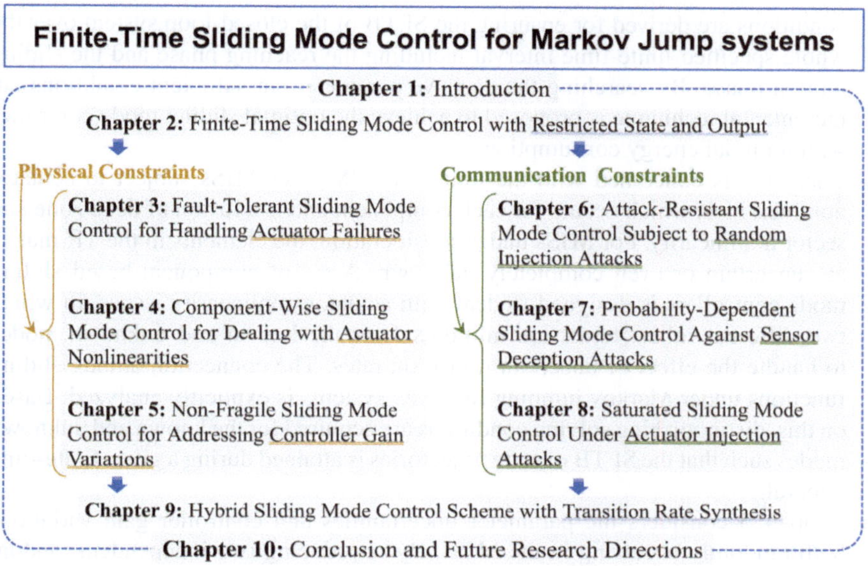

Fig. 1.6 The organization structure of this book

- In this chapter introduces the research backgrounds and research status of MJSs, finite-time control in MJSs, SMC in MJSs, and the pivotal assumptions, definitions, and lemmas.
- Chapter 2 proposes two basic research frameworks for SMC under restricted states and outputs, respectively, to ensure the SFTB of MJSs with completely known TRs and IO-SFTS of MJSs with partially known TRs. (1) Focusing on restricted states for MJSs with completely known TRs, a sliding mode controller is constructed by using the given time parameter, which can effectively drive the state trajectories onto the specified sliding surface during a specified finite (possibly *short*) time interval. By utilizing the proposed partitioning strategy, the SFTB over *reaching phase* and *sliding motion phase* are, respectively, analyzed during the specified finite time interval. (2) When further considering restricted outputs for MJSs with partially known TRs, an extended definition of IO-SFTS is given to reduce the conservativeness arising from zero initial condition $x(0) = 0$. A parameter-related sliding mode control strategy is proposed to eliminate the effect of incomplete transition rates, and sufficient conditions are derived to ensure the IO-SFTS of the closed-loop systems over the finite-time interval including reaching phase and sliding motion phase.
- Chapter 3 studies the finite-time SMC method for MJSs with actuator faults. The considered actuator faults manifest as incomplete executions of the control signals. In virtue of the available bound information of actuator faults, a fault-tolerant sliding mode controller is designed such that state trajectories are driven to the sliding surface before the given short time. The fault-parameter-based sufficient

conditions are derived for ensuring the SFTB of the closed-loop system over the whole specified finite-time interval including the reaching phase and the sliding motion phase. By searching the decisive parameters in sufficient conditions, an incremental technique is proposed to achieve the optimal sliding mode controller with minimal energy consumption.

- Chapter 4 is concerned with the finite-time SMC for MJSs subject to actuator nonlinearities, which occurs for each component and is a fusion of dead zone and sector nonlinearity. For MJSs under consideration, the elements in the TR matrix are uncertain or even completely unknown. A set of component-based sliding mode controllers is designed to deal with actuator nonlinearity issues, in which two different robust terms are introduced for the known and unknown modes to handle the effect of uncertain transition rates. The connection among sliding functions under Markov jumping for SMC systems is explicitly analyzed. Based on this, distinguishing solving conditions are acquired for the known and unknown modes such that the SFTB of state trajectories is attained during a given finite-time interval.
- Chapter 5 considers the parameter uncertainties and controller gain variations both randomly occurring in the MJSs. By introducing two independent random variables obeying Bernoulli distribution, the random characteristics of parameter uncertainties and controller gain variations are described. A mode-dependent sliding surface is constructed, and the non-fragile SMC scheme is synthesized to achieve the reachability of the specified sliding surface in a finite time interval. Furthermore, the SFTB over both the reaching and sliding stages are ensured simultaneously under sufficient conditions via skillful derivation, based on which the sliding gains and controller gains are co-designed.
- Chapter 6 deals with FDI attacks in MJSs happening during C/A channel. It is assumed that the adversaries may inject false data in a probabilistic way into the transmitted control signals via C/A channel. Meanwhile, there may exist randomly occurring uncertainties and peak-bounded external disturbances. By employing the upper-bound information of FDI attacks, a sliding mode controller is designed, which can not only robustly confront the influences of attacks but also drive state trajectories onto the specified sliding surface during a given finite-time interval. A key feature is that a set of mode-dependent sufficiently small scalars are introduced into some coupled Lyapunov inequalities such that the feasible solutions are easily obtained for the SFTB of the closed-loop systems.
- Chapter 7 develops security SMC design for MJSs suffering from both channel fading and deception attacks over the S/C channel. The multiplicative fading model is introduced to describe the fading phenomenon in the unpredictable communication network, and the channel coefficients are mutually independent random variables that take values in certain intervals. Besides, the stochastically occurring deception attacks are modeled by a random process following the Bernoulli distribution. Under the effect of channel fading and deception attacks, a sliding surface is constructed dependent on fading probability, which can weaken the impacts of channel fading. A security SMC law is proposed by means of the occurrence probability and the upper bound of the attack deceptive information, which sup-

presses the stochastically occurring deception attacks and enables the specified sliding surface to be arrived within the given finite time interval. Furthermore, by adopting the partitioning strategy, sufficient conditions are derived for the stochastic finite-time boundedness over two stages of SMC, i.e., the reaching phase and sliding phase.

- Chapter 8 addresses both physical and communication constraints in MJSs, that is, the actuator saturation and injection attack. The injection attacks are supposed to be unknown but real-time occurring, which can be referred to as a nonlinear function. With the help of the learning capability of neural networks, the nonlinear upper bound of injection attacks is analyzed and estimated, and the security control strategy based on neural networks is proposed to improve the reliability of control. Considering the transient performance implementation, a saturation-parameter-dependent sliding mode controller is proposed to guarantee the reachability of the specified sliding surface within a given finite time, in which the adaptive rules of neural network parameters are designed to reduce the effects of the considered attack. The reaching and sliding motion stages are analyzed, respectively, and sufficient conditions are achieved to guarantee the SFTB of the closed-loop system.
- Chapter 9 investigates a collaborative control framework for MJSs by synthesizing the SMC and TR matrix to achieve the SFTB, breaking through the assumption that TRs are priori. First, an existing condition on the TR matrix is deduced to guarantee the SFTB of the unforced MJSs, and a hybrid design scheme integrating the TR matrix with the sliding mode controller is synthesized to ensure the SFTB over both the reaching phase and sliding motion phase. Second, further considering the effects of actuator saturation, the saturated-lower-bound matrix is utilized to construct the sliding mode controller for the guarantee of reachability before a given time despite actuator saturation. Under this SMC law, the collaborative design scheme of the TR matrix and sliding mode controller gains is provided, establishing SFTB conditions over the whole given finite-time interval. For the double deterministic parameters in the proposed design scheme, a gridding search algorithm is presented to solve the optimal combination of sliding mode controller gains and TR matrix.
- Chapter 10 concludes this book and gives some future research directions.

References

1. Zhang, L., Ning, Z., Shi, Y.: Analysis and synthesis for a class of stochastic switching systems against delayed mode switching: a framework of integrating mode weights. Automatica **99**, 99–111 (2019)
2. Lin, Z., Liu, J., Wu, Q., et al.: Mixed H_2/H_∞ pitch control of wind turbine with a Markovian jump model. Int. J. Control **91**(1), 156–169 (2018)
3. Luenberger D.G., Arbel A.: Singular dynamic Leontief systems. Econ.: J. Econ. Soc., 991–995 (1977)
4. Zhang, L., Yang, T., Shi, P., et al.: Analysis and Design of Markov Jump Systems with Complex Transition Probabilities. Springer, Berlin, Germany (2016)

5. Yao, J., Zhai, C., Wang, G.: Stabilization of delayed Markovian jump systems with sampled controllers. J. Frankl. Inst. **361**(17), 107194 (2024)
6. da Silva, L.P., Oliveira, R.C.: H_2 detector-based state feedback control of continuous-time MJLS subject to uncertain transition rate matrices. J. Frankl. Inst. **361**(15), 107123 (2024)
7. Qi, W., Li, R., Zong, G., Yan, H., Wu, Z.G., Cheng, J.: Asynchronous dynamic output feedback control for discrete nonlinear networked semi-Markov jump models with cyber attacks and applications. IEEE Trans. Autom. Sci. Eng. (2024). https://doi.org/10.1109/TASE.2024.3441545
8. Zhou, X., Wang, Y., Ji, Z.: Event-triggered output feedback H_∞ control for Markovian jump networked control systems with signal quantization. Int. J. Robust Nonlinear Control **34**(9), 5910–5928 (2024)
9. Liu, J.J., Ogura, M., Li, Q., Lam, J.: Learning-based stabilization of Markov jump linear systems. Neurocomputing **586**, 127618 (2024)
10. Yao, J., Zhai, C., Wang, G.: Stabilization of delayed Markovian jump systems with sampled controllers. J. Frankl. Inst. **361**(17), 107194 (2024)
11. Wang, L., Long, Y., Li, T., Park, J.H.: MPC-based asynchronous attack tolerant control for uncertain Markov jump cyber–physical systems. IEEE Trans. Cybern. **54**(12), 7489–7500 (2024)
12. Wu, T., Ding, Y., Zhang, S., Liu, Y., Yang, J., Zhang, L.: Efficient model predictive control for Markov jump nonlinear systems via controllable sets and sum-of-square programming. IEEE Control Syst. Lett. **8**, 382–387 (2024)
13. Yao, X., Wen, X.: Composite hierarchical hybrid anti-disturbance control for Markovian jump systems with event-triggered disturbance. Syst. Control Lett. **185**, 105734 (2024)
14. Wang, D., Xu, S.: Composite anti-disturbance control for nonlinear hidden Markov jump systems under replay attacks: A dynamic output-feedback method. IEEE Trans. Cybern. **54**(11), 7038–7047 (2024)
15. Cheng, J., Wu, J., Yan, H., Wu, Z.G., Zhang, D.: Neural network-based anti-disturbance control of nonhomogeneous Markov switching systems. IEEE Trans. Circuits Syst.-Express Briefs **71**(6), 3151–3155 (2024)
16. Xiong, L., Song, R., Huang, S., Ban, C., Li, P., et al.: Markov jump system modeling and control of inverter-fed remote area weak grid via quantized sliding mode. IEEE Trans. Power Syst. (2024). https://doi.org/10.1109/TPWRS.2024.3494857
17. Cheng, J., Xu, J., Yan, H., Wu, Z.G., Qi, W.: Neural network-based sliding mode control for semi-Markov jumping systems with singular perturbation. IEEE Trans. Cybern. (2024). https://doi.org/10.1109/TCYB.2024.3481870
18. Xu, T., Niu, Y., Chen, B., Deng, F.: Protocol-based asynchronous sliding mode control for Markovian jump systems under random sampling. Nonlinear Dyn. **113**, 697–710 (2025)
19. Krasovskii, N.N., Lidskii, E.A.: Analysis design of controller in systems with random attributes-Part 2. Autom. Remote Control **22**, 1141–1146 (1961)
20. Mahmoud, M., Jiang, J., Zhang, Y.: Active Fault Tolerant Control Systems: Stochastic Analysis and Synthesis. Springer, New York, USA (2003)
21. Oliveira, A.M., Costa, O.L.V.: An iterative approach for the discrete-time dynamic control of Markov jump linear systems with partial information. Int. J. Robust Nonlinear Control **30**(2), 495–511 (2020)
22. Xiong, J., Lam, J., Gao, H., et al.: On robust stabilization of Markovian jump systems with uncertain switching probabilities. Automatica **41**(5), 897–903 (2005)
23. de Souza, C.E.: Robust stability and stabilization of uncertain discrete-time Markovian jump linear systems. IEEE Trans. Autom. Control **51**(5), 836–841 (2006)
24. Zhang, L., Boukas, E.K., Lam, J.: Analysis and synthesis of Markov jump linear systems with time-varying delays and partially known transition probabilities. IEEE Trans. Autom. Control **53**(10), 2458–2464 (2008)
25. Sehgal, V.K.: Markovian models based stochastic communication in networks-in-package. IEEE Trans. Parallel Distrib. Syst. **26**(10), 2806–2821 (2014)

26. Feng, J.E., Lam, J., Shu, Z.: Stabilization of Markovian systems via probability rate synthesis and output feedback. IEEE Trans. Autom. Control **55**(3), 773–777 (2010)
27. Wang, G.: Stabilization of singular Markovian jump systems via switching probability rate design. In: Proceedings of the 31st Chinese Control Conference, pp. 2732–2726 (2012)
28. Wang, G., Zhang, P.: Mode-independent H_∞ control of singular Markovian jump systems via switching probability rate design. In: Proceedings of the 32nd Chinese Control Conference, pp. 1325–1329 (2013)
29. Zhang, P., Cao, J., Wang, G.: Mode-independent guaranteed cost control of singular Markovian delay jump systems with switching probability rate design. Int. J. Innov. Comput. Inf. Control **10**(4), 1291–1303 (2014)
30. Song, J., Niu, Y., Lam, J., et al.: A hybrid design approach for output feedback exponential stabilization of Markovian jump systems. IEEE Trans. Autom. Control **63**(5), 1404–1417 (2018)
31. Li, Q., Zhang, J., Lin, P.: Adaptive event-triggered fault detection for nonlinear positive semi-Markov jump systems. Int. J. Adapt. Control Signal Process. **36**(11), 2677–2700 (2022)
32. Lu, D., Zhang, X., Lan, Y., Wang, X.: Fault detection and isolation for discrete-time Markovian jump systems with generally bounded transition probabilities: a zonotope-based method. Trans. Inst. Meas. Control **43**(13), 2948–2959 (2021)
33. Long, Y., Cheng, Y., Li, T., Bai, W., Chen, K., Bai, L.: Asynchronous frequency-dependent fault detection for nonlinear Markov jump systems under wireless fading channels. IEEE Trans. Cybern. **52**(12), 13598–13608 (2021)
34. Niu, Y., Wang, X.: Sliding mode control design for uncertain delay systems with partial actuator degradation. Inter. J. Syst. Sci. **40**(4), 403–409 (2009)
35. Chen, B., Niu, Y., Zou, Y.: Adaptive sliding mode control for stochastic Markovian jumping systems with actuator degradation. Automatica **49**(6), 1748–1754 (2013)
36. Ran, G., Shu, Z., Lam, H.K., Liu, J., Li, C.: Dissipative tracking control of nonlinear Markov jump systems with incomplete transition probabilities: a multiple-event-triggered approach. IEEE Trans. Fuzzy Syst. **31**(7), 2389–2400 (2022)
37. Ma, H., Jia, Y.: H_∞ control for discrete-time nonlinear Markov jump systems with multiplicative noise and sector constraint. Int. J. Robust Nonlinear Control **24**(16), 2347–2364 (2014)
38. Zhang, Q., Yan, H., Tian, Y., Qi, P.: Asynchronous control for discrete-time singular Markovian jump systems with actuator saturation. IEEE Trans. Circuits Syst. II-Express Briefs **71**(8), 3890–3894 (2024)
39. Qi, W., Park, J.H., Cheng, J., Kao, Y., Gao, X.: Anti-windup design for stochastic Markovian switching systems with mode-dependent time-varying delays and saturation nonlinearity. Nonlinear Anal.: Hybrid Syst. **26**, 201–211 (2017)
40. Zhang, J., Liu, D., Ma, Y., Yu, P.: Non-fragile H_∞ memory sampled-data state-feedback control for continuous-time nonlinear Markovian jump fuzzy systems with time-varying delay. Inf. Sci. **577**, 214–233 (2021)
41. Li, X., Li, Y., Cheng, J., Shi, K., Qiu, K.: Static output feedback LFC for semi-Markov type interconnected multi-area power systems: a non-fragile PI strategy. Int. J. Robust Nonlinear Control **34**(15), 10437–10453 (2024)
42. Pasqualetti, F., Dörfler, F., Bullo, F.: Attack detection and identification in cyber-physical systems. IEEE Trans. Autom. Control **58**(11), 2715–2729 (2013)
43. Ding, D., Han, Q.L., Xiang, Y., et al.: A survey on security control and attack detection for industrial cyber-physical systems. Neurocomputing **275**, 1674–1683 (2018)
44. Chen, X., Wang, Y.: Event-triggered attack-tolerant tracking control design for networked nonlinearcontrol systems under DoS jamming attacks. Sci. China Inf. Sci. **63**(5), 150207 (2020)
45. Zhao, S., Zhang, H., Xiong, L., Wen, S., Cao, J., Zhang, Y.: Resilient adaptive event-triggered synchronization control of piecewise-homogeneous Markov jump delayed neural networks under aperiodic DoS attacks. Int. J. Robust Nonlinear Control **34**(3), 1493–1521 (2024)
46. Zhang, Y., Wu, Z.G.: Asynchronous control of Markov jump systems under aperiodic DoS attacks. IEEE Trans. Circuits Syst. II– Express Briefs **70**(2), 685–689 (2023)

47. Tan, Y., Liu, Q., Liu, J., Xie, X., Fei, S.: Observer-based security control for interconnected semi-Markovian jump systems with unknown transition probabilities. IEEE Trans. Cybern. **52**(9), 9013–9025 (2022)
48. Cao, Z., Niu, Y., Park, J.H., Zhu, K.: Attack-resistant sliding mode control for Markov jump systems with redundant channels: a novel two-layer mapping approach. IEEE Trans. Autom. Sci. Eng. (2024). https://doi.org/10.1109/TASE.2024.3379192
49. Zhang, D., Wang, Q.G., Feng, G., et al.: A survey on attack detection, estimation and control of industrial cyber-physical systems. ISA Trans. **116**, 1–16 (2021)
50. Chen, B., Niu, Y., Zou, Y.: Security control for Markov jump system with adversarial attacks and unknown transition rates via adaptive sliding mode technique. J. Frankl. Inst. **356**(6), 3333–3352 (2019)
51. Ding, D., Tang, Z., Wang, Y., Ji, Z.: Secure synchronization of complex networks under deception attacks against vulnerable nodes. Appl. Math. Comput. **399**, 126017 (2021)
52. Yang, H., Han, H., Yin, S., Han, H., Wang, P.: Sliding mode-based adaptive resilient control for Markovian jump cyber-physical systems in face of simultaneous actuator and sensor attacks. Automatica **142**, 110345 (2022)
53. Tan, C., Gao, C., Peng, J., Xie, X., Wang, Y.: Neural network-based security control for TS fuzzy system with cooperative event-triggered mechanism. IEEE Trans. Fuzzy Syst. **32**(8), 4633–4645 (2024)
54. Anbalagan, P., Jeong, J.H., Joo, Y.H.: Secure reachable set synthesis and reliable event-driven mechanism for TS fuzzy semi-Markovian networked control systems against deception attacks. IEEE Trans. Fuzzy Syst. **32**(7), 3928–3942 (2024)
55. Zhang, Z., Shen, H., Su, L.: H_∞/Passive synchronization of semi-Markov jump neural networks subject to hybrid attacks via an activation function division approach. J. Syst. Sci. Complex. **37**(3), 1023–1036 (2024)
56. Kamenkov, G.: On stability of motion over a finite interval of time (in Russian). J. Appl. Math. Mech. **17**, 529–540 (1953)
57. Lebedev, A.: On stability of motion during a given interval of time (in Russian). J. Appl. Math. Mech. **18**, 139–148 (1954)
58. Dorato, P.: Short time stability in linear time-varying systems. IRE Int. Conv. Rec. **4**, 83–87 (1961)
59. Weiss, L., Infante, E.F.: On the stability of systems defined over a finite-time interval. Natl. Acad. Sci. USA, 44–48 (1965)
60. Weiss, L., Infante, E.F.: Finite-time stability under perturbing forces and on product spaces. IEEE Trans. Autom. Control **2**(2), 54–59 (1965)
61. Michel, A., Wu, S.: Stability of discrete-time systems over a finite interval of time. Int. J. Control **9**, 679–694 (1969)
62. Michel, A., Porter, D.: Practical stability and finite-time stability of discontinuous systems. IEEE Trans. Circuit Theory **19**(2), 123–129 (1972)
63. Dorato, P., Abdallah, C.T., Famularo, D.: Robust finite-time stability design via linear matrix inequalities. In: 36th IEEE Conference on Decision and Control, pp. 1305–1306 (1997)
64. Bhat, S.P., Bernstein, D.S.: Finite-time stability of continuous autonomous systems. SIAM J. Control Optim. **38**(3), 751–766 (2000)
65. Amato, F., Ariola, M., Dorato, P.: Finite-time control of linear systems subject to parametric uncertainties and disturbances. Automatica **37**(9), 1459–1463 (2001)
66. Amato, F., Ariola, M., Cosentino, C.: Finite-time stabilization via dynamic output feedback. Automatica **42**(2), 337–342 (2006)
67. Amato, F., Ariola, M.: Finite-time control of discrete-time linear systems. IEEE Trans. Autom. Control **50**(5), 724–729 (2005)
68. Amato, F., Ambrosino, R., Cosentino, C., et al.: Input-output finite time stabilization of linear systems. Automatica **46**(9), 1558–1562 (2010)
69. Amato, F., Carannante, G., Tommasi, G.D., et al.: Input-output finite-time stability of linear systems: necessary and sufficient conditions. IEEE Trans. Autom. Control **57**(12), 3051–3063 (2012)

70. Tan, F., Zhou, B., Duan, G.: Finite-time stabilization of linear time-varying systems by piecewise constant feedback. Automatica **68**, 277–285 (2016)
71. Rotondo, D., Nejjari, F., Puig, V.: Dilated LMI characterization for the robust finite time control of discrete-time uncertain linear systems. Automatica **63**, 16–20 (2013)
72. He, S., Liu, F.: Finite-time state feedback stabilization of Markov jump systems (in China). Control Decis. **24**(1), 91–95 (2009)
73. Ran, G., Liu, J., Li, C., Chen, H., Han, C.: Finite-time filtering for fuzzy nonlinear semi-Markov jump systems with deception attacks and aperiodical transmission. J. Frankl. Inst. **360**(16), 12576–12597 (2023)
74. Tian, G.: Finite-time H_∞ control for stochastic Markovian jump systems with time-varying delay and generally uncertain transition rates. Int. J. Syst. Sci. **52**(12), 2446–2459 (2021)
75. Song, J., Niu, Y., Zou, Y.: Asynchronous output feedback control of time-varying Markovian jump systems within a finite-time interval. J. Frankl. Inst. **354**(15), 6747–6765 (2017)
76. Li, F., Ni, Z., Su, L., Xia, J., Shen, H.: Passivity-based finite-region control of 2-D hidden Markov jump Roesser systems with partial statistical information. Nonlinear Anal.: Hybrid Syst. **51**, 101433 (2024)
77. Orlov, Y.: Finite time stability and robust control synthesis of uncertain switched systems. SIAM J. Control Optim. **43**(4), 1253–1271 (2006)
78. Lin, X., Du, H., Li, S.: Finite-time boundedness and L_2-gain analysis for switched delay systems with norm-bounded disturbance. Appl. Math. Comput. **217**(12), 5982–5993 (2011)
79. Zong, G., Wang, R., Zheng, W.X., et al.: Finite-time H_∞ control for discrete-time switched nonlinear systems with time delay. Int. J. Robust Nonlinear Control **25**(6), 914–936 (2015)
80. Zhang, L., Wang, S., Karimi, H.R., et al.: Robust finite-time control of switched linear systems and application to a class of servomechanism systems. IEEE/ASME Trans. Mechatron. **20**(5), 2476–2485 (2015)
81. ElBsat, M.N., Yaz, E.E.: Robust and resilient finite-time bounded control discrete-time uncertain nonlinear systems. Automatica **49**(7), 2292–2296 (2013)
82. Wu, H., Feng, S.: Guaranteed-cost finite-time fuzzy control for temperature-constrained nonlinear coupled heat-ODE systems. IEEE Trans. Syst. Man Cybern.: Syst. (2017)
83. Zhang, G., Wang, W.: Finite-region stability and boundedness for discrete 2D Fornasini-Marchesini second models. Int. J. Syst. Sci. **48**(4), 778–787 (2017)
84. Zhang, G., Trentelman, H.L., Wang, W., et al.: Input-output finite-region stability and boundedness for discrete 2D Fornasini-Marchesini models. Syst. Control Lett. **99**, 9–16 (2017)
85. Wang, Y., Xu, S., Ahn, C.K.: Almost sure finite-time control for Markovian jump systems under asynchronous switching with applications: a sliding mode approach. IEEE Trans. Circuits Syst. I: Regul. Pap. **69**(9), 3726–3735 (2022)
86. Qi, W., Zong, G., Karimi, H.R.: Finite-time observer-based sliding mode control for quantized semi-Markov switching systems with application. IEEE Trans. Ind. Inform. **16**(2), 1259–1271 (2019)
87. Chen, H., Zong, G., Zhao, X.: Asynchronous secure control of fuzzy MJSs with time delays: a multi-level probabilistic event-triggered mechanism. IEEE Trans. Autom. Sci. Eng. (2024). https://doi.org/10.1109/TASE.2024.3421530
88. Cheng, P., Wang, H., Stojanovic, V., Liu, F., He, S., Shi, K.: Dissipativity-based finite-time asynchronous output feedback control for wind turbine system via a hidden Markov model. Int. J. Syst. Sci. **53**(15), 3177–3189 (2022)
89. Bandyopadhyay, B., Janardhanan, S.: Discrete-Time Sliding Mode Control: A Multirate Output Feedback Approach, vol. 323. Springer, Berlin, Germany (2006)
90. Bartolini, G., Fridman, L., Pisano, A., et al.: Modern Sliding Mode Control Theory: New Perspectives and Applications, vol. 375. Springer, Berlin, Germany (2008)
91. Milosavljevic, D.: General conditions for existence of a quasi-sliding mode on the switching hyperplane in discrete variable structure systems. Autom. Remote Control **46**, 307–314 (1985)
92. Bandyopadhyay, B., Behera, A.K.: Event-Triggered Sliding Mode Control: A New Approach to Control System Design, vol. 139. Springer, Switzerland (2018)

93. Niu, Y., Ho, D.W.C., Wang, X.: Sliding mode control for Itô stochastic systems with Markovian switching. Automatica **43**(10), 1784–1790 (2007)
94. Song, J., Niu, Y., Lam, H.K., et al.: Asynchronous sliding mode control of singularly perturbed semi-Markovian jump systems: application to an operational amplifier circuit. Automatica **118**, 109026 (2020)
95. Chen, B., Niu, Y., Zou, Y.: Adaptive sliding mode control for stochastic Markovian jumping systems with actuator degradation. Automatica **49**(6), 1748–1754 (2013)
96. Wei, Y., Park, J.H., Qiu, J., et al.: Sliding mode control for semi-Markovian jump systems via output feedback. Automatica **81**, 133–141 (2017)
97. Chen, B., Niu, Y., Zou, Y.: Sliding mode control for stochastic Markovian jumping systems subject to successive packet losses. J. Frankl. Inst. **351**(4), 2169–2184 (2014)
98. Hu, J., Zhang, P., Kao, Y., et al.: Sliding mode control for Markovian jump repeated scalar nonlinear systems with packet dropouts: the uncertain occurrence probabilities case. Appl. Math. Comput. **362**, 124574 (2019)
99. Nie, L., Chen, D., Hu, J.: Quantized sliding mode control for networked Markovian jump systems under round-robin protocol: the output feedback case. Int. J. Control Autom. Syst. **19**, 2674–2686 (2021)
100. Liu, M., Zhang, L., Shi, P., et al.: Sliding mode control of continuous-time Markovian jump systems with digital data transmission. Automatica **80**, 200–209 (2017)
101. Chen, B., Niu, Y., Liu, H.: Input-to-state stabilization of stochastic Markovian jump systems under communication constraints: genetic algorithm-based performance optimization. IEEE Trans. Cybern. **52**(10), 10379–10392 (2022)
102. Liu, L., Ma, L., Zhang, J., et al.: Sliding mode control for nonlinear Markovian jump systems under denial-of-service attacks. IEEE/CAA J. Autom. Sinica **7**(6), 1638–1648 (2019)
103. Chen, B., Niu, Y.: Sliding mode switched control for Markovian jumping systems subject to intermittent DoS attacks. Int. J. Robust Nonlinear Control **32**(3), 1545–1560 (2022)
104. Li, M., Chen, Y., Liu, Y.: Sliding-mode secure control for jump cyber-physical systems with malicious attacks. J. Frankl. Inst. **358**(7), 3424–3440 (2021)
105. Chen, B., Zou, Y., Niu, Y.: Dynamic event-triggered sliding mode security control for Markovian jump systems: learning-based iteration optimization method. Int. J. Robust Nonlinear Control **32**(5), 2500–2517 (2022)
106. Skorokhod, A.V.: Asymptotic Methods in the Theory of Stochastic Differential Equation. American Mathematical Society (1989)
107. Ma, H., Jia, Y.: Input-output finite-time mean square stabilisation of stochastic systems with Markovian jump. Int. J. Syst. Sci. **45**(3), 325–336 (2014)
108. Song, J., Niu, Y., Zou, Y.: Finite-time stabilization via sliding mode control. IEEE Trans. Autom. Control **62**(3), 1478–1483 (2017)
109. Song, J., Niu, Y., Zou, Y.: Finite-time sliding mode control synthesis under explicit output constraint. Automatica **65**, 111–114 (2016)
110. Chen, B., Huang, J., Niu, Y.: Sliding mode control for Markovian jumping systems with actuator nonlinearities. Int. J. Syst. Sci. **43**(4), 656–664 (2012)

Chapter 2
Finite-Time Sliding Mode Control with Restricted States and Outputs

2.1 Introduction

Markov jump systems (MJSs) have been widely applied in the fields of economics, robotics and wireless communications in recent years, since MJSs can effectively model many practice systems experiencing abrupt changes in their structures, in which the switching among various modes are dominated by a Markov process [1]. Many control methods are employed to stabilize MJSs [2–4], among which sliding mode control (SMC) has been extended to MJSs and shows effective robustness to external disturbances [6–11]. The SMC problem for Itô stochastic systems with Markov switching has been considered in [6], which has established the connections among the designed sliding surfaces corresponding to every mode. In [8], the common integral sliding surface has been chosen to remove the random switching phenomenon for sliding surface. It is worthy of noting that the aforementioned works were based on Lyapunov stability, that is, the dynamic behavior of the controlled system was considered within a sufficiently long (in principle infinite) time interval.

Actually, many practical system may be considered only during a short finite-time interval, which concerns the quantitative behavior of the state/output variables over an assigned finite-time interval [13–15]. Such attained transient performance is called finite-time boundedness (FTB), which means that the system state/output (weighted) norm does not exceed an assigned threshold during the specified time interval $[0, T]$. According to different behaviors/objectives that needs to be constrained, stochastic finite-time boundedness (SFTB) concerns the transient behavior of states for MJSs and input-output stochastic finite-time stabilization input-output stochastic finite-time stabilization (IO-SFTS) concerns the transient behavior of outputs for MJSs.

In addition, the complete knowledge on transition rates (TRs) may be expensive or time-consuming obtained, or even unavailable due to the limited techniques. Accordingly, MJSs with uncertain or unknown TRs have attracted increasing attention [10–12]. With the partially known TRs, the finite set of modes has been divided

Z. Cao and Y. Niu, *Finite-Time Sliding Mode Control for Markov Jump Systems*,
Studies in Systems, Decision and Control 603,
https://doi.org/10.1007/978-981-96-7651-4_2

into two parts in [10], i.e., the one with known TR and the other with unknown one, for which two different robust terms were introduced. In [11], the incomplete TRs have been further considered with unknown case, in which each TR could be known, completely unknown or only its boundary information was known.

In this chapter, a finite-time SMC framework is proposed for MJSs involving SFTB under state constraints and IO-SFTS under outputs constraints. Specifically, the research is divided into two parts.

- When considering state constraints, a parameter-related sliding mode controller is constructed to drive the state trajectories onto the specified integral sliding surface strictly before the assigned time T. By utilizing the proposed partition strategy, the SFTB over reaching phase and sliding motion phase are respectively analyzed during the specified finite time interval.
- When considering output constraints, the problem of incomplete TR matrix are involved. An extended definition on IO-SFTS will be considered, in which we replace the zero initial condition $x(0) = 0$ by an output constraint of the initial condition, which may effectively avoid the limitation and conservativeness arising from zero initial condition. A segmented SMC law is designed according to the diagonal elements of TRs to guarantee the reachability of the specified sliding surface. Meanwhile, the sufficient conditions are derived to ensure the IO-SFTS over the reaching phase and the sliding motion phase despite partially known TRs.

2.2 SFTB Under Restricted States

Consider the MJS defined on a complete probability space $(\Omega, \mathscr{F}, \Pr)$ as in system (1.6) without parameter uncertainties $\Delta A_i(t) \equiv 0$, in which the matched nonlinear perturbation $f_i(x(t))$ and unmatched external disturbance $w(t)$ satisfy Assumptions 1.3 and 1.4, respectively. The transition probabilities (TPs) are given as in (1.7) with the assumptions that all TRs are completely available.

For the specified finite interval $[0, T]$, the objective of this section is to propose the finite-time SMC framework of MJS (1.6) ($\Delta A_i(t) \equiv 0$) for SFTB performance, including: (1) guarantee the reachability of sliding surface in a finite time $T^* < T$; and (2) maintain SFTB for the whole time interval $[0, T]$.

2.2.1 Parameter-Related Controller Design

In this subsection, we will design a suitable controller to drive the system trajectories onto the specified sliding surface. To this end, a mode-independent sliding function is firstly designed as follows:

$$s(t) = Gx(t), \tag{2.1}$$

where matrix $G \in \mathbb{R}^{n \times n}$ is chosen such that GB_i is nonsingular for all $i \in \mathbb{S}$.

In virtue of the sliding function (2.1), we design the following SMC law:

$$u(t) = K_i x(t) - \rho_i(t)\text{sgn}(s(t)), \tag{2.2}$$

where the mode-dependent gain $K_i \in \mathbb{R}^{m \times n}$ is constant matrix to be designed later, and the robust term $\rho_i(t)$ is given as

$$\rho_i(t) \triangleq \varrho_i + \mu_i \delta + \gamma_i \|x(t)\|, \tag{2.3}$$

with

$$\mu_i \triangleq \|(GB_i)^{-1} GD_i\|, \ \gamma_i \triangleq \tilde{\gamma}_i + \|K_i\|, \tag{2.4}$$

$$\tilde{\gamma}_i \triangleq \|(GB_i)^{-1} GA_i\| + \frac{1}{2}\|\sum_{j=1}^{N} \lambda_{ij}(GB_j)^{-1}G\| + \epsilon_i, \tag{2.5}$$

and an adjustable parameter $\varrho_i > 0$ satisfying

$$\varrho_i \geq \frac{\lambda_{\max}\left((GB_i)^{-1}\right)}{T - \kappa}\|Gx(0)\| \tag{2.6}$$

with κ being a given sufficient small constant with $0 < \kappa < T$.

Remark 2.1 According to the mode-independent sliding function (2.1), the sliding surface $s(t) = 0$ is also mode-independent. A mode-independent sliding surface may avoid the switching frequently from one mode to another. Additionally, due to the mode-independence of $s(t)$, the term γ_i in (2.3) is chosen as a constant, leading to ease of implementation of the controller design.

Remark 2.2 It should be pointed out that the designed SMC law (2.2) depends on the assigned time T via the selection criterion (2.3) and (2.6). The selection (2.6) on ϱ_i in this section is different from some existing works, for example, in [7, 9], similar parameter was potentially chosen zero due to no finite-time requirement; in [6, 8], similar parameter was selected arbitrarily for ensuring the finite-time reachability. This is just the key feature of the proposed finite-time SMC method. Moreover, it should be pointed out that the scalar κ plays a key role to guarantee $T^* < T$ strictly, which means that the existence of the sliding mode phase during the interval $[T^*, T]$ can be ensured in the strict sense. This is the main difference from the related results in [16, 17] with $T^* \leq T$.

Theorem 2.1 *For a given finite time T, consider the MJS (1.6) ($\Delta A_i(t) \equiv 0$) with the sliding function (2.1). The SMC law (2.2)–(2.6) can ensure that the state trajectories arrive at the sliding surface $s(t) = 0$ before a finite time T^* with $T^* < T$ and remain there in the subsequent time.*

Proof Choose the Lyapunov function as

$$V_1(x(t), i) = s^{\mathrm{T}}(t)(GB_i)^{-1}s(t). \tag{2.7}$$

We obtain the infinitesimal operator as

$$\mathscr{L}V_1(x(t), i) = 2s^{\mathrm{T}}(t)(GB_i)^{-1}G\dot{x}(t) + s^{\mathrm{T}}(t)\left(\sum_{j=1}^{N}\lambda_{ij}(GB_j)^{-1}\right)s(t). \tag{2.8}$$

From (2.2), one further has

$$\begin{aligned}
&\mathscr{L}V_1(x(t), i)\\
\le & -2s^{\mathrm{T}}(t)\rho_i(t)\mathrm{sgn}(s(t)) + 2\|s(t)\|\|f_i(x(t))\|\\
&+ 2\|s(t)\|\left(\|(GB_i)^{-1}GA_i\| + \|K_i\|\right)\|x(t)\|\\
&+ 2\|s(t)\|\|(GB_i)^{-1}GD_i\|\|w(t)\| + s^{\mathrm{T}}(t)\left(\sum_{j=1}^{N}\lambda_{ij}(GB_j)^{-1}\right)s(t).
\end{aligned} \tag{2.9}$$

Subsisting (2.3) and (2.4) into (2.9) and noting $\|s(t)\| \le |s(t)|$, we can get

$$\begin{aligned}
&\mathscr{L}V_1(x(t), i)\\
\le & -2(\varrho_i + \mu_i\delta + \gamma_i\|x(t)\|)|s(t)| + 2\|s(t)\|\|f_i(x(t))\|\\
&+ 2\|s(t)\|\left(\|(GB_i)^{-1}GA_i\| + \|K_i\|\right)\|x(t)\|\\
&+ 2\|s(t)\|\|(GB_i)^{-1}GD_i\|\|w(t)\| + \|s(t)\|\left(\sum_{j=1}^{N}\lambda_{ij}(GB_j)^{-1}G\right)\|x(t)\|\\
\le & -2\varrho_i\|s(t)\|.
\end{aligned} \tag{2.10}$$

Accordingly, we obtain that the state trajectories will be driven onto the specified sliding surfaces $s(x(t), i) = 0$ by the SMC law (2.2)–(2.6).

The following will further give the reaching instant T^*. Recalling (2.7), one has $V_1(x(t), i) \le \lambda_{\max}\left((GB_i)^{-1}\right)\|s(t)\|^2$, Thus, we obtain

$$\mathscr{L}V_1(x(t), i) \le -\frac{2\varrho_i}{\sqrt{\lambda_{\max}\left((GB_i)^{-1}\right)}}\sqrt{V_1(x(t), i)}. \tag{2.11}$$

Integrating (2.11) from 0 to t and then taking expectation yield

$$\mathbb{E}\left\{\sqrt{V_1(x(t), i)}\right\} - \sqrt{V_1(x(0), i)} \le -\frac{\varrho_i}{\sqrt{\lambda_{\max}\left((GB_i)^{-1}\right)}}t, \tag{2.12}$$

Obviously, there exists an instant $T^* > 0$ as

$$T^* = \frac{\sqrt{\lambda_{\max}\left((GB_i)^{-1}\right)}}{\varrho_i}\sqrt{V_1(x(0), i)}. \tag{2.13}$$

such that for $t \geq T^*$, $\mathbb{E}\{V_1(x(t), i)\} = 0$, from which the reachability of sliding surface is attained (in the mean square).

From $V_1(x(0), i) \leq \lambda_{\max}\left((GB_i)^{-1}\right)\|s(0)\|^2$ and $s(0) = Gx(0)$, we can get

$$T^* \leq \frac{\lambda_{\max}\left((GB_i)^{-1}\right)}{\varrho_i}\|Gx(0)\| \leq T - \kappa < T. \tag{2.14}$$

This means that the state trajectories of MJS (1.6) ($\Delta A_i(t) \equiv 0$) can be globally driven onto the specified sliding surface $s(t) = 0$ before the finite time T^* with $T^* < T$. This completes the proof. □

2.2.2 SFTB Analysis

We shall give sufficient conditions based on linear matrix inequalities (LMIs) in the following theorem to guarantee SFTB over the whole finite-time interval $[0, T]$.

Theorem 2.2 *For MJS* (1.6) ($\Delta A_i(t) \equiv 0$), *given parameters* (c_1, c_2, T, R, δ) *and a positive scalar* η_i, *if there exist matrices* $\mathcal{P}_i > 0$, $\mathcal{Q}_i > 0$, $\mathcal{W}_i > 0$, *a real matrix* $\mathcal{L}_i$ *and positive scalars* c^*, ξ_i, ϑ ($\forall i \in \mathbb{S}$), *satisfying the following LMIs:*

$$\begin{bmatrix} \Omega_{11i} & \Omega_{12i} \\ * & \Omega_{22i} \end{bmatrix} < 0, \tag{2.15}$$

$$\begin{bmatrix} \Psi_{11i} & \Psi_{12i} \\ * & \Psi_{22i} \end{bmatrix} < 0, \tag{2.16}$$

$$\begin{bmatrix} -e^{-\eta_i T}\mathcal{Q}_i + 6mn\eta_i\mathcal{W}_i & \sqrt{6m\eta_i}\tilde{\gamma}_i\mathcal{P}_i \\ * & -I \end{bmatrix} < 0, \tag{2.17}$$

$$\begin{bmatrix} -\mathcal{W}_i & \mathcal{L}_i^{\mathrm{T}} \\ * & -I \end{bmatrix} < 0, \tag{2.18}$$

$$c_1 < c^* < c_2, \tag{2.19}$$

$$0 < \mathcal{P}_i < \vartheta R^{-1}, \tag{2.20}$$

$$-\tfrac{e^{-\eta_i T}c_2}{2}\vartheta + c^* + \delta^2 T\eta_i\vartheta < 0, \tag{2.21}$$

$$\begin{bmatrix} -\frac{1}{2}e^{-\eta_i T}c^* + 3mT\varrho_i^2\eta_i + \delta^2 T(1 + 3m\mu_i^2)\eta_i & \sqrt{c_1} \\ * & -\vartheta \end{bmatrix} < 0, \tag{2.22}$$

where

$$\begin{aligned}
&\Omega_{11i} \triangleq \mathrm{He}(\mathcal{P}_i A_i^{\mathrm{T}} + \mathcal{L}_i^{\mathrm{T}} B_i^{\mathrm{T}}) + (\lambda_{ii} - \eta_i)P_i + Q_i, \\
&\Psi_{11i} \triangleq \mathrm{He}(\mathcal{P}_i \bar{A}_i^{\mathrm{T}}) + (\lambda_{ii} - \eta_i)P_i, \\
&\Omega_{12i} \triangleq \begin{bmatrix} D_i & -B_i & \xi_i B_i & \epsilon_i \mathcal{P}_i & \mathcal{M}_i \end{bmatrix}, \ \Psi_{12i} \triangleq \begin{bmatrix} \bar{D}_i & \mathcal{M}_i \end{bmatrix}, \\
&\Omega_{22i} \triangleq -\mathrm{diag}\{\eta_i I, \eta_i I, \xi_i I, \xi_i I, \mathcal{N}\}, \ \Psi_{22i} \triangleq -\mathrm{diag}\{\eta_i I, \mathcal{N}\}, \\
&\mathcal{M}_i \triangleq [\sqrt{\lambda_{i1}}\mathcal{P}_i \cdots \sqrt{\lambda_{ij}}\mathcal{P}_i \cdots \sqrt{\lambda_{iN}}\mathcal{P}_i], \ j = 1, 2, \cdots, N, j \neq i, \\
&\mathcal{N} \triangleq \mathrm{diag}\{\mathcal{P}_1, \cdots, \mathcal{P}_j, \cdots, \mathcal{P}_N\}, \ j = 1, 2, \cdots, N, j \neq i,
\end{aligned}$$

the closed-loop system is SFTB with respect to $\aleph_{[0,T]}$, *meanwhile, the controller gains in* (2.1) *are given as* $K_i = \mathcal{L}_i \mathcal{P}_i^{-1}$, $(i \in \mathbb{S})$.

Proof The following proof includes three steps, by which the SFTB over the whole finite-time interval $[0, T]$ is attained.

► **Step 1**. SFTB over reaching phase for $t \in [0, T^*]$.
Substituting the SMC law (2.2)–(2.6) into the MJS (1.6) with $\Delta A_i(t) \equiv 0$, the closed-loop MJS can be obtained during the reaching phase $[0, T^*]$ as follows:

$$\dot{x}(t) = \tilde{A}_i x(t) + D_i w(t) + B_i f_i(x(t)) - B_i \rho_{si}(t) \tag{2.23}$$

with $\tilde{A}_i \triangleq A_i + B_i K_i$ and $\rho_{si}(t) \triangleq \rho_i(t)\mathrm{sgn}(s(t))$.

Choose the following Lyapunov function

$$V_2(x(t), i) = x^{\mathrm{T}}(t) P_i x(t) + \int_0^t x^{\mathrm{T}}(\tau) Q x(\tau) \mathrm{d}\tau. \tag{2.24}$$

Denote $\tilde{P}_i \triangleq \sum_{j=1}^N \lambda_{ij} P_j$. We obtain the infinitesimal operator as

$$\begin{aligned}
&\mathscr{L}V_2(x(t), i) \\
&= x^{\mathrm{T}}(t)(\mathrm{He}(\tilde{A}_i^{\mathrm{T}} P_i) + \tilde{P}_i + Q)x(t) + w^{\mathrm{T}}(t) D_i^{\mathrm{T}} P_i x(t) + x^{\mathrm{T}}(t) P_i D_i w(t) \\
&\quad - \rho_{si}^{\mathrm{T}}(t) B_i^{\mathrm{T}} P_i x(t) - x^{\mathrm{T}}(t) P_i B_i \rho_{si}(t) + 2x^{\mathrm{T}}(t) P_i B_i f_i(x(t)).
\end{aligned} \tag{2.25}$$

For $\xi_i > 0$, we obtain

$$2x^{\mathrm{T}}(t) P_i B_i f_i(x(t)) \leq \xi_i x^{\mathrm{T}}(t) P_i B_i B_i^{\mathrm{T}} P_i x(t) + \xi_i^{-1} \epsilon_i^2 x^{\mathrm{T}}(t) x(t). \tag{2.26}$$

Putting (2.26) into (2.25) derives

$$\begin{aligned}
&\mathscr{L}V_2(x(t), i) \\
&\leq x^{\mathrm{T}}(t) \left(\mathrm{He}(\tilde{A}_i^{\mathrm{T}} P_i) + \tilde{P}_i + Q + \xi_i P_i B_i B_i^{\mathrm{T}} P_i + \xi_i^{-1} \epsilon_i^2 I \right) x(t) \\
&\quad - \rho_{si}^{\mathrm{T}}(t) B_i^{\mathrm{T}} P_i x(t) - x^{\mathrm{T}}(t) P_i B_i \rho_{si}(t)
\end{aligned}$$

$$+ w^{\mathrm{T}}(t)D_i^{\mathrm{T}}P_ix(t) + x^{\mathrm{T}}(t)P_iD_iw(t). \tag{2.27}$$

Define an auxiliary function as follows:

$$J_1(x(t),i) = \mathscr{L}V_2(x(t),i) - \eta_iV_2(x(t),i) - \eta_iw^{\mathrm{T}}(t)w(t) - \eta_i\rho_{si}^{\mathrm{T}}(t)\rho_{si}(t). \tag{2.28}$$

Then, by Schur Complement (Lemma 1.3), it can be shown that $J_1(x(t),i) + \eta_i\int_0^t x^{\mathrm{T}}(\tau)Qx(\tau)\mathrm{d}\tau < 0$ can be ensured by the following condition:

$$\begin{bmatrix} \mathrm{He}(\tilde{A}_i^{\mathrm{T}}P_i) + Q + \tilde{P}_i - \eta_iP_i & P_iD_i & -P_iB_i & \xi_iP_iB_i & \epsilon_iI \\ * & -\eta_iI & 0 & 0 & 0 \\ * & * & -\eta_iI & 0 & 0 \\ * & * & * & -\xi_iI & 0 \\ * & * & * & * & -\xi_iI \end{bmatrix} < 0. \tag{2.29}$$

Furthermore, $J_1(x(t),i) < 0$ can be ensured by $J_1(x(t),i) + \eta_i\int_0^t x^{\mathrm{T}}(\tau)Qx(\tau)\mathrm{d}\tau < 0$, namely, $J_1(x(t),i)$ can be ensured by (2.29). Taking the expectation on the inequality $J_1(x(t),i) < 0$ yields

$$\mathbb{E}\{\mathscr{L}V_2(x(t),i)\} < \eta_i\mathbb{E}\{V_2(x(t),i)\} + \eta_iw^{\mathrm{T}}(t)w(t) + \eta_i\rho_{si}^{\mathrm{T}}(t)\rho_{si}(t). \tag{2.30}$$

According to Mean Inequality, $\rho_{si}(t)$ satisfies from (2.3)

$$\rho_{si}^{\mathrm{T}}(t)\rho_{si}(t) = m\rho_i^{\mathrm{T}}(t)\rho_i(t) \le 3m\varrho_i^2 + 3m\gamma_i^2\|x(t)\|^2 + 3m\mu_i^2\delta^2, \tag{2.31}$$

which will be used in the following derivation.

Multiplying both sides of expression (2.30) by $e^{-\eta_it}$ and integrating the obtained inequality from 0 to t with $t \in [0, T^*]$, it follows from (2.29) that

$$\begin{aligned} &e^{-\eta_it}\mathbb{E}\{V_2(x(t),i)\} \\ <&\mathbb{E}\{V_2(x(0),r_0)\} + \eta_i\int_0^t e^{-\eta_i\tau}\left(w^{\mathrm{T}}(\tau)w(\tau) + \rho_{si}^{\mathrm{T}}(\tau)\rho_{si}(\tau)\right)\mathrm{d}\tau \\ \le&\bar{\sigma}c_1 + 3mT\varrho_i^2\eta_i + \delta^2T(1+3m\mu_i^2)\eta_i + \int_0^t x^{\mathrm{T}}(\tau)\left(3m\gamma_i^2\eta_iI\right)x(\tau)\mathrm{d}\tau \end{aligned} \tag{2.32}$$

with $\bar{\sigma} \triangleq \max_{i\in\mathbb{S}}\{\lambda_{\max}(R^{-1/2}P_iR^{-1/2})\}$.

On the other hand, it follows from (2.24) that

$$\begin{aligned} e^{-\eta_it}\mathbb{E}\{V_2(x(t),i)\} \ge{}& e^{-\eta_iT}\underline{\sigma}\mathbb{E}\{x^{\mathrm{T}}(t)Rx(t)\} \\ &+ \int_0^t x^{\mathrm{T}}(\tau)(e^{-\eta_iT}Q)x(\tau)\mathrm{d}\tau \end{aligned} \tag{2.33}$$

with $\underline{\sigma} \triangleq \min_{i\in\mathbb{S}}\{\lambda_{\min}(R^{-1/2}P_iR^{-1/2})\}$.

It yields from (2.32) and (2.33) that

$$3m\gamma_i^2\eta_i I < e^{-\eta_i T} Q, \tag{2.34}$$
$$e^{-\eta_i T}\underline{\sigma}\mathbb{E}\{x^{\mathrm{T}}(t)Rx(t)\} \leq \bar{\sigma}c_1 + 3mT\varrho_i^2\eta_i + \delta^2 T(1 + 3m\mu_i^2)\eta_i. \tag{2.35}$$

Further, if there exists a scalar c^* satisfying

$$c_1 < c^* < c_2, \tag{2.36}$$
$$\frac{\bar{\sigma}c_1+3mT\varrho_i^2\eta_i+\delta^2 T(1+3m\mu_i^2)\eta_i}{\underline{\sigma}} < e^{-\eta_i T}c^*, \tag{2.37}$$

we have from (2.35) that $\mathbb{E}\{x^{\mathrm{T}}(t)Rx(t)\} < c^*$ for all $t \in [0, T^*]$. Thus, according to Definition 1.5, the closed-loop system (2.23) is SFTB with respect to $\aleph_{[0,T^*]}$. It is concluded that if the inequalities (2.29), (2.34), (2.36) and (2.37) hold, the SFTB of the closed-loop system (2.23) can be ensured with respect to $\aleph_{[0,T^*]}$.

► **Step 2**. SFTB over sliding motion phase for $t \in [T^*, T]$.
According to the SMC theory, when the system trajectories reach onto the sliding surface $s(t) = 0$, we have the following equivalent controller:

$$u_{eq}(t) = -(GB_i)^{-1}GA_i x(t) - (GB_i)^{-1}GD_i w(t) - f_i(x(t)). \tag{2.38}$$

By substituting (2.38) into system (1.6) with $\Delta A_i(t) \equiv 0$, we get the following sliding mode dynamics:

$$\dot{x}(t) = \bar{A}_i x(t) + \bar{D}_i w(t) \tag{2.39}$$

with $\bar{A}_i \triangleq (I - B_i(GB_i)^{-1}G)A_i$ and $\bar{D}_i \triangleq (I - B_i(GB_i)^{-1}G)D_i$.

Choose the following Lyapunov function

$$V_3(x(t), i) = x^{\mathrm{T}}(t)P_i x(t). \tag{2.40}$$

We have

$$\begin{aligned}\mathscr{L}V_3(x(t), i) &= x^{\mathrm{T}}(t)\left(\mathrm{He}(\bar{A}_i^{\mathrm{T}}P_i) + \tilde{P}_i\right)x(t) \\ &\quad + w^{\mathrm{T}}(t)\bar{D}_i^{\mathrm{T}}P_i x(t) + x^{\mathrm{T}}(t)P_i\bar{D}_i w(t).\end{aligned} \tag{2.41}$$

Define an auxiliary function as follows:

$$J_2(x(t), i) = \mathscr{L}V_3(x(t), i) - \eta_i V_3(x(t), i) - \eta_i w^{\mathrm{T}}(t)w(t). \tag{2.42}$$

Then, by applying Schur Complement (Lemma 1.3), it can be shown that $J_2(x(t), i) < 0$ can be ensured by the following condition:

$$\begin{bmatrix} \mathrm{He}(\bar{A}_i^{\mathrm{T}}P_i) + \tilde{P}_i - \eta_i P_i & P_i\bar{D}_i \\ * & -\eta_i I \end{bmatrix} < 0. \tag{2.43}$$

Taking the expectation of $J_2(x(t), i) < 0$ yields

$$\mathbb{E}\{\mathscr{L}V_3(x(t), i)\} < \eta_i \mathbb{E}\{V_3(x(t), i)\} + \eta_i w^{\mathrm{T}}(t)w(t). \tag{2.44}$$

It is ensured from Step 1 that the resultant closed-loop system is SFTB with respect to $\aleph_{[0,T^*]}$ over the reaching phase, which implies that the initial condition of sliding motion phase at time T^* satisfies $\mathbb{E}\{x^{\mathrm{T}}(T^*)Rx(T^*)\} < c^*$. Multiplying both sides of expression (2.44) by $e^{-\eta_i t}$ and integrating the obtained inequality from T^* to t with $t \in [T^*, T]$, it follows from (2.43) that

$$\begin{aligned} & e^{-\eta_i t}\mathbb{E}\{V_3(x(t), i)\} \\ & < \mathbb{E}\{V_3(x(T^*), r_{T^*})\} + \eta_i \int_{T^*}^{\mathrm{T}} e^{-\eta_i \tau} w^{\mathrm{T}}(\tau)w(\tau)\mathrm{d}\tau \\ & < \bar{\sigma}c^* + \delta^2 T\eta_i. \end{aligned} \tag{2.45}$$

Besides, one has

$$e^{-\eta_i t}\mathbb{E}\{V_3(x(t), i)\} \geq e^{-\eta_i T}\underline{\sigma}\mathbb{E}\{x^{\mathrm{T}}(t)Rx(t)\}. \tag{2.46}$$

Putting together (2.45) and (2.46), we have

$$\mathbb{E}\{x^{\mathrm{T}}(t)Rx(t)\} \leq \frac{\bar{\sigma}c^* + \delta^2 T\eta_i}{e^{-\eta_i T}\underline{\sigma}}. \tag{2.47}$$

Further, if there exists a scalar c^* satisfying

$$c_1 < c^* < c_2, \tag{2.48}$$

$$\frac{\bar{\sigma}c^* + \delta^2 T\eta_i}{\underline{\sigma}} < e^{-\eta_i T}c_2, \tag{2.49}$$

we have from (2.47) that $\mathbb{E}\{x^{\mathrm{T}}(t)Rx(t) < c_2$ for all $t \in [T^*, T]$. Thus, according to *Definition* 1.5, the sliding mode dynamics (2.39) are SFTB with respect to $\aleph_{[T^*,T]}$. It is concluded that if the inequalities (2.43), (2.48) and (2.49) hold, the SFTB of the sliding mode dynamics (2.39) can be ensured with respect to $\aleph_{[T^*,T]}$.

► **Step 3**. Integration of the results from Steps 1 and 2.
According to the conclusions in Steps 1 and 2, if the inequalities (2.29), (2.34), (2.36), (2.37), (2.43) and (2.49) hold, the SFTB of the overall closed-loop system (2.39) can be ensured with respect to $\aleph_{[0,T]}$.

By pre- and post-multiplying (2.29) with $\mathrm{diag}\{P_i^{-1}, I, I, I, I\}$, pre- and post-multiplying (2.43) with $\mathrm{diag}\{P_i^{-1}, I\}$ and then utilizing Schur Complement (Lemma 1.3), it is easily shown that the conditions (2.29) and (2.43) can be ensured by (2.15) and (2.16), where $\mathcal{P}_i \triangleq P_i^{-1}$, $\mathcal{Q}_i \triangleq \mathcal{P}_i Q \mathcal{P}_i$, $\mathcal{L}_i \triangleq K_i \mathcal{P}_i$.

Denote $\mathcal{W}_i \triangleq \zeta_i \mathcal{P}_i \mathcal{P}_i$ for $\zeta_i > 0$. Then, $K_i^{\mathrm{T}}K_i < \zeta_i I$ can be ensured by (2.18). Recalling $\gamma_i = \tilde{\gamma} + \|K_i\|$, the inequality (2.34) can be ensured by

$$6m\tilde{\gamma}_i^2\eta_i I + 6m\|K_i\|^2\eta_i I < e^{-\eta_i T}Q, \tag{2.50}$$

which can be further ensured by (2.17), combining with $K_i^{\mathrm{T}}K_i < \zeta_i I$. As such, the inequality (2.34) can be guaranteed by conditions (2.17) and (2.18).

By condition (2.20) and the fact, $\lambda_{\max}(R^{1/2}\mathcal{P}_i R^{1/2}) = 1/\lambda_{\min}(R^{-1/2}P_i R^{-1/2})$, one has

$$\underline{\sigma} < \frac{1}{2}, \bar{\sigma} > \frac{1}{\vartheta}.$$

Thus, it is easily shown that the inequality (2.37) can be ensured by (2.20) and (2.22), and the inequality (2.49) can be ensured by (2.20) and (2.21). This completes the proof. □

Remark 2.3 In the previous work [16] on finite-time SMC, the gain of robust term in the controller is dependent on the known constants. Differently, the corresponding gain γ_i in (2.4) is dependent on not only the known constants but also the norm of designed matrix K_i, which brings difficulty in deriving solving scheme (as shown in (2.50)). To remove this obstacle, we skillfully construct constraint condition $K_i^{\mathrm{T}}K_i < \zeta_i I$ and utilize the relationship $\|K_i\|^2 = \mathrm{tr}(K_i^{\mathrm{T}}K_i) \leq n\lambda_{\max}(K_i^{\mathrm{T}}K_i)$ to obtain solvable conditions (2.17) and (2.18).

2.2.3 Simulation

Consider a two-mode MJS (1.6) ($\Delta A_i(t) \equiv 0$) with parameters as follows:
Subsystem 1:

$$A_1 = \begin{bmatrix} -5 & -2 \\ 1 & -3 \end{bmatrix}, \quad B_1 = \begin{bmatrix} 0.4 \\ 0.5 \end{bmatrix}, \quad D_1 = \begin{bmatrix} 0.03 \\ 0.01 \end{bmatrix},$$

Subsystem 2:

$$A_2 = \begin{bmatrix} -1 & -2 \\ 0 & 0 \end{bmatrix}, \quad B_2 = \begin{bmatrix} 0.1 \\ 0.4 \end{bmatrix}, \quad D_2 = \begin{bmatrix} 0.02 \\ 0.03 \end{bmatrix},$$

and

$$G = \begin{bmatrix} 2 & 10 \end{bmatrix}, \quad R = \begin{bmatrix} 0.1 & 0 \\ 0 & 0.1 \end{bmatrix}, \quad x(0) = \begin{bmatrix} -4 \\ 3 \end{bmatrix},$$

$$f_i(x(t)) = \epsilon_i \sin\left(10\sqrt{x_1^2 + x_2^2}\right), \quad \epsilon_1 = 0.5, \ \epsilon_2 = 0.1$$

$$w(t) = \delta\cos(t), \ \delta = 0.01.$$

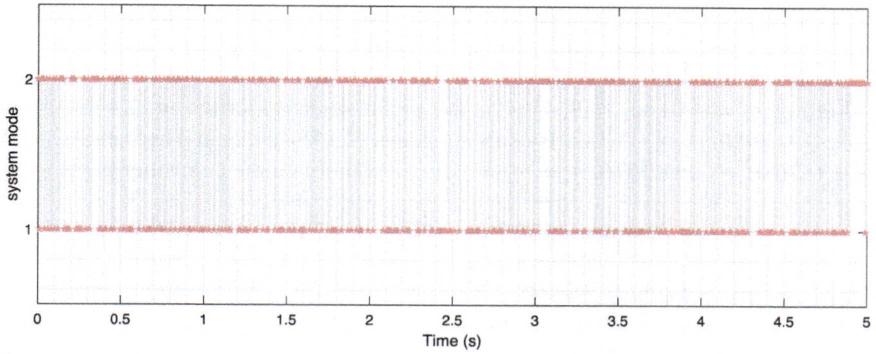

Fig. 2.1 A possible system mode transition sequence

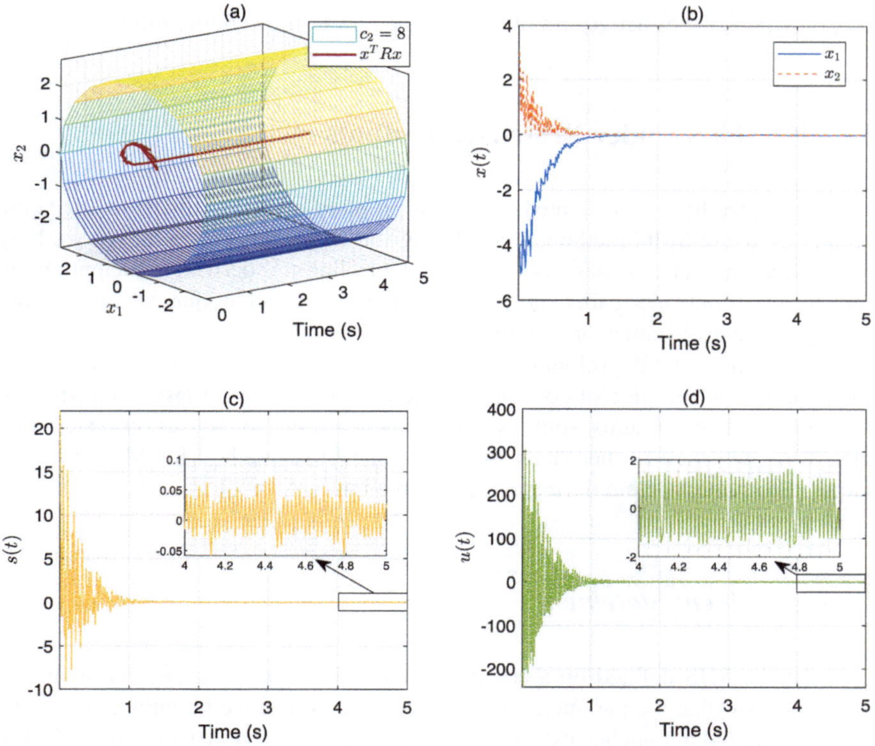

Fig. 2.2 Simulation results of controlled system: **a** evolution of weighted system state variable $x^{\mathrm{T}}(t)Rx(t)$; **b** state responses $x(t)$; **c** sliding variable $s(t)$; **d** control signal $u(t)$

The TR matrix is given as

$$\Lambda = \begin{bmatrix} -100 & 100 \\ 66.67 & -66.67 \end{bmatrix}.$$

Letting $c_1 = 3.01$, $c_2 = 8$, $T = 5$, the scalar selection criterion (2.6) is given as $\varrho_1 \geq 0.7586$, $\varrho_2 \geq 0.6387$. By solving LMIs (2.15)–(2.22) in Theorem 2.2 with $\varrho_1 = 0.7686$, $\varrho_2 = 1.0576$, $\eta_1 = \eta_2 = 0.1$, we can obtain the controller gains

$$K_1 = \begin{bmatrix} -8.4695 & -10.7738 \end{bmatrix}, \ K_2 = \begin{bmatrix} -0.7561 & -4.6656 \end{bmatrix},$$

and the segmentation boundary $c^* = 7.0909$.

The simulations are undergone by using the Euler discretization approach with initial state $x(0) = [-4\ 4]^{\mathrm{T}}$ and the stepsize as $\Delta t = 0.01$s. The simulation results are shown in Figs. 2.1 and 2.2, where Fig. 2.1 shows a possible mode transition sequence. It is exhibited in Fig. 2.2 that the weighted state $x^{\mathrm{T}}(t)Rx(t)$ does not exceed $c_2 = 8$ in closed-loop case, which means the controlled system is SFTB with respect to $\aleph_{[0,5]}$. This illustrates the effectiveness of the designed finite-time sliding mode controller.

2.3 IO-SFTS Under Restricted Outputs

In Sect. 2.2, we have investigated the SFTB problem of MJSs subject to state constraints by using SMC method. Considering the characteristics of SMC, we have designed a parameter-related sliding mode controller and derived sufficient conditions to solve controller gains via "partition–integration" technique, which provides a complete control framework for finite-time SMC.

In the above SFTB problem, the transient performance of interest are state-oriented, that is, the state of controlled system should satisfy transient constraints. Except for state constraints, some practical engineering also focuses on system outputs. Thus, we shall further study the finite-time SMC method for MJSs to solve output constraint problem during assigned time interval $[0, T]$.

2.3.1 Problem Statement

Consider the MJS defined on a complete probability space $(\Omega, \mathscr{F}, \mathrm{Pr})$ as in system (1.6), in which the parameter uncertainty $\Delta A_i(t)$, matched nonlinear perturbation $f_i(x(t))$ and unmatched external disturbance $w(t)$ satisfy Assumptions 1.2–1.4, respectively.

In Sect. 2.2, the TRs are assumed to be completely known, which places great demands on the probability statistics mechanism. In many actual scenarios, the TRs might not be completely available. In this section, the TRs of the jumping process $\{r_t\}_{t\geq 0}$ are further considered to be partially known with TPs generated by (1.7).

In partially known TR matrix Λ, there are three types of elements: (1) exactly known elements; (2) uncertain elements with known lower and upper bounds; and (3) completely unknown elements. For instance, for system (1.6) with three operation modes, the TR matrix may be described by

$$\Lambda = \begin{bmatrix} \lambda_{11} & ? & ? \\ ? & ? & \lambda_{23} \\ ? & \varsigma_{32} & \varsigma_{33} \end{bmatrix},$$

where λ_{ij} $(i, j = 1, 2, 3)$ and "?" represent the completely known elements and completely unknown elements, respectively, and ς_{ij} $(i, j = 1, 2, 3)$ denote the elements with known lower and upper bounds ($\underline{\varsigma}_{ij} \le \varsigma_{ij} \le \bar{\varsigma}_{ij}$). Accordingly, the elements in the ith row of matrix Λ may be divided into the following three aspects:

$$\mathbb{S}_{k1}^i \triangleq \{j : \lambda_{ij} \text{ is exactly known}\},$$
$$\mathbb{S}_{k2}^i \triangleq \{j : \lambda_{ij} \text{ is uncertain with known lower and upper bounds}\},$$
$$\mathbb{S}_{uk}^i \triangleq \{j : \text{ no information available for } \lambda_{ij}\}.$$

Apparently, the completely known elements may be treated as a special case of boundary known ones. Define $\mathbb{S}_k^i \triangleq \mathbb{S}_{k1}^i + \mathbb{S}_{k2}^i = \{j : \underline{\lambda}_{ij} \le \lambda_{ij} \le \bar{\lambda}_{ij}\}$, then $\mathbb{S} = \mathbb{S}_k^i + \mathbb{S}_{uk}^i$. If $\mathbb{S}_k^i \neq \emptyset$, it can be further described as

$$\mathbb{S}_k^i \triangleq \{\mathcal{K}_1^i, \mathcal{K}_2^i, \ldots, \mathcal{K}_l^i, \ldots, \mathcal{K}_{m_i}^i\},$$

where $\mathcal{K}_l^i$ $(l = 1, 2, \ldots, m_i)$ denotes the site of the lth known element in the ith row of matrix Λ, and m_i represents the number of known elements in the ith row of matrix Λ.

Considering the improved definition of IO-SFTS during the given interval $[0, T]$, the aim of this section is to propose a finite-time SMC framework of the system (1.6) for IO-SFTS performance including: (1) design an appropriate sliding mode controller to achieve the reachability in a finite time $T^* < T$; (2) present a reliable solving algorithm to ensure the IO-SFTS during both reaching phase within $[0, T^*]$ and sliding motion phase within $[T^*, T]$ despite incomplete TRs and parameter uncertainties.

2.3.2 Segmented Controller Design

First of all, some abbreviations are introduced for convenience as $\lambda_k^i = -\sum_{j\in\mathbb{S}_k^i} \lambda_{ij}$, $\underline{\lambda}_k^i = -\sum_{j\in\mathbb{S}_k^i} \underline{\lambda}_{ij}$. Then, we construct a mode-dependent sliding function as follows:

$$s(x(t), i) = G_i x(t), \tag{2.51}$$

with G_i being chosen such that $G_i B_i$ is nonsingular for each $i \in \mathbb{S}$, which can be attained by choosing $G_i = B_i^{\mathrm{T}} X_i$ with $X_i > 0$ due to the full column rank of B_i.

Remark 2.4 It should be pointed out that the constructed sliding function (2.1) in Sect. 2.1 is mode-independent while the sliding function (2.51) is mode-dependent. The sliding gain is a key factor for control performance implementation, and thus to discretionarily choose an common sliding gain for all modes is quite hard for designers/engineers. Although a mode-independent sliding surface may avoid the switching frequently from one mode to another, it may also reduce the freedom of control design. In order to remove this obstacle, the mode-dependent sliding gains $G_i = B_i^{\mathrm{T}} X_i$ $(i \in \mathbb{S})$ are considered in this section, which can be determined by choosing $X_i > 0$ $(i \in \mathbb{S})$.

For the specified finite-time interval $[0, T]$, a suitable sliding mode controller should be designed to drive the states trajectories onto the specified sliding surfaces $s(x(t), i) = 0$ before a finite instant T^* with $T^* < T$ and then are maintained there during the interval $[T^*, T]$. To this end, we design the following piecewise SMC law:

$$u(t) = \begin{cases} K_i x(t) - \rho_i(t)\mathrm{sgn}(s(x(t), i)), & s(x(t), i) \neq 0 \\ K_i x(t), & s(x(t), i) = 0, \end{cases} \tag{2.52}$$

with

$$\rho_i(t) \triangleq \varrho_i + \gamma_i \|x(t)\| + \mu_i \delta + \frac{\vartheta_i \|x(t)\|^2}{2\|s(x(t), i)\|}, \tag{2.53}$$

$$\gamma_i \triangleq \|K_i\| + \tilde{\gamma}_i, \tag{2.54}$$

$$\tilde{\gamma}_i \triangleq \|(G_i B_i)^{-1} G_i M_i\| \|N_i\| + \|(G_i B_i)^{-1} G_i A_i\| + \epsilon_i, \tag{2.55}$$

$$\mu_i \triangleq \|(G_i B_i)^{-1} G_i D_i\|, \tag{2.56}$$

$$\vartheta_i \triangleq \begin{cases} \| \sum_{j \in \mathbb{S}_k^i} \bar{\lambda}_{ij} G_j^{\mathrm{T}} (G_j B_j)^{-1} G_j \| + \|\underline{\lambda}_k^i G_j^{\mathrm{T}} (G_j B_j)^{-1} G_j\|, i \in \mathbb{S}_k^i, j \in \mathbb{S}_{uk}^i, \\ \| \sum_{j \in \mathbb{S}_k^i} \bar{\lambda}_{ij} G_j^{\mathrm{T}} (G_j B_j)^{-1} G_j \| + \|\underline{\lambda}_k^i X_i\|, i \in \mathbb{S}_{uk}^i, j \in \mathbb{S}_{uk}^i, j \neq i, \end{cases} \tag{2.57}$$

$$\varrho_i \geq \frac{\lambda_{\max}[(G_i B_i)^{-1}]}{T - \kappa} \|G_i x(0)\|, \tag{2.58}$$

where $\varrho_i > 0$ is a adjustable parameter dependent on assigned time T, κ is a given sufficient small constant with $0 < \kappa < T$, and $\mathcal{X}_i$ will be designed in Theorem 2.3.

In the sequel, it will be shown that the SMC law (2.52)–(2.58) can ensure the reachability of the specified sliding surfaces $s(x(t), i) = 0$ during the interval $[0, T^*]$ with $T^* < T$.

Theorem 2.3 *For a given finite time $T > 0$, if exist matrices $\mathcal{X}_i > 0$ satisfying the following inequalities:*

$$G_i^{\mathrm{T}} (G_i B_i)^{-1} G_i \geq \mathcal{X}_i, i \in \mathbb{S}_{uk}^i, \tag{2.59}$$

$$G_j^{\mathrm{T}} (G_j B_j)^{-1} G_j \leq \mathcal{X}_i, j \in \mathbb{S}_{uk}^i, i \neq j. \tag{2.60}$$

the state trajectories of the MJS (1.6) *can be driven onto the sliding surface* $s(x(t), i) = 0$ *by the SMC law* (2.52)–(2.58) *during the interval* $[0, T^*]$ *with* $T^* < T$ *and remain there in the subsequent time.*

Proof Choose the Lyapunov function as

$$V_1(x(t), i) = s^{\mathrm{T}}(x(t), i)(G_i B_i)^{-1} s(x(t), i). \tag{2.61}$$

Then, we obtain the infinitesimal operator as

$$\begin{aligned}
&\mathscr{L}V_1(x(t), i)\\
&= 2s^{\mathrm{T}}(x(t), i)(G_i B_i)^{-1} G_i(A_i + \Delta A_i(t))x(t) + 2s^{\mathrm{T}}(x(t), i)u(t)\\
&\quad + 2s^{\mathrm{T}}(x(t), i) f_i(x(t)) + 2s^{\mathrm{T}}(x(t), i)(G_i B_i)^{-1} G_i D_i w(t)\\
&\quad + \sum_{j\in\mathbb{S}} \lambda_{ij} s^{\mathrm{T}}(x(t), j)(G_j B_j)^{-1} s(x(t), j)\\
&= 2s^{\mathrm{T}}(x(t), i)K_i x(t) + 2s^{\mathrm{T}}(x(t), i)(G_i B_i)^{-1} G_i(A_i + \Delta A_i(t))x(t)\\
&\quad - 2s^{\mathrm{T}}(x(t), i)\rho_i(t)\mathrm{sgn}(s(x(t), i)) + 2s^{\mathrm{T}}(x(t), i) f_i(x(t))\\
&\quad + 2s^{\mathrm{T}}(x(t), i)(G_i B_i)^{-1} G_i D_i w(t)\\
&\quad + x^{\mathrm{T}}(t)\Big(\sum_{j\in\mathbb{S}} \lambda_{ij} G_j^{\mathrm{T}}(G_j B_j)^{-1} G_j\Big)x(t).
\end{aligned} \tag{2.62}$$

From (2.62), we can see that the term $x^{\mathrm{T}}(t)\Big(\sum_{j\in\mathbb{S}} \lambda_{ij} G_j^{\mathrm{T}}(G_j B_j)^{-1} G_j\Big)x(t)$ is unknown due to some unavailable TRs λ_{ij}. We shall consider two possible cases: $i \in \mathbb{S}_k^i$ and $i \in \mathbb{S}_{uk}^i$, respectively, that is, the diagonal element known or unknown.

► **Case 1**. The diagonal element is known ($i \in \mathbb{S}_k^i$).
In this case, due to the following fact

$$\lambda_k^i > 0, \quad \sum_{j\in\mathbb{S}_{uk}^i} \frac{\lambda_{ij}}{\lambda_k^i} = 1, \tag{2.63}$$

we have

$$\begin{aligned}
&\sum_{j\in\mathbb{S}} \lambda_{ij} G_j^{\mathrm{T}}(G_j B_j)^{-1} G_j\\
&= \sum_{j\in\mathbb{S}_k^i} \lambda_{ij} G_j^{\mathrm{T}}(G_j B_j)^{-1} G_j + \sum_{j\in\mathbb{S}_{uk}^i} \lambda_{ij} G_j^{\mathrm{T}}(G_j B_j)^{-1} G_j\\
&\le \sum_{j\in\mathbb{S}_k^i} \bar{\lambda}_{ij} G_j^{\mathrm{T}}(G_j B_j)^{-1} G_j + \lambda_k^i \sum_{j\in\mathbb{S}_{uk}^i} \frac{\lambda_{ij}}{\lambda_k^i} G_j^{\mathrm{T}}(G_j B_j)^{-1} G_j.
\end{aligned} \tag{2.64}$$

Putting (2.64) into (2.62), one can obtain

$$\mathscr{L}V_1(x(t), i)$$

$$
\begin{aligned}
&\leq \sum_{j\in\mathbb{S}^i_{uk}} \frac{\lambda_{ij}}{\lambda^i_k}\Big(2s^{\mathrm{T}}(x(t),i)K_i x(t) + 2s^{\mathrm{T}}(x(t),i)(G_iB_i)^{-1}G_iM_iF_i(t)N_ix(t) \\
&- 2s^{\mathrm{T}}(x(t),i)\rho_i(t)\mathrm{sgn}(s(x(t),i)) + 2s^{\mathrm{T}}(x(t),i)f_i(x(t)) \\
&+ 2s^{\mathrm{T}}(x(t),i)(G_iB_i)^{-1}G_iD_iw(t) + x^{\mathrm{T}}(t)\Big(\sum_{j\in\mathbb{S}^i_k}\bar{\lambda}_{ij}G_j^{\mathrm{T}}(G_jB_j)^{-1}G_j\Big)x(t) \\
&+ x^{\mathrm{T}}(t)(\underline{\lambda}^i_k G_j^{\mathrm{T}}(G_jB_j)^{-1}G_j)x(t)\Big).
\end{aligned}
\tag{2.65}
$$

Then, substituting the control law (2.52)–(2.58) into (2.65) and noting $\|s(x(t),i)\| \leq |s(x(t),i)|$, we have

$$
\begin{aligned}
&\mathscr{L}V_1(x(t),i) \\
&\leq \sum_{j\in\mathbb{S}^i_{uk}} \frac{\lambda_{ij}}{\lambda^i_k}\Big(2\|s(x(t),i)\|\Big(\|K_i\|\|x(t)\| + \|(G_iB_i)^{-1}G_iM_i\|\|N_i\|\|x(t)\| \\
&+ \epsilon_i\|x(t)\| + \|(G_iB_i)^{-1}G_iD_i\|\|w(t)\|\Big) + \|x(t)\|\Big\|\sum_{j\in\mathbb{S}^i_k}\bar{\lambda}_{ij}G_j^{\mathrm{T}}(G_jB_j)^{-1}G_j\Big\|\|x(t)\| \\
&+ \|x(t)\|\|\underline{\lambda}^i_k G_j^{\mathrm{T}}(G_jB_j)^{-1}G_j\|\|x(t)\| - 2\|s(x(t),i)\|\rho_i(t)\Big) \\
&\leq -2\varrho_i\|s(x(t),i)\| < 0.
\end{aligned}
\tag{2.66}
$$

► **Case 2**. The diagonal element is unknown ($i \in \mathbb{S}^i_{uk}$).
In this case, if there exist a positive definite matrix $\mathcal{X}_i$ satisfying (2.59) and (2.60), noting the fact that $\lambda^i_k = \lambda_{ii} + \sum_{j\in\mathbb{S}^i_{uk}, j\neq i}\lambda_{ij} < 0$, one has

$$
\begin{aligned}
&\sum_{j\in\mathbb{S}}\lambda_{ij}G_j^{\mathrm{T}}(G_jB_j)^{-1}G_j \\
&= \sum_{j\in\mathbb{S}^i_k}\lambda_{ij}G_j^{\mathrm{T}}(G_jB_j)^{-1}G_j + \lambda_{ii}G_i^{\mathrm{T}}(G_iB_i)^{-1}G_i \\
&\quad + \sum_{j\in\mathbb{S}^i_{uk}, j\neq i}\lambda_{ij}G_j^{\mathrm{T}}(G_jB_j)^{-1}G_j \\
&\leq \sum_{j\in\mathbb{S}^i_k}\bar{\lambda}_{ij}G_j^{\mathrm{T}}(G_jB_j)^{-1}G_j + \lambda_{ii}\mathcal{X}_i + \sum_{j\in\mathbb{S}^i_{uk}, j\neq i}\lambda_{ij}\mathcal{X}_i \\
&\leq \sum_{j\in\mathbb{S}^i_k}\bar{\lambda}_{ij}G_j^{\mathrm{T}}(G_jB_j)^{-1}G_j + \underline{\lambda}^i_k\mathcal{X}_i,
\end{aligned}
\tag{2.67}
$$

Putting (2.67) into (2.62) derives

$$
\begin{aligned}
&\mathscr{L}V_1(x(t),i)\\
\le&2s^{\mathrm{T}}(x(t),i)K_ix(t)+2s^{\mathrm{T}}(x(t),i)(G_iB_i)^{-1}G_iM_iF_i(t)N_ix(t)\\
&-2s^{\mathrm{T}}(x(t),i)\rho_i(t)\mathrm{sgn}(s(x(t),i))+2s^{\mathrm{T}}(x(t),i)f_i(x(t))\\
&+2s^{\mathrm{T}}(x(t),i)(G_iB_i)^{-1}G_iD_iw(t)\\
&+x^{\mathrm{T}}(t)\Big(\sum_{j\in\mathbb{S}_k^i}\bar{\lambda}_{ij}G_j^{\mathrm{T}}(G_jB_j)^{-1}G_j\Big)x(t)+x^{\mathrm{T}}(t)(\underline{\lambda}_k^iX_i)x(t).
\end{aligned}
\tag{2.68}
$$

Similar to the derivation procedure in **Case 1**, we have

$$
\mathscr{L}V_1(x(t),i)\le-2\varrho_i\|s(x(t),i)\|<0,
\tag{2.69}
$$

which, together with (2.66) means that the state trajectories can be driven onto the specified switching surfaces $s(x(t),i)=0$ by the SMC law (2.52)–(2.58) for all $i\in\mathbb{S}$.

The following will further derive the reaching instant T^*. We obtain from (2.61):

$$
V_1(x(t),i)\le\lambda_{\max}((G_iB_i)^{-1})\|s(x(t),i)\|^2,
\tag{2.70}
$$

which, together with (2.69), yields:

$$
\mathscr{L}V_1(x(t),i)\le-\frac{2\varrho_i}{\sqrt{\lambda_{\max}((G_iB_i)^{-1})}}\sqrt{V_1(x(t),i)}.
\tag{2.71}
$$

Integrating (2.71) from 0 to t and then taking the expectation give

$$
\mathbb{E}\left\{\sqrt{V_1(x(t),i)}\right\}-\sqrt{V_1(x(0),i)}\le-\varrho_it.
\tag{2.72}
$$

Thus, it is obtained that there exists an instant T^* satisfying

$$
T^*=\frac{\sqrt{\lambda_{\max}((G_iB_i)^{-1})}}{\varrho_i}\sqrt{V_1(x(0),i)},
\tag{2.73}
$$

such that for $t\ge T^*$, $\mathbb{E}\{V_1(x(t),i)\}=0$, consequently, the sliding surface $s(x(t),i)=0$ is attained (in the mean square).

Note that one has from (2.70)

$$
T^*\le\frac{\lambda_{\max}((G_iB_i)^{-1})}{\varrho_i}\|s(x(0),i)\|,
\tag{2.74}
$$

which, together with the fact $s(x(0), i) = G_i x(0)$, yields

$$T^* \leq \frac{\lambda_{\max}((G_i B_i)^{-1})}{\varrho_i} \|G_i x(0)\|. \tag{2.75}$$

For the rule (2.58) on ϱ_i, we can ensure $T^* < T$, which means that the state trajectories of system (1.6) will be driven onto the specified sliding surface $s(x(t), i) = 0$ in finite time T^* with $T^* < T$. This completes the proof. □

Remark 2.5 It is noted that the sliding variables $s(x(t), i)$ $(i \in \mathbb{S})$ are mode-dependent, which will make the weak infinitesimal of $V_1(x(t), i)$ in (2.61) different from the one of the function as (2.7) with mode-independent variable $s(t)$. In fact, its key feature for $\mathscr{L}V_1(x(t), i)$ is the last term in (2.62), which was simply dealt in most existing works as [18, 19] with the expression $s^T(x(t), i) \sum_{j\in\mathbb{S}} \lambda_{ij}(G_j B_j)^{-1} s(x(t), i)$. In the above proof, we modify this derivation on reachability of the mode-dependent sliding surface, which provides an improvement of SMC theory for MJSs.

2.3.3 IO-SFTS Analysis

Similar to the SFTB analysis in Sect. 2.2, there are two phases for the output trajectories of SMC systems: the reaching phase within $[0, T^*]$ and the sliding motion phase within $[T^*, T]$. Accordingly, we still adopt "partition–integration" technique to analyze the IO-SFTS of the controlled system. For convenience, some abbreviations are introduced as follows $P_k^i = \sum_{j\in\mathbb{S}_k^i, j\neq i} \lambda_{ij} P_j$, $\bar{P}_k^i = \sum_{j\in\mathbb{S}_k^i, j\neq i} \bar{\lambda}_{ij} P_j$.

(1) **IO-SFTS over Reaching Phase within** $[0, T^*]$.
During the reaching phase within $[0, T^*]$, the system states are outside of the sliding surface, i.e., $s(x(t), i) \neq 0$. By substituting SMC law (2.52) into the MJS (1.6), we obtain the following closed-loop system:

$$\begin{cases} \dot{x}(t) = (\bar{A}_i + \Delta A_i(t))x(t) - B_i \rho_{si}(t) + B_i f_i(x(t)) + D_i w(t), \\ y(t) = C_i x(t), \end{cases} \tag{2.76}$$

where $\bar{A}_i \triangleq A_i + B_i K_i$ and $\rho_{si}(t) \triangleq \rho_i(t)\text{sgn}(s(x(t), i))$.

The following gives the sufficient condition of the IO-FTS with respect to $\Re_{[0,T^*]}$.

Theorem 2.4 *For the MJS* (2.76) *with the given parameters* $(\beta_1, \beta_2, T, R, \delta)$ *and a feasible scalar* $\varepsilon_i > 0$, *if there exist positive constants* ξ_i, ζ_i, β^* *and matrices* $P_i > 0$, $Q > 0$ *and* $W_i > 0$ $(\forall i \in \mathbb{S})$, *satisfying the following matrix inequalities:*

$$i \in \mathbb{S}_k^i, \begin{bmatrix} \Phi_{1i} & \Omega_{12i} \\ * & \Omega_{22i} \end{bmatrix} < 0, \tag{2.77}$$

$$i \in \mathbb{S}_{uk}^i, \begin{cases} \begin{bmatrix} \Phi_{2i} & \Omega_{12i} \\ * & \Omega_{22i} \end{bmatrix} < 0, \\ W_i < P_i, \ P_j < W_i, \ j \in \mathbb{S}_{uk}^i, \ j \neq i, \end{cases} \tag{2.78}$$

$$4\varepsilon_i \left(\gamma_i^2 + \frac{\vartheta_i^2}{tr(G_i^{\mathrm{T}} G_i)} \right) I < e^{-\varepsilon_i T} Q, \tag{2.79}$$

$$\beta_1 < \beta^* < \beta_2, \tag{2.80}$$

$$\frac{\bar{\varkappa}\beta_1 + \varepsilon_i \delta^2 T(1 + 4\mu_i^2) + 4\varepsilon_i \varrho_i^2 T}{\underline{\varkappa}} < e^{-\varepsilon_i T} \beta^*, \tag{2.81}$$

where

$$\Phi_{1i} \triangleq \begin{bmatrix} \mathrm{He}(P_i \bar{A}_i) + Q - \varepsilon_i P_i + \bar{P}_k^i + \bar{\lambda}_{ii} P_i + \underline{\lambda}_k^i P_j & P_i D_i & -P_i B_i \\ * & -\varepsilon_i I & 0 \\ * & * & -\varepsilon_i I \end{bmatrix}, \ j \in \mathbb{S}_{uk}^i,$$

$$\Phi_{2i} \triangleq \begin{bmatrix} \mathrm{He}(P_i \bar{A}_i) + Q - \varepsilon_i P_i + \bar{P}_k^i + \underline{\lambda}_k^i W_i & P_i D_i & -P_i B_i \\ * & -\varepsilon_i I & 0 \\ * & * & -\varepsilon_i I \end{bmatrix},$$

$$\Omega_{12i} \triangleq \begin{bmatrix} \xi_i P_i M_i & N_i^{\mathrm{T}} & \zeta_i P_i B_i & \alpha_i I \\ 0 & 0 & 0 & 0 \\ 0 & 0 & 0 & 0 \end{bmatrix}, \quad \Omega_{22i} \triangleq -\mathrm{diag}\{\xi_i I, \xi_i I, \zeta_i I, \zeta_i I\},$$

$$\bar{\varkappa} \triangleq \max_{i \in \mathbb{S}} \left\{ \lambda_{\max} \left((C_i^{\mathrm{T}} R C_i)^{-\frac{1}{2}} P_i (C_i^{\mathrm{T}} R C_i)^{-\frac{1}{2}} \right) \right\},$$

$$\underline{\varkappa} \triangleq \min_{i \in \mathbb{S}} \left\{ \lambda_{\min} \left((C_i^{\mathrm{T}} R C_i)^{-\frac{1}{2}} P_i (C_i^{\mathrm{T}} R C_i)^{-\frac{1}{2}} \right) \right\}.$$

the closed-loop system (2.76) *is IO-SFTS with respect to* $\Re_{[0,T^*]}$.

Proof For the following Lyapunov functional candidate:

$$V_2(x(t), i) = x^{\mathrm{T}}(t) P_i x(t) + \int_0^t x^{\mathrm{T}}(\tau) Q x(\tau) \mathrm{d}\tau, \tag{2.82}$$

we obtain its infinitesimal operator as

$$\begin{aligned} &\mathscr{L} V_2(x(t), i) \\ &= \dot{x}^{\mathrm{T}}(t) P_i x(t) + x^{\mathrm{T}}(t) P_i \dot{x}(t) + x^{\mathrm{T}}(t) \sum_{j \in \mathbb{S}} \lambda_{ij} P_j x(t) + x^{\mathrm{T}}(t) Q x(t) \\ &= x^{\mathrm{T}}(t) \left(\bar{A}_i^{\mathrm{T}} P_i + P_i \bar{A}_i + \sum_{j \in \mathbb{S}} \lambda_{ij} P_j + Q \right) x(t) + 2x^{\mathrm{T}}(t) P_i B_i f_i(x(t)) \\ &\quad + 2x^{\mathrm{T}}(t) P_i \Delta A_i(t) x(t) + w^{\mathrm{T}}(t) D_i^{\mathrm{T}} P_i x(t) + x^{\mathrm{T}}(t) P_i D_i w(t) \\ &\quad - \rho_{si}^{\mathrm{T}}(t) B_i^{\mathrm{T}} P_i x(t) - x^{\mathrm{T}}(t) P_i B_i \rho_{si}(t). \end{aligned} \tag{2.83}$$

For $\xi_i > 0$ and $\zeta_i > 0$, one has

$$2x^{\mathrm{T}}(t)P_i\Delta A_i(t)x(t) \le \xi_i x^{\mathrm{T}}(t)P_iM_iM_i^{\mathrm{T}}P_ix(t) + \xi_i^{-1}x^{\mathrm{T}}(t)N_i^{\mathrm{T}}N_ix(t), \quad (2.84)$$
$$2x^{\mathrm{T}}(t)P_iB_if(x(t),t,i) \le \zeta_i x^{\mathrm{T}}(t)P_iB_iB_i^{\mathrm{T}}P_ix(t) + \zeta_i^{-1}\epsilon_i^2x^{\mathrm{T}}(t)x(t). \quad (2.85)$$

Then, we obtain

$$\begin{aligned}
&\mathscr{L}V_2(x(t),i)\\
\le& x^{\mathrm{T}}(t)(\mathrm{He}(P_i\bar{A}_i) + \sum_{j\in\mathbb{S}}\lambda_{ij}P_j + Q + \xi_iP_iM_iM_i^{\mathrm{T}}P_i + \xi_i^{-1}N_i^{\mathrm{T}}N_i\\
&+ \zeta_iP_iB_iB_i^{\mathrm{T}}P_i + \zeta_i^{-1}\epsilon_i^2I)x(t) - \rho_{si}^{\mathrm{T}}(t)B_i^{\mathrm{T}}P_ix(t) - x^{\mathrm{T}}(t)P_iB_i\rho_{si}(t)\\
&+ w^{\mathrm{T}}(t)D_i^{\mathrm{T}}P_ix(t) + x^{\mathrm{T}}(t)P_iD_iw(t). \quad (2.86)
\end{aligned}$$

Define an auxiliary function as follows:

$$J_1(x(t),i) = \mathscr{L}V_2(x(t),i) - \varepsilon_iV_2(x(t),i) - \varepsilon_iw^{\mathrm{T}}(t)w(t) - \varepsilon_i\rho_{si}^{\mathrm{T}}(t)\rho_{si}(t). \quad (2.87)$$

It is easily shown that $J_1(x(t),i) + \int_0^{\mathrm{T}} x^{\mathrm{T}}(\tau)Qx(\tau)\mathrm{d}\tau < 0$ can be ensured by the following inequality:

$$\begin{bmatrix}
\Psi_i & P_iD_i & -P_iB_i & \xi_iP_iM_i & N_i^{\mathrm{T}} & \zeta_iP_iB_i & \alpha_iI\\
* & -\varepsilon_iI & 0 & 0 & 0 & 0 & 0\\
* & * & -\varepsilon_iI & 0 & 0 & 0 & 0\\
* & * & * & -\xi_iI & 0 & 0 & 0\\
* & * & * & * & -\xi_{1i}I & 0 & 0\\
* & * & * & * & * & -\zeta_iI & 0\\
* & * & * & * & * & * & -\zeta_iI
\end{bmatrix} < 0, \quad (2.88)$$

where $\Psi_i \triangleq \mathrm{He}(P_i\bar{A}_i) + \sum_{j\in\mathbb{S}}\lambda_{ij}P_j + Q - \varepsilon_iP_i$.

We can see that the term $\sum_{j\in\mathbb{S}}\lambda_{ij}P_j$ in above matrix Ψ_i can not be calculated directly due to some unavailable TRs λ_{ij}. In order to obtain a feasible condition, we shall consider two possible cases: $i \in \mathbb{S}_k^i$ and $i \in \mathbb{S}_{uk}^i$, respectively, that is, the diagonal element known or unknown.

► **Case 1**. The diagonal element is known ($i \in \mathbb{S}_k^i$).
In this case, due to the fact in (2.63) we have

$$\begin{aligned}
\sum_{j\in\mathbb{S}}\lambda_{ij}P_j &= \sum_{j\in\mathbb{S}_k^i}\lambda_{ij}P_j + \sum_{j\in\mathbb{S}_{uk}^i}\lambda_{ij}P_j = P_k^i + \lambda_{ii}P_i + \lambda_k^i\sum_{j\in\mathbb{S}_{uk}^i}\frac{\lambda_{ij}}{\lambda_k^i}P_j\\
&\le \bar{P}_k^i + \bar{\lambda}_{ii}P_i + \underline{\lambda}_k^i\sum_{j\in\mathbb{S}_{uk}^i}\frac{\lambda_{ij}}{\lambda_k^i}P_j. \quad (2.89)
\end{aligned}$$

According to (2.88), one has

$$\sum_{j\in\mathbb{S}^i_{uk}} \frac{\lambda_{ij}}{\lambda^i_k} \left[\begin{array}{ccc|c} \mathrm{He}(P_i\bar{A}_i) + Q - \varepsilon_i P_i + \bar{P}^i_k + \bar{\lambda}_{ii} P_i + \underline{\lambda}^i_k P_j & P_i D_i & -P_i B_i & \\ * & -\varepsilon_i I & 0 & \Omega_{12i} \\ * & * & -\varepsilon_i I & \\ \hline & * & & \Omega_{22i} \end{array}\right] < 0, \tag{2.90}$$

which can be ensured by (2.77).

► **Case 2**. The diagonal element is unknown ($i \in \mathbb{S}^i_{uk}$).
In this case, if there exist a positive definite matrix W_i satisfying (2.78), noting $\lambda^i_k = \lambda_{ii} + \sum_{j\in\mathbb{S}^i_{uk}, j\neq i} \lambda_{ij} < 0$, one has

$$\begin{aligned} \sum_{j\in\mathbb{S}} \lambda_{ij} P_j &= \sum_{j\in\mathbb{S}^i_k} \lambda_{ij} P_j + \lambda_{ii} P_i + \sum_{j\in\mathbb{S}^i_{uk}, j\neq i} \lambda_{ij} P_j \\ &< P^i_k + \lambda_{ii} W_i + \sum_{j\in\mathbb{S}^i_{uk}, j\neq i} \lambda_{ij} W_i \\ &\leq P^i_k + \underline{\lambda}^i_k W_i, \end{aligned} \tag{2.91}$$

By means of expression (2.88), we have

$$\left[\begin{array}{ccc|c} \mathrm{He}(P_i\bar{A}_i) + Q - \varepsilon_i P_i + \bar{P}^i_k + \underline{\lambda}^i_k W_i & P_i D_i & -P_i B_i & \\ * & -\varepsilon_i I & 0 & \Omega_{12i} \\ * & * & -\varepsilon_i I & \\ \hline & * & & \Omega_{22i} \end{array}\right] < 0, \tag{2.92}$$

which can be ensured by the first condition of (2.78).

From Cases 1 and 2, we can conclude that (2.88) can be ensured by (2.77) or (2.78). Moreover, taking the expectation of $J_1(x(t), i) < 0$ yields

$$\mathbb{E}\{\mathscr{L}V_2(x(t), i)\} < \varepsilon_i \mathbb{E}\{V_2(x(t), i)\} + \varepsilon_i w^{\mathrm{T}}(t) w(t) + \varepsilon_i \rho^{\mathrm{T}}_{si}(t) \rho_{si}(t). \tag{2.93}$$

Besides, according to the following expressions

$$\|x(t)\|^2 = x^{\mathrm{T}}(t)x(t) = tr(x^{\mathrm{T}}(t)x(t)), \tag{2.94}$$

$$\begin{aligned} \|s(x(t), i)\|^2 &= s^{\mathrm{T}}(x(t), i)s(x(t), i) \\ &= tr(x^{\mathrm{T}}(t)G^{\mathrm{T}}_i G_i x(t)) \\ &= tr(x(t)x^{\mathrm{T}}(t))tr(G^{\mathrm{T}}_i G_i), \end{aligned} \tag{2.95}$$

it can obtained that $\rho_{si}(t)$ satisfies the following inequality:

$$\begin{aligned}\rho_{si}^{\mathrm{T}}(t)\rho_{si}(t) &= m\rho_i^{\mathrm{T}}(t)\rho_i(t) \leq 4m\varrho_i^2 + 4m\gamma_i^2\|x(t)\|^2 \\ &\quad + 4m\mu_i^2\delta^2 + \frac{4m\vartheta_i^2}{tr(G_i^{\mathrm{T}}G_i)}\|x(t)\|^2. \end{aligned} \tag{2.96}$$

Multiplying both sides of expression (2.93) by $e^{-\varepsilon_i t}$ and integrating the obtained inequality from 0 to t with $t \in [0, T^*]$, we have

$$\begin{aligned} &e^{-\varepsilon_i t}\mathbb{E}\{V_2(x(t), i)\} \\ &< \mathbb{E}\{V_2(x(0), r_0)\} + \varepsilon_i \int_0^t e^{-\varepsilon_i \tau} w^{\mathrm{T}}(\tau)w(\tau)\mathrm{d}\tau + \varepsilon_i \int_0^t e^{-\varepsilon_i \tau}\rho_{si}^{\mathrm{T}}(\tau)\rho_{si}(\tau)\mathrm{d}\tau \\ &\leq \bar{\varkappa}\beta_1 + \delta^2 T(1 + 4m\mu_i^2)\varepsilon_i + 4mT\varepsilon_i\varrho_i^2 \\ &\quad + 4m\varepsilon_i\left(\gamma_i^2 + \frac{\vartheta_i^2}{tr(G_i^{\mathrm{T}}G_i)}\right)\int_0^t x^{\mathrm{T}}(\tau)x(\tau)\mathrm{d}\tau. \end{aligned} \tag{2.97}$$

On the other hand, it follows from (2.82) that

$$\begin{aligned} &e^{-\varepsilon_i t}\mathbb{E}\{V_2(x(t), i)\} \\ &\geq e^{-\varepsilon_i t}\mathbb{E}\{x^{\mathrm{T}}(t)P_i x(t)\} + \int_0^t x^{\mathrm{T}}(\tau)(e^{-\varepsilon_i t}Q)x(\tau)\mathrm{d}\tau \\ &\geq e^{-\varepsilon_i T}\underline{\varkappa}\mathbb{E}\{y^{\mathrm{T}}(t)Ry(t)\} + \int_0^t x^{\mathrm{T}}(\tau)(e^{-\varepsilon_i T}Q)x(\tau)\mathrm{d}\tau. \end{aligned} \tag{2.98}$$

By condition (2.79), it yields from (2.97) and (2.98) that

$$\mathbb{E}\{y^{\mathrm{T}}(t)Ry(t)\} \leq \frac{\bar{\varkappa}\beta_1 + \varepsilon_i\delta^2 T(1 + 4\mu_i^2) + 4\varepsilon_i\varrho_i^2 T}{e^{-\varepsilon_i T}\underline{\varkappa}}. \tag{2.99}$$

By considering (2.80), (2.81), it follows from (2.99) that $\mathbb{E}\{y^{\mathrm{T}}(t)Ry(t)\} < \beta^*$ for all $t \in [0, T^*]$. Thus, according to Definition 1.6, the resulting closed-loop system (2.76) is IO-SFTS with respect to $\Re_{[0,T^*]}$. This completes the proof. □

(2) **IO-SFTS over Sliding Motion Phase within** $[T^*, T]$
According to SMC theory, when the state trajectories maintain in the sliding surface $s(x(t), i) = 0$ for each $i \in \mathbb{S}$ during the interval $[T^*, T]$, we have the following equivalent controller:

$$u_{eq}(t) = -f_i(x(t)) - (G_i B_i)^{-1}G_i[(A_i + \Delta A_i(t))x(t) + D_i w(t)]. \tag{2.100}$$

By substituting (2.100) into the model (1.6), we get the following sliding mode dynamics:

$$\begin{cases} \dot{x}(t) = \big(\tilde{A}_i + (I - \tilde{G}_i)\Delta A_i(t)\big)x(t) + (I - \tilde{G}_i)D_i w(t), \\ y(t) = C_i x(t). \end{cases} \tag{2.101}$$

where $\tilde{A}_i \triangleq (I - \tilde{G}_i)A_i$ and $\tilde{G}_i \triangleq B_i(G_i B_i)^{-1}G_i$.

In the following theorem, the sufficient condition is derived to guarantee the IO-SFTS with respect to $\Re_{[}T^*, T]$.

Theorem 2.5 *For the given parameters* $(\beta_1, \beta_2, T, R, \delta)$ *and feasible scalar* $\varepsilon_i > 0$, *consider the sliding mode dynamic* (2.101) *with incomplete TRs. If there exist positive constants* ϕ_i, ψ_i, β^* *and matrices* $P_i > 0$ *and* $W_i > 0$ $(\forall i \in \mathbb{S})$, *satisfying the following matrix inequalities:*

$$i \in \mathbb{S}_k^i, \begin{bmatrix} \Upsilon_{1i} & \Gamma_{12i} \\ * & \Gamma_{22i} \end{bmatrix} < 0, \tag{2.102}$$

$$i \in \mathbb{S}_{uk}^i, \begin{cases} \begin{bmatrix} \Upsilon_{2i} & \Gamma_{12i} \\ * & \Gamma_{22i} \end{bmatrix} < 0, \\ W_i < P_i, \ P_j < W_i, \ j \in \mathbb{S}_{uk}^i, \ j \neq i, \end{cases} \tag{2.103}$$

$$\beta_1 < \beta^* < \beta_2, \tag{2.104}$$

$$\frac{\bar{\varkappa}\beta^* + \varepsilon_i \delta^2 T}{\underline{\varkappa}} < e^{-\varepsilon_i T}\beta_2, \tag{2.105}$$

where

$$\Upsilon_{1i} \triangleq \begin{bmatrix} \mathrm{He}(P_i\tilde{A}_i) - \varepsilon_i P_i + \bar{P}_k^i + \bar{\lambda}_{ii}P_i + \underline{\lambda}_k^i P_j & P_i D_i \\ * & -\varepsilon_i I \end{bmatrix}, \ j \in \mathbb{S}_{uk}^i,$$

$$\Upsilon_{2i} \triangleq \begin{bmatrix} \mathrm{He}(P_i\tilde{A}_i) - \varepsilon_i P_i + \bar{P}_k^i + \underline{\lambda}_k^i W_i & P_i D_i \\ * & -\varepsilon_i I \end{bmatrix},$$

$$\Gamma_{12i} \triangleq \begin{bmatrix} \xi_i P_i M_i & N_i^{\mathrm{T}} & \phi_i P_i B_i (G_i B_i)^{-1} G_i M_i & N_i^{\mathrm{T}} & \psi_i P_i B_i (G_i B_i)^{-1} G_i & 0 \\ 0 & 0 & 0 & 0 & 0 & D_i^{\mathrm{T}} \\ 0 & 0 & 0 & 0 & 0 & 0 \end{bmatrix},$$

$$\Gamma_{22i} \triangleq -\mathrm{diag}\{\xi_i I, \xi_i I, \phi_i I, \phi_i I, \psi_i I, \psi_i I\}.$$

the sliding mode dynamics (2.101) *are IO-SFTS with respect to* $\Re_{[T^*,T]}$.

Proof Choose the following Lyapunov functional candidate:

$$V_3(x(t), i) = x^{\mathrm{T}}(t)P_i x(t). \tag{2.106}$$

We obtain the infinitesimal operator as

$$\mathscr{L}V_3(x(t), i) = \dot{x}^{\mathrm{T}}(t)P_i x(t) + x^{\mathrm{T}}(t)P_i\dot{x}(t) + x^{\mathrm{T}}(t)\sum_{j\in\mathbb{S}}\lambda_{ij}P_j x(t)$$

$$
\begin{aligned}
&= x^{\mathrm{T}}(t)\left(\tilde{A}_i^{\mathrm{T}} P_i + P_i \tilde{A}_i + \sum_{j\in\mathbb{S}} \lambda_{ij} P_j\right) x(t) + 2x^{\mathrm{T}}(t) P_i \Delta A_i(t) x(t) \\
&\quad - 2x^{\mathrm{T}}(t) P_i \tilde{G}_i \Delta A_i(t) x(t) + w^{\mathrm{T}}(t) D_i^{\mathrm{T}} P_i x(t) + x^{\mathrm{T}}(t) P_i D_i w(t) \\
&\quad - 2x^{\mathrm{T}}(t) P_i \tilde{G}_i D_i w(t).
\end{aligned} \tag{2.107}
$$

For $\phi_i > 0$ and $\psi_i > 0$, one has

$$
-2x^{\mathrm{T}}(t) P_i \tilde{G}_i \Delta A_i(t) x(t) \leq \phi_i x^{\mathrm{T}}(t) P_i \tilde{G}_i M_i M_i^{\mathrm{T}} \tilde{G}_i^{\mathrm{T}} P_i x(t) + \phi_i^{-1} x^{\mathrm{T}}(t) N_i^{\mathrm{T}} N_i x(t), \tag{2.108}
$$

$$
-2x^{\mathrm{T}}(t) P_i \tilde{G}_i D_i w(t) \leq \psi_i x^{\mathrm{T}}(t) P_i \tilde{G}_i \tilde{G}_i^{\mathrm{T}} P_i x(t) + \psi_i^{-1} w^{\mathrm{T}}(t) D_i^{\mathrm{T}} D_i w(t). \tag{2.109}
$$

Define an auxiliary function as follows:

$$
J_2(x(t), i) = \mathscr{L} V_3(x(t), i) - \varepsilon_i V_3(x(t), i) - \varepsilon_i w^{\mathrm{T}}(t) w(t). \tag{2.110}
$$

It is easily shown that $J_2(x(t), i) < 0$ can be ensured by the following inequality:

$$
\begin{bmatrix}
\mathrm{He}(P_i \tilde{A}_i) + \sum_{j\in\mathbb{S}} \lambda_{ij} P_j - \varepsilon_i P_i & P_i D_i & \xi_i P_i M_i & N_i^{\mathrm{T}} & \phi_i P_i \tilde{G}_i M_i & N_i^{\mathrm{T}} & \psi_i P_i \tilde{G}_i & 0 \\
* & -\varepsilon_i I & 0 & 0 & 0 & 0 & 0 & D_i^{\mathrm{T}} \\
* & * & -\xi_i I & 0 & 0 & 0 & 0 & 0 \\
* & * & * & -\xi_i I & 0 & 0 & 0 & 0 \\
* & * & * & * & -\phi_i I & 0 & 0 & 0 \\
* & * & * & * & * & -\phi_i I & 0 & 0 \\
* & * & * & * & * & * & -\psi_i I & 0 \\
* & * & * & * & * & * & * & -\psi_i I
\end{bmatrix} < 0. \tag{2.111}
$$

Similar to the derivation procedure of Theorem 2.4, we shall consider the following two cases.

► **Case 1**. The diagonal element is known ($i \in \mathbb{S}_k^i$).
In this case, according to (2.63)–(2.89), one has

$$
\sum_{j\in\mathbb{S}_{uk}^i} \frac{\lambda_{ij}}{\lambda_k^i} \left[\begin{array}{cc|c}
\mathrm{He}(P_i \tilde{A}_i) - \varepsilon_i P_i + \bar{P}_k^i + \bar{\lambda}_{ii} P_i + \underline{\lambda}_k^i P_j & P_i D_i & \Gamma_{12i} \\
* & -\varepsilon_i I & \\
\hline
* & & \Gamma_{22i}
\end{array}\right] < 0, \tag{2.112}
$$

which can be ensured by (2.102).

► **Case 2**. The diagonal element is unknown ($i \in \mathbb{S}_{uk}^i$).
In this case, according to (2.91), we have

$$
\left[\begin{array}{cc|c}
\mathrm{He}(P_i \tilde{A}_i) - \varepsilon_i P_i + \bar{P}_k^i + \underline{\lambda}_k^i W_i & P_i D_i & \Lambda_{12i} \\
* & -\varepsilon_i I & \\
\hline
* & & \Lambda_{22i}
\end{array}\right] < 0, \tag{2.113}
$$

which can be ensured by (2.103).

From Cases 1 and 2, we can conclude that (2.111) can be ensured by (2.112) or (2.113). Then, taking the expectation of $J_2(x(t), i) < 0$ yields

$$\mathbb{E}\{\mathscr{L}V_3(x(t), i)\} < \varepsilon_i \mathbb{E}\{V_3(x(t), i)\} + \varepsilon_i w^{\mathrm{T}}(t)w(t). \tag{2.114}$$

By multiplying both sides of expression (2.114) with $e^{-\varepsilon_i t}$ and integrating the obtained inequality from T^* to t with $t \in [T^*, T]$, we have

$$\begin{aligned} e^{-\varepsilon_i t}\mathbb{E}\{V_3(x(t), i)\} &< e^{-\varepsilon_i T^*}\mathbb{E}\{V_3(x(T^*), r_{T^*})\} + \varepsilon_i \int_{T^*}^{t} e^{-\varepsilon_i \tau} w^{\mathrm{T}}(\tau)w(\tau)\mathrm{d}\tau \\ &< \bar{\varkappa}\mathbb{E}\{y^{\mathrm{T}}(T^*)Ry(T^*)\} + \varepsilon_i \delta^2 T \\ &< \bar{\varkappa}\beta^* + \varepsilon_i \delta^2 T. \end{aligned} \tag{2.115}$$

In addition, it follows from (2.106) that

$$e^{-\varepsilon_i t}\mathbb{E}\{V_3(x(t), i)\} \geq e^{-\varepsilon_i T}\underline{\varkappa}\mathbb{E}\{y^{\mathrm{T}}(t)Ry(t)\}. \tag{2.116}$$

According to (2.115) and (2.116), we have

$$\mathbb{E}\{y^{\mathrm{T}}(t)R)y(t)\} < \frac{\bar{\varkappa}\beta^* + \varepsilon_i \delta^2 T}{e^{-\varepsilon_i T}\underline{\varkappa}}. \tag{2.117}$$

If there exists a scalar β^* satisfying (2.104), (2.105), we have from (2.117) that $\mathbb{E}\{y^{\mathrm{T}}(t)R)y(t)\} < \beta_2$ for all $t \in [T^*, T]$. Thus, according to Definition 1.6, the sliding mode dynamics (2.101) is IO-SFTS with respect to $\Re_{[T^*, T]}$. This completes the proof. □

2.3.4 Solving Algorithm for Controller Gains

It is worthy of pointing out that the control gains K_i ($i \in \mathbb{S}$) should be designed to ensure the conditions in Theorems 2.4 and 2.5 simultaneously. However, it is noted that the matrix inequalities in Theorems 2.4 and 2.5 are nonlinear, e.g., $\|K_i\|^2$, $\bar{\varkappa}\beta^*$, which are difficult to solve. In the sequel, we shall give sufficient conditions based on LMIs, which can guarantee that the conditions in Theorems 2.4 and 2.5 hold simultaneously.

Theorem 2.6 *For the given parameters* $(\beta_1, \beta_2, T, R, \delta)$ *and a feasible scalar* $\varepsilon_i > 0$, *consider the MJS* (2.76) *with incomplete TRs. If there exist constants* $\xi_i > 0$, $\zeta_i > 0$, $\phi_i > 0$, $\psi_i > 0$, $\beta^* > 0$, $\eta_i > 0$ *and matrices* $\mathcal{P}_i > 0$, $\mathcal{W}_i > 0$, $\mathcal{V}_i > 0$, $\mathcal{U}_i > 0$ *and* $\mathcal{L}_i$ ($\forall i \in \mathbb{S}$), *satisfying the following matrix inequalities:*

$$i \in \mathbb{S}_k^i, \begin{bmatrix} \Pi_{11i} & \Sigma_{12i} & \Pi_{13i} \\ * & \Sigma_{22i} & 0 \\ * & * & \Pi_{33i} \end{bmatrix} < 0, \tag{2.118}$$

$$i \in \mathbb{S}_{uk}^i, \begin{cases} \begin{bmatrix} \mathcal{F}_{11i} & \Sigma_{12i} & \mathcal{F}_{13i} \\ * & \Sigma_{22i} & 0 \\ * & * & \mathcal{F}_{33i} \end{bmatrix} < 0, \\ \mathcal{W}_i < \mathcal{P}_i, \\ \begin{bmatrix} -\mathcal{W}_i & \mathcal{P}_i \\ * & -\mathcal{P}_j \end{bmatrix} < 0, \ j \in \mathbb{S}_{uk}^i, j \neq i, \end{cases} \tag{2.119}$$

$$\begin{bmatrix} -\mathcal{V}_i & \mathcal{L}_i^{\mathrm{T}} \\ * & -I \end{bmatrix} < 0, \tag{2.120}$$

$$\begin{bmatrix} -e^{-\varepsilon_i T}\mathcal{U}_i + 8n\varepsilon_i \mathcal{V}_i & 2\vartheta_i\sqrt{\frac{\varepsilon_i}{tr(G_i^{\mathrm{T}}G_i)}}\mathcal{P}_i & 4\varepsilon_i\tilde{\gamma}_i\mathcal{P}_i \\ * & -I & 0 \\ * & * & -2\varepsilon_i I \end{bmatrix} < 0, \tag{2.121}$$

$$\beta_1 < \beta^* < \beta_2, \tag{2.122}$$

$$\eta_i(C_i^{\mathrm{T}}RC_i)^{-1} < \mathcal{P}_i < 2(C_i^{\mathrm{T}}RC_i)^{-1}, \tag{2.123}$$

$$\begin{bmatrix} -\frac{1}{2}e^{-\varepsilon_i T}\beta^* + \varepsilon_i\delta^2 T(1+4\mu_i^2) + 4\varepsilon_i\varrho_i^2 T & \sqrt{\beta_1} \\ * & -\eta_i \end{bmatrix} < 0, \tag{2.124}$$

$$-\frac{1}{2}e^{-\varepsilon_i T}\beta_2\eta_i + \beta^* + \varepsilon_i\delta^2 T\eta_i < 0, \tag{2.125}$$

where

$$\Sigma_{12i} \triangleq \begin{bmatrix} -B_i & 0 & \xi_i M_i & 0 & \mathcal{P}_i N_i^{\mathrm{T}} & 0 & \zeta_i B_i & 0 & \epsilon_i\mathcal{P}_i & 0 & 0 \\ 0 & \xi_i M_i & 0 & \mathcal{P}_i N_i^{\mathrm{T}} & 0 & \phi_i B_i(GB_i)^{-1}GM_i & 0 & \mathcal{P}_i N_i^{\mathrm{T}} & 0 & \psi_i B_i(GB_i)^{-1}G & 0 \\ 0 & 0 & 0 & 0 & 0 & 0 & 0 & 0 & 0 & 0 & 0 \\ 0 & 0 & 0 & 0 & 0 & 0 & 0 & 0 & 0 & 0 & D_i^{\mathrm{T}} \end{bmatrix},$$

$$\Sigma_{22i} \triangleq -\mathrm{diag}\{\varepsilon_i I, \xi_i I, \xi_i I, \xi_i I, \xi_i I, \phi_i I, \zeta_i I, \phi_i I, \zeta_i I, \psi_i I, \psi_i I\},$$

$$\Pi_{11i} \triangleq \begin{bmatrix} \mathrm{He}(A_i\mathcal{P}_i + B_i\mathcal{L}_i) + (\bar{\lambda}_{ii} - \varepsilon_i)\mathcal{P}_i + \mathcal{U}_i & 0 & D_i & 0 \\ 0 & \mathrm{He}(\tilde{A}_i\mathcal{P}_i) + (\bar{\lambda}_{ii} - \varepsilon_i)\mathcal{P}_i & 0 & D_i \\ * & * & -\varepsilon_i I & 0 \\ * & * & * & -\varepsilon_i I \end{bmatrix},$$

$$\Pi_{13i} \triangleq \begin{bmatrix} \sqrt{\underline{\lambda}_k^i}\mathcal{P}_i & 0 & M_{1k}^i & 0 \\ 0 & \sqrt{\underline{\lambda}_k^i}\mathcal{P}_i & 0 & M_{1k}^i \\ 0 & 0 & 0 & 0 \\ 0 & 0 & 0 & 0 \end{bmatrix}, \quad \Pi_{33i} \triangleq -\mathrm{diag}\{\mathcal{P}_j, \mathcal{P}_j, N_{1k}^i, N_{1k}^i\},$$

$$M_{1k}^i \triangleq [\sqrt{\bar{\lambda}_{i\mathcal{K}_1^i}}\mathcal{P}_i, \ldots, \sqrt{\bar{\lambda}_{i\mathcal{K}_{m_i}^i}}\mathcal{P}_i], \quad N_{1k}^i \triangleq [\mathcal{P}_{\mathcal{K}_1^i}, \ldots, \mathcal{P}_{\mathcal{K}_{m_i}^i}], \quad \mathcal{K}_l^i \neq i,$$

$$\mathcal{F}_{11i} \triangleq \begin{bmatrix} \mathrm{He}(A_i\mathcal{P}_i + B_i\mathcal{L}_i) + \underline{\lambda}_k^i\mathcal{W}_i - \varepsilon_i\mathcal{P}_i + \mathcal{U}_i & 0 & D_i & 0 \\ 0 & \mathrm{He}(\tilde{A}_i\mathcal{P}_i) + \underline{\lambda}_k^i\mathcal{W}_i - \varepsilon_i\mathcal{P}_i & 0 & D_i \\ * & * & -\varepsilon_i I & 0 \\ * & * & * & -\varepsilon_i I \end{bmatrix},$$

$$\mathcal{F}_{13i} \triangleq \begin{bmatrix} M_{2k}^i & 0 \\ 0 & M_{2k}^i \\ 0 & 0 \\ 0 & 0 \end{bmatrix}, \quad \mathcal{F}_{33i} \triangleq -\mathrm{diag}\{N_{2k}^i, N_{2k}^i\},$$

$$M_{2k}^i \triangleq [\sqrt{\bar{\lambda}_{i\mathcal{K}_1^i}}\mathcal{P}_i, \ldots, \sqrt{\bar{\lambda}_{i\mathcal{K}_{n_i}^i}}\mathcal{P}_i], \quad N_{2k}^i \triangleq [\mathcal{P}_{\mathcal{K}_1^i}, \ldots, \mathcal{P}_{\mathcal{K}_{n_i}^i}], \quad \mathcal{K}_l^i \neq i,$$

the closed-loop system (2.76) is IO-SFTS with respect to $\Re_{[0,T]}$ over the whole finite-time interval $[0, T]$, meanwhile, the control gains in (2.52) are given as $K_i = \mathcal{L}_i\mathcal{P}_i^{-1}$ $(i \in \mathbb{S})$.

Proof By taking the conditions (2.77)–(2.81) in Theorem 2.4 and conditions (2.102)–(2.104) in Theorem 2.5 simultaneously, we have the following inequalities:

$$i \in \mathbb{S}_k^i, \begin{bmatrix} \Theta_{1i} & \Xi_{12i} \\ * & \Xi_{22i} \end{bmatrix} < 0, \tag{2.126}$$

$$i \in \mathbb{S}_{uk}^i, \begin{cases} \begin{bmatrix} \Theta_{2i} & \Xi_{12i} \\ * & \Xi_{22i} \end{bmatrix} < 0, \\ W_i < P_i, \; P_j < W_i, \; j \in \mathbb{S}_{uk}^i, \; j \neq i, \end{cases} \tag{2.127}$$

$$4\varepsilon_i\left(\gamma_i^2 + \frac{\vartheta_i^2}{tr(G_i^{\mathrm{T}}G_i)}\right)I < e^{-\varepsilon_i T}Q, \tag{2.128}$$

$$\beta_1 < \beta^* < \beta_2, \tag{2.129}$$

$$\frac{\bar{\varkappa}\beta_1 + \varepsilon_i\delta^2T(1 + 4\mu_i^2) + 4\varepsilon_i\varrho_i^2T}{\underline{\varkappa}} < e^{-\varepsilon_i T}\beta^*, \tag{2.130}$$

$$\frac{\bar{\varkappa}\beta^* + \varepsilon_i\delta^2T}{\underline{\varkappa}} < e^{-\varepsilon_i T}\beta_2, \tag{2.131}$$

where

$$\Theta_{1i} \triangleq \begin{bmatrix} \mathrm{He}(P_i\bar{A}_i)+Q-\varepsilon_i P_i+\bar{P}_k^i+\bar{\lambda}_{ii}P_i+\underline{\lambda}_k^i P_j & 0 & P_iD_i & 0 \\ * & \mathrm{He}(P_i\tilde{A}_i)-\varepsilon_i P_i+\bar{P}_k^i+\bar{\lambda}_{ii}P_i+\underline{\lambda}_k^i P_j & 0 & P_iD_i \\ * & * & -\varepsilon_i I & 0 \\ * & * & * & -\varepsilon_i I \end{bmatrix},$$

$$\Theta_{2i} \triangleq \begin{bmatrix} \mathrm{He}(P_i\bar{A}_i)+Q-\varepsilon_i P_i+\bar{P}_k^i+\underline{\lambda}_k^i W_i & 0 & P_iD_i & 0 \\ * & \mathrm{He}(P_i\tilde{A}_i)-\varepsilon_i P_i+\bar{P}_k^i+\underline{\lambda}_k^i W_i & 0 & P_iD_i \\ * & * & -\varepsilon_i I & 0 \\ * & * & * & -\varepsilon_i I \end{bmatrix},$$

$$\Xi_{12i} \triangleq \left[\begin{matrix} -P_iB_i & 0 & \xi_iP_iM_i & 0 & N_i^{\mathrm{T}} & 0 \\ 0 & \xi_iP_iM_i & 0 & N_i^{\mathrm{T}} & 0 & \phi_iP_iB_i(G_iB_i)^{-1}G_iM_i \\ 0 & 0 & 0 & 0 & 0 & 0 \\ 0 & 0 & 0 & 0 & 0 & 0 \end{matrix}\right.$$
$$\left.\begin{matrix} \alpha_i I & 0 & 0 & & \\ 0 & N_i^{\mathrm{T}} & 0 & \psi_iP_iB_i(G_iB_i)^{-1}G_i & 0 \\ 0 & 0 & 0 & 0 & 0 \\ 0 & 0 & 0 & 0 & D_i^{\mathrm{T}} \end{matrix}\right],$$

$$\Xi_{22i} \triangleq -\mathrm{diag}\{\varepsilon_i I, \xi_i I, \xi_i I, \xi_i I, \xi_i I, \phi_i I, \zeta_i I, \phi_i I, \zeta_i I, \psi_i I, \psi_i I\}.$$

Denote $\mathcal{P}_i \triangleq P_i^{-1}$, $\mathcal{W}_i \triangleq \mathcal{P}_i W_i \mathcal{P}_i$, $\mathcal{L}_i \triangleq K_i \mathcal{P}_i$, $\mathcal{U}_i \triangleq \mathcal{P}_i Q \mathcal{P}_i$ and $\mathcal{V}_i \triangleq \alpha_i \mathcal{P}_i \mathcal{P}_i$. It is easily shown that the inequalities (2.126) and (2.127) are ensured by (2.118) and (2.119).

Notice that γ_i is dependent on K_i, that is, γ_i is not a priori. From (2.54), we have $\gamma_i^2 \leq 2\tilde{\gamma}_i^2 + 2\|K_i\|^2$. Thus, inequality (2.128) can be ensured by

$$8\varepsilon_i \|K_i\|^2 I + 8\varepsilon_i \tilde{\gamma}_i^2 I + \frac{4\varepsilon_i \vartheta_i^2}{tr(G_i^{\mathrm{T}} G_i)} I < e^{-\varepsilon_i T} Q. \tag{2.132}$$

From condition (2.120), we can obtain $K_i^{\mathrm{T}} K_i < \alpha_i I$, that is, $\|K_i\|^2 < n\alpha_i$. Therefore, (2.132) holds if the following inequality holds:

$$8n\varepsilon_i \alpha_i I + 8\varepsilon_i \tilde{\gamma}_i^2 I + \frac{4\varepsilon_i \vartheta_i^2}{tr(G_i^{\mathrm{T}} G_i)} I < e^{-\varepsilon_i T} Q, \tag{2.133}$$

which can be ensured by (2.121).

On the other hand, by the condition (2.123) and the following fact

$$\lambda_{\max}\{(C_i^{\mathrm{T}} R C_i)^{\frac{1}{2}} \mathcal{P}_i (C_i^{\mathrm{T}} R C_i)^{\frac{1}{2}}\} = \frac{1}{\lambda_{\min}\{(C_i^{\mathrm{T}} R C_i)^{-\frac{1}{2}} P_i (C_i^{\mathrm{T}} R C_i)^{-\frac{1}{2}}\}}, \tag{2.134}$$

one has $\frac{1}{2} < \underline{\varkappa}$, $\bar{\varkappa} < \frac{1}{\eta_i}$. Thus, it is easily shown that the inequalities (2.129)–(2.131) can be ensured by (2.122)–(2.125). This completes the proof. □

Remark 2.6 It is noted that sufficient conditions are given in the Theorem 2.6 to ensure the IO-SFTS for a class of MJSs with incomplete TRs. Obviously, when TRs are completely known, the lower bound of an element is equal to the upper bound ($\underline{\lambda}_{ij} = \bar{\lambda}_{ij}$). Accordingly, by utilizing $\underline{\lambda}_{ij} = \bar{\lambda}_{ij} = \lambda_{ij}$, the conditions in Theorem 2.6 can also guarantee the IO-SFTS for MJSs with completely known TRs. When all TRs are completely unknown, that is, $\lambda_k^i = 0$, the conditions (2.118)–(2.125) will reduce to (2.119)–(2.125) with $\mathcal{P}_1 = \mathcal{P}_2 = \mathcal{P}_3$ and $\mathcal{W}_1 = \mathcal{W}_2 = \mathcal{W}_3$.

2.3.5 Simulation

In this section, a modified example from a typical DC chopper circuit (Fig. 2.3) is utilized to illustrate the applicability of the theoretical results.

It is assumed that the "V" in Fig. 2.3 is always turned on so that the invariable average output voltage can be achieved. In the load circuit, the jumping among capacitors 1, 2 and 3 is stochastic and obeys Markov process. Based on Kirchhoff's laws, taking the voltage $u_2(t)$ of the capacitor and current $i(t)$ of the inductor in the load network as state variables (see details in [20, 21]), the system can be modeled as:

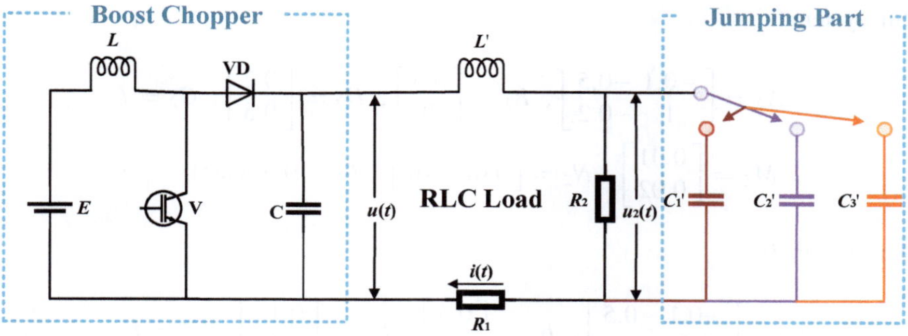

Fig. 2.3 Boost chopper with RLC load

$$\begin{cases} \frac{\mathrm{d}i(t)}{\mathrm{d}t} = -\frac{1}{L'}R_1 i(t) - \frac{1}{L'}u_2(t) + \frac{1}{L'}u(t), \\ \frac{\mathrm{d}u_2(t)}{\mathrm{d}t} = \frac{1}{C_j'}i(t) - \frac{1}{C_j'R_2}u_2(t), \quad j = 1, 2, 3. \end{cases} \tag{2.135}$$

Defining $x(t) \triangleq [i(t) \;\; u_2(t)]^{\mathrm{T}}$, we have

$$\dot{x}(t) = A_j x(t) + B_j u(t), \tag{2.136}$$

where

$$A_j \triangleq \begin{bmatrix} -\frac{1}{L'}R_1 & -\frac{1}{L'} \\ \frac{1}{C_j'} & -\frac{1}{C_j'R_2} \end{bmatrix}, \quad B_j \triangleq \begin{bmatrix} \frac{1}{L'} \\ 0 \end{bmatrix}, \quad j = 1, 2, 3.$$

The parameters are given in Table 2.1, by which one has subsystems:
Subsystem 1:

$$A_1 = \begin{bmatrix} -0.1 & -0.5 \\ 2 & -0.4 \end{bmatrix}, \; B_1 = \begin{bmatrix} 0.5 \\ 0 \end{bmatrix}, \; D_1 = \begin{bmatrix} 0.1 \\ 0.2 \end{bmatrix}, \; C_1 = I,$$

$$M_1 = \begin{bmatrix} 0.01 \\ 0.01 \end{bmatrix}, \; N_1 = \begin{bmatrix} 0.02 \; 0.02 \end{bmatrix}, \; F_1 = 0.5\sin(t),$$

Table 2.1 Parameters in the circuit system

Parameter	Value	Parameter	Value
L'	2 H	C_1'	0.5 F
R_1	0.2 Ω	C_2'	1 F
R_2	5 Ω	C_3'	2 F

Subsystem 2:

$$A_2 = \begin{bmatrix} -0.1 & -0.5 \\ 1 & -0.2 \end{bmatrix}, \; B_2 = \begin{bmatrix} 0.5 \\ 0 \end{bmatrix}, \; D_2 = \begin{bmatrix} 0.2 \\ 0.3 \end{bmatrix}, \; C_2 = I$$
$$M_2 = \begin{bmatrix} 0.01 \\ 0.02 \end{bmatrix}, \; N_2 = \begin{bmatrix} 0.02 \; 0.01 \end{bmatrix}, \; F_2 = 0.5\cos(t),$$

Subsystem 3:

$$A_3 = \begin{bmatrix} -0.1 & -0.5 \\ 0.5 & -0.1 \end{bmatrix}, \; B_3 = \begin{bmatrix} 0.5 \\ 0 \end{bmatrix}, \; D_3 = \begin{bmatrix} 0.1 \\ 0.2 \end{bmatrix}, \; C_3 = I$$
$$M_3 = \begin{bmatrix} 0.01 \\ 0.01 \end{bmatrix}, \; N_3 = \begin{bmatrix} 0.01 \; 0.01 \end{bmatrix}, \; F_3 = 0.1(\sin(t) + \cos(t)),$$

and $w(t) = 0.5\sin(2t)$ with $\delta = 0.5$, $f_i(x(t)) = 0.1\sin(\sqrt{x_1^2 + x_2^2 + x_3^2})$ with $\epsilon_i = 0.1$, $i = 1, 2, 3$. It is assumed that the TR matrix is given as

$$\Lambda = \begin{bmatrix} -0.3 & ? & ? \\ ? & ? & 0.1 \\ ? & \varsigma_{32} & ? \end{bmatrix},$$

where ς_{32} is unknown but with known bound, that is, $0.35 \le \varsigma_{32} \le 1.05$, and those terms denoted by "?" are completely known. According to the given TR matrix, we can obtain $\mathbb{S}_k^1 = \{1\}$, $\mathbb{S}_{uk}^1 = \{2, 3\}$, $\mathbb{S}_k^2 = \{3\}$, $\mathbb{S}_{uk}^2 = \{1, 2\}$, $\mathbb{S}_k^3 = \{2\}$, $\mathbb{S}_{uk}^3 = \{1, 3\}$.

Now, we objective is to design an SMC law to achieve the IO-SFTS with $\beta_1 = 3.5$, $\beta_2 = 4$, $T = 5$ and $R = \text{diag}\{1, 1\}$. The parameters ϱ_i and μ_i in (2.54) are given as $\varrho_1 = \varrho_2 = \varrho_3 = 0.21$, $\mu_1 = 0.2$, $\mu_2 = 0.4$, $\mu_3 = 0.2$. By solving LMIs (2.118)–(2.125) in Theorem 2.6 for the chosen matrices $X_i = \text{diag}\{16, 1\}$ $(i = 1, 2, 3)$, we obtain

$$K_1 = 10^{-3} \times \begin{bmatrix} -0.5924 \; -0.0023 \end{bmatrix}, \; K_2 = \begin{bmatrix} -0.0064 \; -0.0007 \end{bmatrix},$$
$$K_3 = \begin{bmatrix} -0.0274 \; 0.0077 \end{bmatrix}.$$

The simulations are undergone by using the Euler discretization approach with initial state $x(0) = [-0.5 \;\; 1.8]^{\mathrm{T}}$ and the stepsize as $\Delta t = 0.01$s.

The simulation results as shown in Fig. 2.4. There are some notable variations in the control signal as shown in Fig. 2.4d, which is just the feature of the SMC method. Nevertheless, from the evolution of weighted system output variable $y^{\mathrm{T}}(t)Ry(t)$ in Fig. 2.4a under the mode jumping in Fig. 2.4b, it can be seen that the proposed SMC law can effectively deal with the effect of incomplete TRs and ensure the IO-SFTS of the resultant closed-loop systems.

In simulation operation, the discretization stepsize Δt will affect the simulation precision, and the smaller the stepsize Δt, the closer the simulation results are to

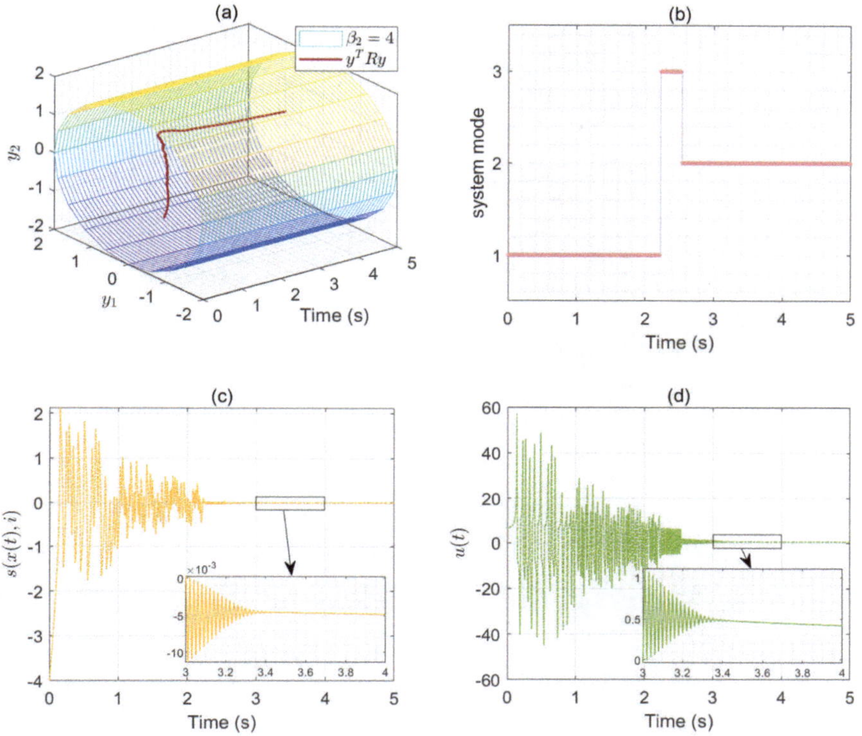

Fig. 2.4 Simulation results of controlled system: **a** evolution of weighted system output variable $y^{\mathrm{T}}(t)Ry(t)$; **b** a possible mode transition sequence; **c** sliding variable $s(x(t), i)$; **d** control signal $u(t)$

the original continuous system results. Based on the above parameters and solution results, Fig. 2.5 depicts the variations of control signals under different discretization stepsizes, from which the smaller the stepsize Δt, the less chatter the control signal. In addition, it is noticed from the reaching time T^* in (2.75) that the adjustable parameters ϱ_i $(i = 1, 2, 3)$ directly affects the arrival time that the sliding surface will be reached. In order to compare the influences of ϱ_i on the reachability, we choose stepsize as $\Delta t = 0.0001$s and obtain the evolutions of sliding variables in Fig. 2.6 under different adjustable parameter ϱ_i. It is shown that for fixed initial condition, the bigger the value of ϱ_i, the faster the sliding surface will be reached.

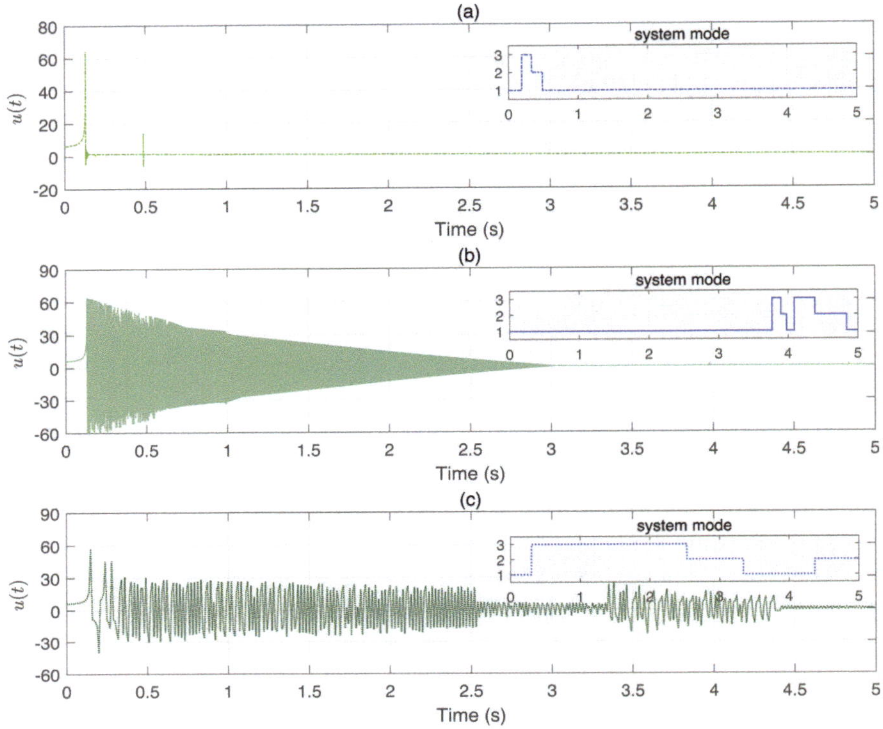

Fig. 2.5 Control signal $u(t)$ under $\varrho_i = 0.21$ $(i = 1, 2, 3)$ with different discretization stepsizes: **a** $\Delta t = 0.0001\text{s}$; **b** $\Delta t = 0.001\text{s}$; **c** $\Delta t = 0.01\text{s}$

2.4 Conclusion

In this chapter, we have investigated the SFTB for MJSs with complete TRs and IO-SFTS for MJSs with incomplete TRs via SMC technique. It is manifested that the reachability with $T^* < T$ for any given short time T can be attained by a scalar selection criterion. Besides, the effects of Markov switching have been considered in the design of both sliding surface and SMC law. By utilizing the "partition–integration" technique, we analyze the SFTB/IO-SFTS during the reaching phase and the sliding motion phase, respectively, and acquire the corresponding algorithms for solving controller gains. It is shown that the designed SMC laws can ensure the SFTB under restricted states and the IO-SFTS under restricted outputs during the given time interval $[0, T]$.

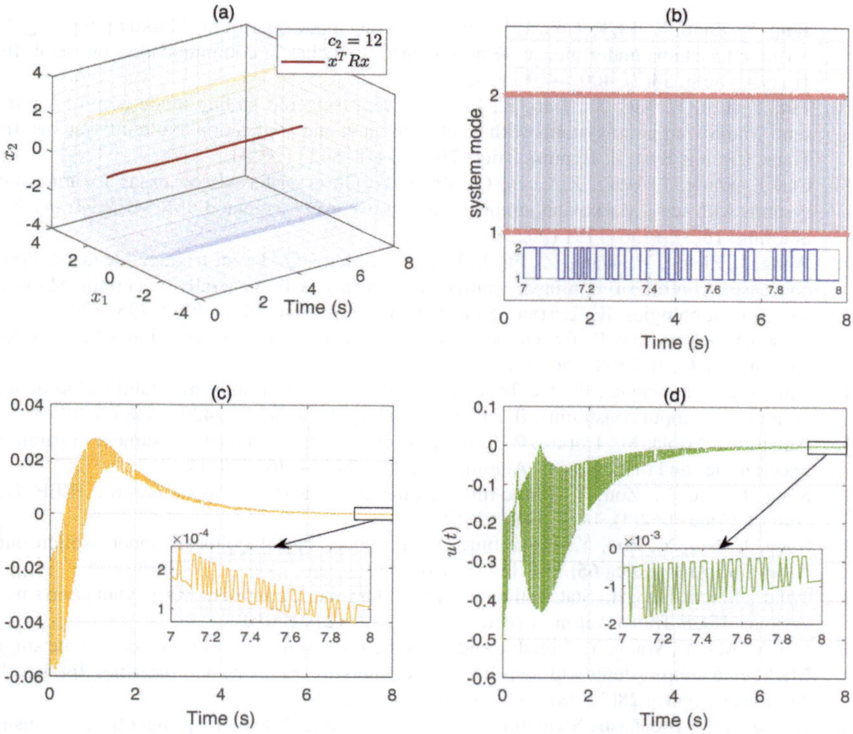

Fig. 2.6 Sliding variable $s(x(t), i)$ under $\Delta t = 0.0001$s with different adjustable parameters ϱ_i $(i = 1, 2, 3)$

References

1. Costa, O.L.V., Fragoso, M.D., Marques, R.P.: Discrete-Time Markovian Jump Linear Systems. Springer, London (2006)
2. Bolzern, P., Colaneri, P., De Nicolao, G.: Almost sure stability of Markov jump linear systems with deterministic switching. IEEE Trans. Autom. Control **58**(1), 209–214 (2013)
3. Zou, Y., Lam, J., Niu, Y., Li, D.: Constrained predictive control synthesis for quantized systems with Markovian data loss. Automatica **55**, 217–225 (2015)
4. Narvaez, A.R.R., Costa, E.F.: Control of continuous-time linear systems with Markov jump parameters in reverse time. IEEE Trans. Autom. Control **65**(5), 2265–2271 (2019)
5. Xu, R., Kao, Y., Gao, M.: Finite-time synchronization of Markovian jump complex networks with generally uncertain transition rates. Trans. Inst. Meas. Control **39**(1), 52–60 (2017)
6. Niu, Y., Ho, D.W.C., Wang, X.: Sliding mode control for Itô stochastic jump systems with Markovian switching. Automatica **43**(10), 1784–1790 (2007)
7. Shen, C., Xu, J., Cheng, J., Yan, H., Cao, J.: Protocol-based SMC for singularly perturbed systems with switching parameters and deception attacks. Inf. Sci. **679**, 121089 (2024)
8. Liu, M., Zhang, L., Shi, P., Zhan, Y.: Sliding mode control of continuous-time Markovian jump systems with digital data transmission. Automatica **80**, 200–209 (2017)

9. Tang, Y., Zhou, X., Li, P., Cao, J., Cheng, J.: Sliding mode control for Markov jump singularly perturbed systems under piecewise homogeneous stochastic communication protocol. IEEE Trans. Cybern. **54**(7), 4002–4013 (2024)
10. Tan, Y., Liu, J., Xie, X., Tian, E., Liu, J.: Event-triggered sliding-mode control for fuzzy semi-Markovian jump systems with dead zone input and application to circuit systems. IEEE Trans. Circuits Syst. II: Express Briefs **70**(9), 3418–3423 (2023)
11. Cai, J., Wang, J., Feng, J., Chen, G., Zhao, Y.: Observer-based consensus for multi-agent systems with semi-Markovian jumping via adaptive event-triggered SMC. IEEE Trans. Netw. Sci. Eng. **10**(3), 1736–1751 (2023)
12. Wang, J., Zhang, H., Wang, Z., Fu, J., Wang, W., Meng, Q.: Event-triggered leader-following consensus control of nonlinear multiagent systems with generally uncertain Markovian switching topologies. IEEE Trans. Syst. Man. Cybern. Syst. **54**(8), 4942–4954 (2024)
13. Amato, F., Ambrosino, R., Cosentino, C.: Input-output finite time stabilization of linear systems. Automatica **46**(9), 1558–1562 (2010)
14. Amato, F., Carannante, G., De Tommasi, G.: Input-output finite-time stabilisation of linear systems with input constraints. IET Control Theory Appl. **8**(14), 1429–1438 (2014)
15. Amato, F., Ariola, M., Dorato, P.: Finite-time control of linear systems subject to parametric uncertainties and disturbances. Automatica **37**(9), 1459–1463 (2001)
16. Song, J., Niu, Y., Zou, Y.: Finite-time stabilization via sliding mode control. IEEE Trans. Autom. Control **62**(3), 1478–1483 (2017)
17. Song, J., Niu, Y., Zou, Y.: Finite-time sliding mode control synthesis under explicit output constraint. Automatica **65**, 111–114 (2016)
18. Wu, L., Shi, P., Gao, H.: State estimation and sliding-mode control of Markovian jump singular systems. IEEE Trans. Autom. Control **55**(5), 1213–1219 (2010)
19. Liu, Y., Ma, Y., Wang, Y.: Reliable sliding mode finite-time control for discrete-time singular Markovian jump systems with sensor fault and randomly occurring nonlinearities. Int. J. Robust Nonlinear Control **28**(2), 381–402 (2018)
20. Boukas, E.K.: Stochastic Switching Systems: Analysis and Design. Springer Science Business Media (2007)
21. Luan, X., Zhao, C., Liu, F.: Finite-time stabilization of switching Markov jump systems with uncertain transition rates. Circuits Syst. Signal Process. **34**(12), 3741–3756 (2015)

Chapter 3
Fault-Tolerant Sliding Mode Control for Handling Actuator Failures

3.1 Introduction

An actuator is a key component in control systems, which converts control signals into mechanical motion. The actuator faults often occur in actual physical systems, which can cause performance deterioration or even instability of the systems. Thus, it is necessary to develop an effective control technique to maintain a receivable stability for the Markov jump systems (MJSs) against actuator failures [1–3]. More recently, the sliding mode control (SMC) technique has witnessed increasing applications in MJSs subject to actuator faults [4–8]. Among them, the on-line estimation mechanism has been proposed in [4] regarding the loss effectiveness of actuators, by which an adaptive sliding mode controller has been designed to attenuate the effect of the actuator degradation. The stabilization problem for MJSs subject to actuator and sensor faults has been also addressed in [8], and a fault tolerant control strategy has been designed by means of a sliding mode observer to estimate the disturbances, actuator and sensor faults simultaneously.

It is worthy of noting that the aforementioned results have been obtained based on Lyapunov stability, that is, the behavior of the controlled system has been considered within a sufficiently long (in principle infinite) time interval. In a lot of industrial process, finite-time boundedness (FTB) is more practical for analyzing the transient behavior within a finite (possibly short) interval. Some interesting results have been obtained on stochastic finite-time boundedness (SFTB) for MJSs [9, 10], while the existing works involving in SFTB cannot be extended to deal with the actuator failures, especially taking the specific structures of MJSs and SMC into consideration. This motivates the study in this chapter. Apparently, when analyzing SFTB of MJSs subject to actuator faults via SMC approaches, the following questions had to be answered, which just make the present research be challenging:

Q1: When the actuator suffers from failures, how to design a sliding mode controller for eliminating the negative effects of actuator faults?

Z. Cao and Y. Niu, *Finite-Time Sliding Mode Control for Markov Jump Systems*, Studies in Systems, Decision and Control 603,
https://doi.org/10.1007/978-981-96-7651-4_3

Q2: For a given finite time T, how to guarantee that the state trajectories can arrive at the designed sliding surface within $[0, T^*]$ with $T^* < T$ and maintain SFTB performance within both $[0, T^*]$ and $[T^*, T]$?

The objective of this chapter is to solve the above questions and then establish a fault-tolerant SMC scheme for MJSs to achieve SFTS performance. The main contributions are highlighted as follows. *(1) In virtue of the specified state constraint domain, the problem of SFTB for a class of MJSs with actuator faults is solved via SMC technique. (2) By means of the achieved fault boundary information, a scalar selection criterion is proposed to develop sliding mode controller for guaranteeing the reachability before the specified finite time. (3) A "partition–integration" technique is introduced to obtain the sufficient conditions for the SFTB of the closed-loop systems over the whole finite-time interval* $[0, T]$.

3.2 Actuator Fault Model

Consider the MJS defined on a complete probability space $(\Omega, \mathscr{F}, \Pr)$ as in system (1.6) with actuator faults, in which the parameter uncertainty $\Delta A_i(t)$, matched nonlinear perturbation $f_i(x(t))$ and unmatched external disturbance $w(t)$ satisfy Assumptions 1.2–1.4, respectively. The transition probabilities (TPs) are generated by (1.7) with the assumptions that all transition rates (TRs) are completely available. Accordingly, the plant can only receive the faulty control signal $u^F(t)$ instead of a complete control signal $u(t)$, as shown in Fig. 3.1.

In this chapter, it is assumed that the actuator faults may happen according to the following model:

$$u^F(t) = (I - \rho)u(t), \tag{3.1}$$

with $\rho \triangleq \text{diag}\{\rho_1, \ldots, \rho_m\}$ satisfying

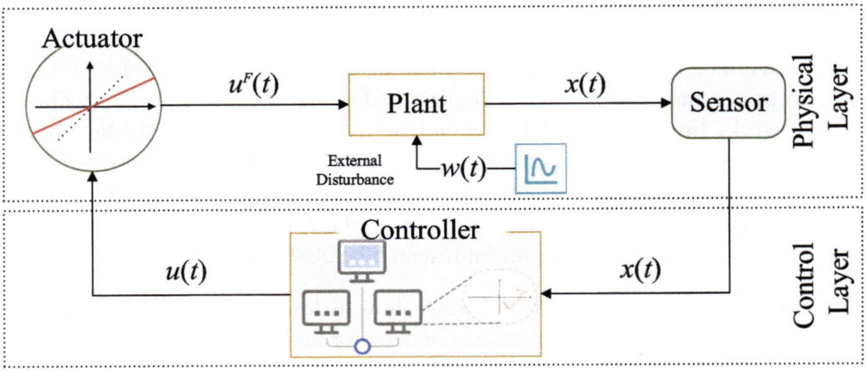

Fig. 3.1 Block diagram of MJSs subject to actuator faults

$$0 \leq \underline{\rho}_k \leq \rho_k \leq \bar{\rho}_k < 1, \qquad k = 1, 2, \ldots, m, \tag{3.2}$$

where the unknown parameter ρ_k $(k = 1, \ldots, m)$ denotes the loss of effectiveness of the kth actuator. Define $\underline{\rho} \triangleq \mathrm{diag}\{\underline{\rho}_1, \ldots, \underline{\rho}_m\}$, $\bar{\rho} \triangleq \mathrm{diag}\{\bar{\rho}_1, \ldots, \bar{\rho}_m\}$.

The system (1.6) subject to actuator faults (3.1) is described as

$$\dot{x}(t) = (A_i + \Delta A_i)x(t) + B_i(I - \rho)u(t) + B_i f_i(x(t)) + D_i w(t). \tag{3.3}$$

Remark 3.1 The actuator model in (3.1) is normal as $\underline{\rho}_k = \bar{\rho}_k = 0$ $(k = 1, 2, \ldots, m)$ and partly faulty as $0 < \underline{\rho}_k \leq \bar{\rho}_k < 1$ $(k = 1, 2, \ldots, m)$. Furthermore, due to different actuator components serving for different control signals, the component-based actuator faults are considered. Hence, the fault model in this chapter is more general.

3.3 Fault-Tolerant Controller Design

In this section, a mode-independent sliding functions will be firstly constructed as follows:

$$s(t) = Gx(t), \tag{3.4}$$

where $G \triangleq \sum_{i=1}^{N} \theta_i B_i^{\mathrm{T}} X_i$ and the matrix X_i is chosen such that GB_i is nonsingular for each $i \in \mathbb{S}$, which can be easily ensured by choosing $X_i > 0$ and $\theta_i > 0$ due to the full column of B_i.

Remark 3.2 In Chap. 2, we note that a mode-independent sliding surface may avoid the switching frequently from one mode to another while a mode-dependent one may increase the freedom of control design. In order to adequately utilize the mode information, we construct a mode-independent sliding function with the sliding gain consisting of the matrices related to all modes, which can be determined by choosing $X_i > 0$ and $\theta_i > 0$ $(i \in \mathbb{S})$.

For the given finite-time interval $[0, T]$, a suitable sliding mode controller should be designed to drive the states trajectories onto the specified sliding surface $s(t) = 0$ during $[0, T^*]$ with $T^* < T$ and maintain there for the rest time interval $[T^*, T]$. To this end, we design the following SMC law:

$$u(t) = -K_i x(t) - \gamma_i \mathrm{sgn}(s(t)), \tag{3.5}$$

with

$$\gamma_i \triangleq \varrho_i + \mu_i \delta, \tag{3.6}$$

$$\mu_i \triangleq \frac{\|(GB_i)^{-1} G D_i\|}{1 - \bar{\rho}_{\max}}, \tag{3.7}$$

and $\varrho_i > 0$ a adjustable parameter to be further described in Theorem 3.1, and $\bar{\rho}_{\max} \triangleq \max\{\bar{\rho}_1, \ldots, \bar{\rho}_m\}$.

3.4 Reachability Under Assigned Convergence Domain

In this section, it will be proven that the SMC law (3.5)–(3.7) can ensure the reachability of the specified sliding surfaces $s(t) = 0$.

Theorem 3.1 *For a given finite time T, the SMC law is designed as* (3.5)–(3.7) *and the adjustable parameters $\varrho_i > 0$ satisfies*

$$\varrho_i \geq \frac{\varkappa_i + \chi_i\sqrt{c_2/\lambda_{\min}(R)}}{1 - \bar{\rho}_{\max}}, \tag{3.8}$$

where

$$\varkappa_i \geq \frac{\lambda_{\max}\left((GB_i)^{-1}\right)}{T - \kappa}\|Gx(0)\|, \tag{3.9}$$

$$\begin{aligned} \chi_i \triangleq &\|(GB_i)^{-1}GA_i\| + \|I - \underline{\rho}\|\|K_i\| + \|(GB_i)^{-1}GM_i\|\|N_i\| \\ &+ \frac{1}{2}\left\|\sum_{j=1}^{N}\lambda_{ij}(GB_j)^{-1}G\right\| + \epsilon_i, \end{aligned} \tag{3.10}$$

then the state trajectories can be driven onto the sliding surface $s(t) = 0$ during the interval $[0, T^*]$ with $T^* < T$ and remain there (in the mean square) in the subsequent time.

Proof Choose the Lyapunov function:

$$V_1(x(t), i) = s^{\mathrm{T}}(t)(GB_i)^{-1}s(t). \tag{3.11}$$

Then, by Definition 1.4, we obtain the infinitesimal operator as

$$\begin{aligned} &\mathscr{L}V_1(x(t), i) \\ &= 2s^{\mathrm{T}}(t)(GB_i)^{-1}(GM_iF_i(t)N_i + GA_i)x(t) \\ &\quad + 2s^{\mathrm{T}}(t)f_i(x(t)) + 2s^{\mathrm{T}}(t)(GB_i)^{-1}GD_iw(t) \\ &\quad + s^{\mathrm{T}}(t)\sum_{j=1}^{N}\lambda_{ij}(GB_j)^{-1}Gx(t) + 2s^{\mathrm{T}}(t)(I - \rho)u(t). \end{aligned} \tag{3.12}$$

Substituting the control law (3.5)–(3.10) into (3.12) and noting $\|s(t)\| \leq |s(t)|$, we can obtain

$$\mathscr{L}V_1(x(t), i) \leq 2\|s(t)\|\big(\chi_i \|x(t)\| - ((1 - \bar{\rho}_{\max})\varrho_i - \varkappa_i) - \varkappa_i\big). \tag{3.13}$$

If the adjustable parameter ϱ_i satisfies the expression (3.8), one has $(1 - \bar{\rho}_{\max})\varrho_i - \varkappa_i \geq 0$ and

$$\frac{(1 - \bar{\rho}_{\max})\varrho_i - \varkappa_i}{\chi_i} \geq \sqrt{\frac{c_2}{\lambda_{\min}(R)}}. \tag{3.14}$$

Now, define the following domains:

$$\Phi_1 \triangleq \{x(t) : \chi_i \|x(t)\| \leq (1 - \bar{\rho}_{\max})\varrho_i - \varkappa_i\}, \tag{3.15}$$

$$\Phi_2 \triangleq \{x(t) : \sqrt{\lambda_{\min}(R)}\|x(t)\| \leq \sqrt{c_2}\}. \tag{3.16}$$

Apparently, one has $\Phi_2 \subseteq \Phi_1$. It will be proven in Theorem 3.2 later that the SFTB of the system (3.3) over the finite-time interval $[0, T]$ can be ensured, that is, one has $x^{\mathrm{T}}(t)Rx(t) < c_2$ in the mean square. That implies that one has $\|x(t)\| \leq \sqrt{c_2/\lambda_{\min}(R)}$ during the interval $[0, T]$ in the mean square. Thus, the state trajectories will remain within the set Φ_1 in the interval $[0, T]$ in the mean square, which yields from (3.13):

$$\mathscr{L}V_1(x(t), i) \leq -2\varkappa_i \|s(t)\|. \tag{3.17}$$

The following will further give the reaching instant T^*. To this end, we obtain from (3.11) $V_1(x(t), i) \leq \lambda_{\max}\left((GB_i)^{-1}\right)\|s(t)\|^2$. Then, one has from (3.17):

$$\mathscr{L}V_1(x(t), i) \leq -\frac{2\varkappa_i}{\sqrt{\lambda_{\max}\left((GB_i)^{-1}\right)}}\sqrt{V_1(x(t), i)}. \tag{3.18}$$

Integrating (3.18) from 0 to t, one can ensure that there exists an instant T^* satisfying

$$T^* = \frac{\sqrt{\lambda_{\max}\left((GB_i)^{-1}\right)}}{\varkappa_i}\sqrt{V_1(x(0), i)}, \tag{3.19}$$

such that for $t \geq T^*$, $\mathbb{E}\{V_1(x(t), i)\} = 0$, consequently, the sliding surface $s(t) = 0$ is attained (in the mean square).

Recalling $V_1(x(0), i) \leq \lambda_{\max}\left((GB_i)^{-1}\right)\|s(t)\|^2$ and the fact $s(0) = Gx(0)$, one has

$$T^* \leq \frac{\lambda_{\max}\left((GB_i)^{-1}\right)}{\varkappa_i}\|Gx(0)\|. \tag{3.20}$$

Thus, from (3.8) and (3.20), one obtains $T^* \leq T - \kappa < T$, which means that the state trajectories of system (3.3) will be driven onto the specified sliding surface $s(t) = 0$ in finite time T^* with $T^* < T$. This complete the proof. □

Remark 3.3 It should be pointed out that the designed SMC law (3.5)–(3.7) depends on the given time T and the bounds of actuator faults $\bar{\rho}_k$ $(k = 1, 2, \ldots, m)$ via the selection on the scalar ϱ_i. Hence, the proposed SMC law can effectively attenuate the effect of actuator faults during the specified finite-time interval. This just answers the first question (**Q1**).

3.5 SFTB Analysis

As is well known, there exist two phases for the state trajectories of SMC systems: the reaching phase within $[0, T^*]$ and the sliding motion phase within $[T^*, T]$. According to partition strategy in Lemma 1.4, the reaching phase within $[0, T^*]$ and the sliding motion phase within $[T^*, T]$ can be connected by parameters c^* and T^*. In this section, by analyzing SFTB problem of the MJSs (3.3) during the reaching phase and the sliding motion phase, respectively, sufficient conditions will be given to guarantee SFTB over the whole finite-time interval $[0, T]$. This answers the second question (**Q2**).

By substituting SMC law (3.5) into (3.3), we obtain the following closed-loop system:

$$\begin{aligned} \dot{x}(t) =& \big(A_i + \Delta A_i(t) - B_i(I - \rho)K_i\big)x(t) + B_i f_i(x(t)) \\ &+ D_i w(t) - B_i(I - \rho)\gamma_{si}(t), \end{aligned} \tag{3.21}$$

where $\gamma_{si}(t) \triangleq \gamma_i(t)\mathrm{sgn}(s(t))$.

3.5.1 *Reaching and Sliding Motion Phases Analysis*

In the following theorem, the sufficient condition is derived via the "partition–integration" technique to guarantee SFTB over the whole finite-time interval $[0, T]$.

Theorem 3.2 *For the given parameters (c_1, c_2, T, R, δ) and a feasible scalar η_i, if there exist positive constants c^*, ϑ, α_i, φ_i, ζ_i, ξ_i, and matrices $\mathcal{P}_i > 0$ and $\mathcal{L}_i$ $(\forall i \in \mathbb{S})$, satisfying the following inequalities:*

$$\begin{bmatrix} \Lambda_{11i} & \Lambda_{12i} \\ * & \Lambda_{22i} \end{bmatrix} < 0, \tag{3.22}$$

$$\begin{bmatrix} \Omega_{11i} & \Omega_{12i} \\ * & \Omega_{22i} \end{bmatrix} < 0, \tag{3.23}$$

$$c_1 < c^* < c_2, \tag{3.24}$$

$$\vartheta R^{-1} < \mathcal{P}_i < R^{-1}, \tag{3.25}$$

$$\begin{bmatrix} -e^{-\eta_i T}c^* + (2\varrho_i^2 T + (1+2\mu_i^2)\delta^2 T)\eta_i & \sqrt{c_1} \\ * & -\vartheta I \end{bmatrix} < 0, \tag{3.26}$$

$$-e^{-\eta_i T}c_2\vartheta + c^* + \eta_i\delta^2 T\vartheta < 0, \tag{3.27}$$

where

$$\begin{aligned}
\Lambda_{11i} &\triangleq \begin{bmatrix} \Xi_{1i} & D_i \\ * & -\eta_i I \end{bmatrix}, \\
\Lambda_{12i} &\triangleq \begin{bmatrix} -B_i(I-\bar{\rho}) & \alpha_i M_i & \mathcal{P}_i N_i^{\mathrm{T}} & \varphi_i B_i & \epsilon_i \mathcal{P}_i I & B_i & \mathcal{L}_i^{\mathrm{T}}\bar{\rho}^{\mathrm{T}} & M_i^k \\ 0 & 0 & 0 & 0 & 0 & 0 & 0 & 0 \end{bmatrix}, \\
\Lambda_{22i} &\triangleq -\mathrm{diag}\{\eta_i I, \alpha_i I, \alpha_i I, \varphi_i I, \varphi_i I, \zeta_i I, \zeta_i I, N_i^k\}, \\
\Xi_{1i} &\triangleq \mathcal{P}_i A_i^{\mathrm{T}} + A_i\mathcal{P}_i - \mathcal{L}_i^{\mathrm{T}} B_i^{\mathrm{T}} - B_i\mathcal{L}_i + (\lambda_{ii}-\eta_i)\mathcal{P}_i, \\
\Omega_{11i} &= \begin{bmatrix} \Xi_{2i} & \Gamma_i D_i \\ * & -\eta_i I \end{bmatrix}, \quad \Omega_{12i} \triangleq \begin{bmatrix} \xi_i\Gamma_i M_i & \mathcal{P}_i N_i^{\mathrm{T}} & M_i^k \\ 0 & 0 & 0 \end{bmatrix}, \\
\Omega_{22i} &\triangleq -\mathrm{diag}\{\xi_i I, \xi_i I, N_i^k\}, \quad \Gamma_i \triangleq I - B_i(GB_i)^{-1}G, \\
\Xi_{2i} &\triangleq \mathcal{P}_i A_i^{\mathrm{T}}\Gamma_i^{\mathrm{T}} + \Gamma_i A_i\mathcal{P}_i + (\lambda_{ii}-\eta_i)\mathcal{P}_i, \\
M_i^k &\triangleq [\sqrt{\lambda_{i1}}\mathcal{P}_i, \sqrt{\lambda_{i2}}\mathcal{P}_i, \ldots, \sqrt{\lambda_{iN}}\mathcal{P}_i], \\
N_i^k &\triangleq \mathrm{diag}\{\mathcal{P}_1, \mathcal{P}_2, \ldots, \mathcal{P}_N\}, i \neq j,
\end{aligned}$$

the closed-loop system (3.21) *is SFTB with respect to* $\aleph_{[0,T]}$. *Moreover, the controller gains in* (3.5) *are given as* $K_i = \mathcal{L}_i\mathcal{P}_i^{-1}$ $(i \in \mathbb{S})$.

Proof We shall analyze the SFTB during the two phases, respectively, over the finite-time interval $[0, T]$.

► **Step 1**. SFTB during reaching phase within $[0, T^*]$.

For the following Lyapunov functional candidate:

$$V_2(x(t), i) = x^{\mathrm{T}}(t)P_i x(t), \tag{3.28}$$

we obtain its infinitesimal operator as

$$\begin{aligned}
&\mathscr{L}V_2(x(t), i) \\
&= x^{\mathrm{T}}(t)\left(\bar{A}_i^{\mathrm{T}}P_i + P_i\bar{A}_i + \sum_{j=1}^{N}\lambda_{ij}P_j\right)x(t) + w^{\mathrm{T}}(t)D_i^{\mathrm{T}}P_i x(t) + x^{\mathrm{T}}(t)P_i D_i w(t) \\
&\quad + 2x^{\mathrm{T}}(t)P_i\Delta A_i(t)x(t) + 2x^{\mathrm{T}}(t)P_i B_i\rho K_i x(t) - \gamma_{si}^{\mathrm{T}}(t)(I-\rho)^{\mathrm{T}}B_i^{\mathrm{T}}P_i x(t) \\
&\quad - x^{\mathrm{T}}(t)P_i B_i(I-\rho)\gamma_{si}(t) + 2x^{\mathrm{T}}(t)P_i B_i f_i(x(t)),
\end{aligned} \tag{3.29}$$

where $\bar{A}_i \triangleq A_i - B_i K_i$.

For $\alpha_i > 0$, $\varphi_i > 0$ and $\zeta_i > 0$, one has

$$2x^{\mathrm{T}}(t)P_i\Delta A_i(t)x(t) \le \alpha_i x^{\mathrm{T}}(t)P_iM_iM_i^{\mathrm{T}}P_ix(t) + \alpha_i^{-1}x^{\mathrm{T}}(t)N_i^{\mathrm{T}}N_ix(t), \quad (3.30)$$
$$2x^{\mathrm{T}}(t)P_iB_if_i(x(t)) \le \varphi_i x^{\mathrm{T}}(t)P_iB_iB_i^{\mathrm{T}}P_ix(t) + \varphi_i^{-1}\epsilon_i^2x^{\mathrm{T}}(t)x(t), \quad (3.31)$$
$$2x^{\mathrm{T}}(t)P_iB_i\rho K_ix(t) \le \zeta_i^{-1}x^{\mathrm{T}}(t)K_i^{\mathrm{T}}\bar{\rho}^{\mathrm{T}}\bar{\rho}K_ix(t) + \zeta_i x^{\mathrm{T}}(t)P_iB_iB_i^{\mathrm{T}}P_ix(t). \quad (3.32)$$

Then, we have

$$\begin{aligned}
&\mathbb{E}\{\mathscr{L}V_2(x(t),i)\} \\
&\le x^{\mathrm{T}}(t)\Bigg(\bar{A}_i^{\mathrm{T}}P_i + P_i\bar{A}_i + \sum_{j=1}^{N}\lambda_{ij}P_j + \varphi_iP_iB_iB_i^{\mathrm{T}}P_i + \varphi_i^{-1}\epsilon_i^2I \\
&\quad + \alpha_iP_iM_iM_i^{\mathrm{T}}P_i + \alpha_i^{-1}N_i^{\mathrm{T}}N_i + \zeta_iP_iB_iB_i^{\mathrm{T}}P_i + \zeta_i^{-1}K_i^{\mathrm{T}}\bar{\rho}^{\mathrm{T}}\bar{\rho}K_i\Bigg)x(t) \\
&\quad + w^{\mathrm{T}}(t)D_i^{\mathrm{T}}P_ix(t) - x^{\mathrm{T}}(t)P_iB_i(I-\rho)\gamma_{si}(t) \\
&\quad + x^{\mathrm{T}}(t)P_iD_iw(t) - \gamma_{si}^{\mathrm{T}}(t)(I-\rho)^{\mathrm{T}}B_i^{\mathrm{T}}P_ix(t).
\end{aligned} \quad (3.33)$$

Define an auxiliary function as

$$J_1(x(t),i) = \mathscr{L}V_2(x(t),i) - \eta_iV_2(x(t),i) - \eta_iw^{\mathrm{T}}(t)w(t) - \eta_i\gamma_{si}^{\mathrm{T}}(t)\gamma_{si}(t).$$

We have $J_1(x(t),i) \le \varpi^{\mathrm{T}}(t)\Psi_{1i}\varpi(t)$ from (3.29)–(3.32) with $\varpi(t) \triangleq [x^{\mathrm{T}}(t)\ w^{\mathrm{T}}(t)\ \gamma_{si}^{\mathrm{T}}(t)]^{\mathrm{T}}$. The following matrix inequality implies $J_1(x(t),i) < 0$ by Schur Complement (Lemma 1.3):

$$\begin{bmatrix}
\Psi_i & P_iD_i & -P_iB_i(I-\bar{\rho}) & \alpha_iP_iM_i & N_i^{\mathrm{T}} & \varphi_iP_iB_i & \epsilon_iI & P_iB_i & K_i^{\mathrm{T}}\bar{\rho}^{\mathrm{T}} \\
* & -\eta_iI & 0 & 0 & 0 & 0 & 0 & 0 & 0 \\
* & * & -\eta_iI & 0 & 0 & 0 & 0 & 0 & 0 \\
* & * & * & -\alpha_iI & 0 & 0 & 0 & 0 & 0 \\
* & * & * & * & -\alpha_iI & 0 & 0 & 0 & 0 \\
* & * & * & * & * & -\varphi_iI & 0 & 0 & 0 \\
* & * & * & * & * & * & -\varphi_iI & 0 & 0 \\
* & * & * & * & * & * & * & -\zeta_iI & 0 \\
* & * & * & * & * & * & * & * & -\zeta_iI
\end{bmatrix} < 0, \quad (3.34)$$

where $\Psi_i \triangleq A_i^{\mathrm{T}}P_i + P_iA_i - K_i^{\mathrm{T}}B_i^{\mathrm{T}}P_i - P_iB_iK_i + \sum_{j=1}^{N}\lambda_{ij}P_j - \eta_iP_i$. Let $\mathcal{P}_i \triangleq P_i^{-1}$ and $\mathcal{L}_i \triangleq K_i\mathcal{P}_i$. It is easily shown that (3.34) can be ensured by (3.22).

Moreover, it follows from $J_1(x(t),i) < 0$ that

$$\mathscr{L}V_2(x(t),i) < \eta_iV_2(x(t),i) + \eta_iw^{\mathrm{T}}(t)w(t) + \eta_i\gamma_{si}^{\mathrm{T}}(t)\gamma_{si}(t). \quad (3.35)$$

Besides, $\gamma_{si}(t)$ satisfies the following inequality:

$$\gamma_{si}^{\mathrm{T}}(t)\gamma_{si}(t) = \gamma_i^{\mathrm{T}}(t)\gamma_i(t) \le 2\varrho_i^2 + 2\mu_i^2w^{\mathrm{T}}(t)w(t). \quad (3.36)$$

Multiplying both sides of expression (3.35) by $e^{-\eta_i t}$ and integrating the obtained inequality from 0 to t with $t \in [0, T^*]$, we have

$$\begin{aligned}
&e^{-\eta_i t}\mathbb{E}\{V_2(x(t), i)\} \\
&< V_2(x(0), r_0) + 2\varrho_i^2\eta_i \int_0^t e^{-\eta_i \tau}\mathrm{d}\tau + (1 + 2\mu_i^2)\eta_i \int_0^t e^{-\eta_i \tau} w^{\mathrm{T}}(\tau)w(\tau)\mathrm{d}\tau \\
&\leq \bar{\sigma}_P c_1 + 2\varrho_i^2 T\eta_i + (1 + 2\mu_i^2)\eta_i\delta^2 T.
\end{aligned} \tag{3.37}$$

On the other hand, it follows from (3.28) that

$$e^{-\eta_i t}\mathbb{E}\{V_2(x(t), i)\} \geq e^{-\eta_i T}\underline{\sigma}_P\mathbb{E}\{x^{\mathrm{T}}(t)Rx(t)\}. \tag{3.38}$$

It yields from (3.37) and (3.38) that

$$\mathbb{E}\{x^{\mathrm{T}}(t)Rx(t)\} \leq \frac{\bar{\sigma}_P c_1 + 2\varrho_i^2 T\eta_i + (1 + 2\mu_i^2)\eta_i\delta^2 T}{e^{-\eta_i T}\underline{\sigma}_P}, \tag{3.39}$$

where

$$\bar{\sigma}_P \triangleq \max_{i \in \mathbb{S}} \left\{\lambda_{\max}\left(R^{-\frac{1}{2}} P_i R^{-\frac{1}{2}}\right)\right\}, \ \underline{\sigma}_P \triangleq \min_{i \in \mathbb{S}} \left\{\lambda_{\min}\left(R^{-\frac{1}{2}} P_i R^{-\frac{1}{2}}\right)\right\}.$$

Further, if there exists scalar $c^* \in (c_1, c_2)$ satisfying

$$\frac{\bar{\sigma}_P c_1 + 2\varrho_i^2 T\eta_i + (1 + 2\mu_i^2)\eta_i\delta^2 T}{\underline{\sigma}_P} < e^{-\eta_i T}c^*, \tag{3.40}$$

we obtain from (3.39) that $\mathbb{E}\{x^{\mathrm{T}}(t)Rx(t)\} < c^*$ for all $t \in [0, T^*]$. Thus, according to Definition 1.5, the closed-loop system (3.21) is SFTB with respect to $\aleph_{[0,T^*]}$. By the condition (3.25), one has $1 < \underline{\sigma}_P$, $\bar{\sigma}_P < \frac{1}{\vartheta}$. Thus, it is easily shown that the inequality (3.40) can be ensured by the inequality (3.26).

► **Step 2**. SFTB over sliding motion phase within $[T^*, T]$.

According to SMC theory, when the state trajectories maintain in the sliding surface $s(t) = 0$, we have the following equivalent controller:

$$\begin{aligned}
u_{eq}(t) = &- (I - \rho)^{-1}(GB_i)^{-1}G\big[(A_i + \Delta A_i(t))x(t) \\
&+ D_i w(t)\big] - (I - \rho)^{-1}f_i(x(t)).
\end{aligned} \tag{3.41}$$

By substituting (3.41) into (3.3), we get

$$\dot{x}(t) = (\tilde{A}_i + \Delta\tilde{A}_i(t))x(t) + \tilde{D}_i w(t), \tag{3.42}$$

where $\tilde{A}_i \triangleq \Gamma_i A_i$, $\Delta\tilde{A}_i(t) \triangleq \Gamma_i\Delta A_i(t)$ and $\tilde{D}_i \triangleq \Gamma_i D_i$.

Choose the Lyapunov functional candidate as (3.28). We obtain the infinitesimal operator as

$$\mathscr{L}V_2(x(t),i) = x^{\mathrm{T}}(t)\left(\mathrm{He}(P_i(\tilde{A}_i + \Delta\tilde{A}_i)) + \sum_{j=1}^{N}\lambda_{ij}P_j\right)x(t) + w^{\mathrm{T}}(t)\tilde{D}_i^{\mathrm{T}}P_ix(t) + x^{\mathrm{T}}(t)P_i\tilde{D}_iw(t). \tag{3.43}$$

For $\xi_i > 0$, one has

$$2x^{\mathrm{T}}(t)P_i\Gamma_i\Delta A_i(t)x(t) \le \xi_i x^{\mathrm{T}}(t)P_i\Gamma_iM_iM_i^{\mathrm{T}}\Gamma_i^{\mathrm{T}}P_ix(t) + \xi_i^{-1}x^{\mathrm{T}}(t)N_i^{\mathrm{T}}N_ix(t). \tag{3.44}$$

Define an auxiliary function as follows:

$$J_2(x(t),i) = \mathscr{L}V_2(x(t),i) - \eta_iV_2(x(t),i) - \eta_iw^{\mathrm{T}}(t)w(t). \tag{3.45}$$

We have $J_2(x(t),i) \le \varsigma^{\mathrm{T}}(t)\Psi_{2i}\varsigma(t)$ from (3.43)–(3.44), where $\varsigma(t) \triangleq [x(t)\ w(t)]^{\mathrm{T}}$. The following matrix inequality implies $J_2(x(t),i) < 0$ by Schur Complement (Lemma 1.3):

$$\begin{bmatrix} A_i^{\mathrm{T}}\Gamma_i^{\mathrm{T}}P_i + P_iA_i\Gamma_i + \sum_{j=1}^{N}\lambda_{ij}P_j - \eta_iP_i & P_i\Gamma_iD_i & \xi_iP_i\Gamma_iM_i & N_i^{\mathrm{T}} \\ * & -\eta_iI & 0 & 0 \\ * & * & -\xi_iI & 0 \\ * & * & * & -\xi_iI \end{bmatrix} < 0. \tag{3.46}$$

It is easily shown that (3.46) can be ensured by (3.23).

By means of $J_2(x(t),i) < 0$, we obtain

$$\mathscr{L}V_2(x(t),i) < \eta_iV_2(x(t),i) + \eta_iw^{\mathrm{T}}(t)w(t). \tag{3.47}$$

It is shown in **Step 1** that the closed-loop system is SFTB with respect to $\aleph_{[0,T^*]}$ over the reaching phase, which implies that the initial condition of sliding motion phase at instant T^* satisfies $\mathbb{E}\{x^{\mathrm{T}}(T^*)Rx(T^*)\} < c^*$. Thus, by multiplying both sides of expression (3.47) by $e^{-\eta_it}$ and integrating the obtained inequality from T^* to t with $t \in [T^*, T]$, we have

$$\begin{aligned} &e^{-\eta_it}\mathbb{E}\{V_2(x(t),i)\} \\ <&e^{-\eta_iT^*}\mathbb{E}\{V_2(x(T^*),r_{T^*})\} + \eta_i\int_{T^*}^{t}e^{-\eta_i\tau}w^{\mathrm{T}}(\tau)w(\tau)\mathrm{d}\tau \\ <&x^{\mathrm{T}}(T^*)P_ix(T^*) + \eta_i\delta^2T \\ \le&\bar{\sigma_P}c^* + \eta_i\delta^2T. \end{aligned} \tag{3.48}$$

In addition, it follows from (3.28) that

$$e^{-\eta_i t}\mathbb{E}\{V_2(x(t),i)\} \geq e^{-\eta_i T}\underline{\sigma}_P\mathbb{E}\{x^{\mathrm{T}}(t)Rx(t)\}. \tag{3.49}$$

Putting together (3.48) and (3.49), we have

$$\mathbb{E}\{x^{\mathrm{T}}(t)Rx(t)\} \leq \frac{\bar{\sigma}_P c^* + \eta_i\delta^2 T}{e^{-\eta_i T}\underline{\sigma}_P}. \tag{3.50}$$

Further, if $c^* \in (c_1, c_2)$ satisfies

$$\frac{\bar{\sigma}_P c^* + \eta_i\delta^2 T}{\underline{\sigma}_P} < e^{-\eta_i T}c_2, \tag{3.51}$$

we obtain from (3.50) that $\mathbb{E}\{x^{\mathrm{T}}(t)Rx(t)\} < c_2$ for all $t \in [T^*, T]$. Moreover, it is easily shown that the inequalities (3.51) can be ensured by (3.27).

Combining **Steps 1 and 2**, it can be shown that the SFTB of closed-loop system (3.21) can be ensured by the conditions (3.22)–(3.27), which completes the proof. □

3.5.2 Incremental-Technique-Based Search Algorithm

It is noted that for solving matrix inequalities (3.24)–(3.27) in Theorem 3.2, the parameters ϱ_i and η_i should be given in advance. That is, if ϱ_i and η_i are prior, matrix inequalities (3.24)–(3.27) can be linear. However, it is seen from (3.8) that ϱ_i is dependent on the norm of the control gain K_i. Since K_i will be obtained only after solving (3.24)–(3.27) in Theorem 3.2, it is difficult to choose the parameter ϱ_i according to expression (3.8).

On the other hand, it can be seen from (3.24), (3.25), and (3.27) that the scalar η_i should be selected in the range of $(0, \frac{1}{T}\ln\frac{c_2}{c_1})$ (The detailed derivation can refer to Ref. [11]). Apparently, a desirable η_i within $(0, \frac{1}{T}\ln\frac{c_2}{c_1})$ should make the obtained control gain matrix K_i to have minimum norm. To this end, a search algorithm for suitable parameters ϱ_i and η_i is given as follows:

Algorithm 3.1 Searching parameters ϱ_i and η_i simultaneously

---

Choose an initial parameter candidate $\varrho_i^p = \frac{\varkappa_i + \chi_{1i}\sqrt{c_2/\lambda_{\min}(R)}}{1-\bar{\rho}_{\max}} + 0.1$, where χ_{1i} is calculated by $\chi_{1i} \triangleq \chi_i - \|I - \underline{\rho}\|\|K_i\|$ with χ_i as in (3.10).

Let $L = \frac{1}{T}\ln\frac{c_2}{c_1}$ and $\varrho_i = \varrho_i^p$, and substitute ϱ_i into linear matrix inequalities (LMIs) (3.22)–(3.27).

for $q = 1$ **to** $\lfloor 100L \rfloor$

  Let $\eta_i^q = 0.01q$ and $\eta_i = \eta_i^q$, and substitute η_i into LMIs (3.22)–(3.27).

  Solve LMIs (3.22)–(3.27).

  if LMIs (3.22)–(3.27) is feasible.

7: Calculate $\|K_i\|$ by $K_i = \mathcal{L}_i \mathcal{P}_i^{-1}$ where $\mathcal{L}_i$ and $\mathcal{P}_i$ is the solution of LMIs (3.22)–(3.27).
8: Let $K_i^q = K_i$.
9: **else**
10: $\|K_i^q\| = \inf$.
11: **end if**
12: **end for**
13: Select $K_i = \min\{\|K_i^1\|, \|K_i^2\|, \ldots, \|K_i^{\lfloor 100L \rfloor}\|\}$. Let the superscript of the minimum be min.
14: Let $\eta_i = \eta_i^{\min}$
15: **if** $\varrho_i^p \geq \frac{\varkappa_i + \chi_i \sqrt{c_2/\lambda_{\min}(R)}}{1-\bar{\rho}_{\max}}$ where $\varkappa_i$ and χ_i are given as in (3.9)–(3.10).
16: $\varrho_i = \varrho_i^p$
17: **else**
18: Set $\varrho_i^p = \varrho_i^p + 0.1$, and repeat steps 2–14 until $\varrho_i^p \geq \frac{\varkappa_i + \chi_i \sqrt{c_2/\lambda_{\min}(R)}}{1-\bar{\rho}_{\max}}$.
19: **end if**

Remark 3.4 Note that ϱ_i is dependent on the determination of controller gain K_i, and the design of controller gain K_i on the condition of ϱ_i. It apparently is difficult to deal with this complex coupling issue. In Chap. 2, this difficulty is solved by introducing a set of constraint condition $K_i^{\mathrm{T}} K_i < \zeta_i I$ and employing corresponding inequality scaling approaches. Although this dealing method can derive solvable conditions, it would bring large conservatism for solving controller gains. In this chapter, we relax the constraint condition $K_i^{\mathrm{T}} K_i < \zeta_i I$ and develop Algorithm 3.1, which further increases flexibility and freedom in the controller design.

3.6 Simulation

In this section, two examples will be given to demonstrate the validity of our design approach. The first numerical example is used to show that the obtained controller can ensure SFTB of the MJSs subject to actuator faults, and the second practical example is derived from an aircraft engine system to illustrate the applicability of the obtained results.

3.6.1 Numerical Example

Consider a two-mode MJS (3.3) with parameters as follows:

Subsystem 1:

$$A_1 = \begin{bmatrix} -0.1 & 0.7 \\ -0.5 & 0.2 \end{bmatrix},\ B_1 = \begin{bmatrix} 0.2 \\ 0.3 \end{bmatrix},\ D_1 = \begin{bmatrix} 0.3 \\ 0.2 \end{bmatrix},\ M_1 = \begin{bmatrix} 0.1 \\ 0.2 \end{bmatrix},\ N_1 = \begin{bmatrix} 0.2 & 0.2 \end{bmatrix},$$

Subsystem 2:

$$A_2 = \begin{bmatrix} -0.5 & 0.2 \\ -0.7 & -0.4 \end{bmatrix}, \; B_2 = \begin{bmatrix} 1.1 \\ -0.2 \end{bmatrix}, \; D_2 = \begin{bmatrix} 0.2 \\ 0.1 \end{bmatrix}, \; M_2 = \begin{bmatrix} 0.2 \\ 0.1 \end{bmatrix}, \; N_2 = \begin{bmatrix} 0.3 & 0.2 \end{bmatrix},$$

and $f_i(x(t)) = 0.02\sin(\sqrt{x_1^2 + x_2^2})(i = 1, 2)$, $w(t) = 0.1e^{-t}$ with $\delta = 0.1$ and $\epsilon_1 = \epsilon_2 = 0.02$. It is assumed that the TR matrix is given as

$$\Lambda = \begin{bmatrix} -0.3 & 0.3 \\ 0.8 & -0.8 \end{bmatrix},$$

and the bounds of actuator faults (3.1) are $\underline{\rho} = 0.4$, $\bar{\rho} = 0.7$.

Set the SFTB parameters as $c_1 = 2$, $c_2 = 12$, $T = 8$ and $R = \text{diag}\{40, 36\}$. By using Algorithm 3.1, the parameters ρ_i and η_i can be obtained as $\varrho_1 = 0.8069$, $\varrho_2 = 0.4501$, $\eta_1 = 0.17$, $\eta_2 = 0.17$, and the parameter μ_i in (3.7) is given as $\mu_1 = 4.0100$, $\mu_2 = 0.5201$. By solving the conditions (3.22)–(3.27), we obtain

$$K_1 = \begin{bmatrix} 1.0297 & 1.6893 \end{bmatrix}, \; K_2 = \begin{bmatrix} 0.9341 & -0.1359 \end{bmatrix},$$

and $c^* = 10.8719$. Then, for the chosen scalars $\theta_1 = \theta_2 = 0.5$ and matrices $X_1 = \text{diag}\{4, 1\}$, $X_2 = \text{diag}\{1.5, 2\}$, the desired sliding function (3.4) and SMC law (3.5) can be obtained as

$$\begin{aligned} s(t) =& [1.2250 \quad -0.0500]x(t), \\ u(t) =& \begin{cases} -[1.0297 \quad 1.6893]x(t) - 1.3260\text{sgn}(s(t)), & \text{under mode 1} \\ -[0.9341 \quad -0.1359]x(t) - 0.5100\text{sgn}(s(t)) & \text{under mode 2.} \end{cases} \end{aligned}$$

Under the above sliding surface $s(t) = 0$ and sliding mode controller, the simulation results with initial state $x(0) = [-0.03 \quad 0.02]^T$ are shown in Fig. 3.2. It is shown in Fig. 3.2(a) that the closed-loop system can be controlled to attain the SFTB effectively in spite of the effect of actuator faults.

3.6.2 Application to Aircraft Engine System

In this section, an example on aircraft engine system will be given to demonstrate the effectiveness of the proposed fault-tolerant SMC method and applicability of the obtained results. Consider the linearized model of the aircraft engine system as a modified version of the example in [5], whose system matrix is taken as follows:

$$A(t) = \begin{bmatrix} -1.46 & 0 \\ 0.1643 + 0.5g(t) & -0.4 + g(t) \end{bmatrix}, \tag{3.52}$$

where $g(t)$ is an uncertain model parameter, whose value is taken as -0.5 or -2 according to a Markov process $r_t \in \{1, 2\}$ with the TR matrix as:

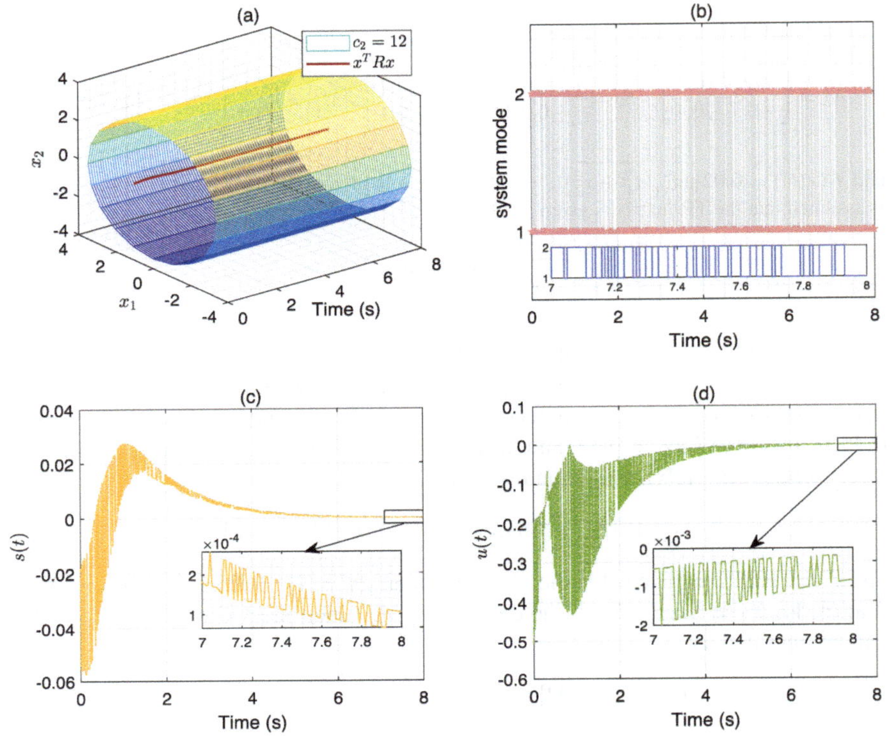

Fig. 3.2 Simulation results of controlled system: **a** evolution of weighted system state variable $x^{\mathrm{T}}(t)Rx(t)$; **b** possible mode transition sequence r_t; **c** sliding variable $s(t)$; **d** control signal $u(t)$

$$\Lambda = \begin{bmatrix} -0.5 & 0.5 \\ 0.7 & -0.7 \end{bmatrix}.$$

Accordingly, we have the following two subsystems:

Subsystem 1:

$$A_1 = \begin{bmatrix} -1.46 & 0 \\ -0.0857 & -0.9 \end{bmatrix},\ B_1 = \begin{bmatrix} 0.05 \\ 0.1 \end{bmatrix},\ D_1 = \begin{bmatrix} 0.02 \\ 0.03 \end{bmatrix},$$

Subsystem 2:

$$A_2 = \begin{bmatrix} -1.46 & 0 \\ -0.8357 & -2.4 \end{bmatrix},\ B_2 = \begin{bmatrix} 0.1 \\ 0.1 \end{bmatrix},\ D_2 = \begin{bmatrix} 0.01 \\ 0.02 \end{bmatrix},$$

and $f_i(x(t)) = 0.02\sin\left(\sqrt{x_1^2 + x_2^2}\right)$ $(i = 1, 2)$, $w(t) = 0.15\mathrm{e}^{-t}$ with $\delta = 0.15$ and $\epsilon_1 = \epsilon_2 = 0.02$. Considering that there exist no parameter uncertainty in the engine system, we have $\Delta A_i = 0$ $(i = 1, 2)$. The bounds of actuator faults (3.1) are set as $\underline{\rho} = 0.5$, $\bar{\rho} = 0.9$.

Now, the objective of this example is to design an SMC law to achieve the SFTB with $c_1 = 2$, $c_2 = 12$, $T = 10$ and $R = \text{diag}\{40, 36\}$. To this end, by using the proposed Algorithm 3.1, the parameters ϱ_i and η_i can be obtained as $\varrho_1 = 0.3102$, $\varrho_2 = 0.5847$, $\eta_1 = 0.11$, $\eta_2 = 0.07$, and the parameters μ_i $(i = 1, 2)$ in (3.7) are given as $\mu_1 = 3.2100$, $\mu_2 = 1.3433$. For the chosen matrices $X_1 = \text{diag}\{4, 1\}$, $X_2 = \text{diag}\{1.5, 2\}$ and scalars $\theta_1 = \theta_2 = 0.5$, by solving LMIs (3.22)–(3.27) in Theorem 3.2, we obtain

$$K_1 = \begin{bmatrix} 0.1388 & 0.1037 \end{bmatrix}, \; K_2 = \begin{bmatrix} -0.0400 & -0.1596 \end{bmatrix},$$

and $c^* = 10.3722$.

The simulation results with initial states $x(0) = [-0.03 \;\; 0.02]^{\mathrm{T}}$ are shown in Figs. 3.3, 3.4 and 3.5. In order to illustrate the adaptation for stochastic mode jumping,

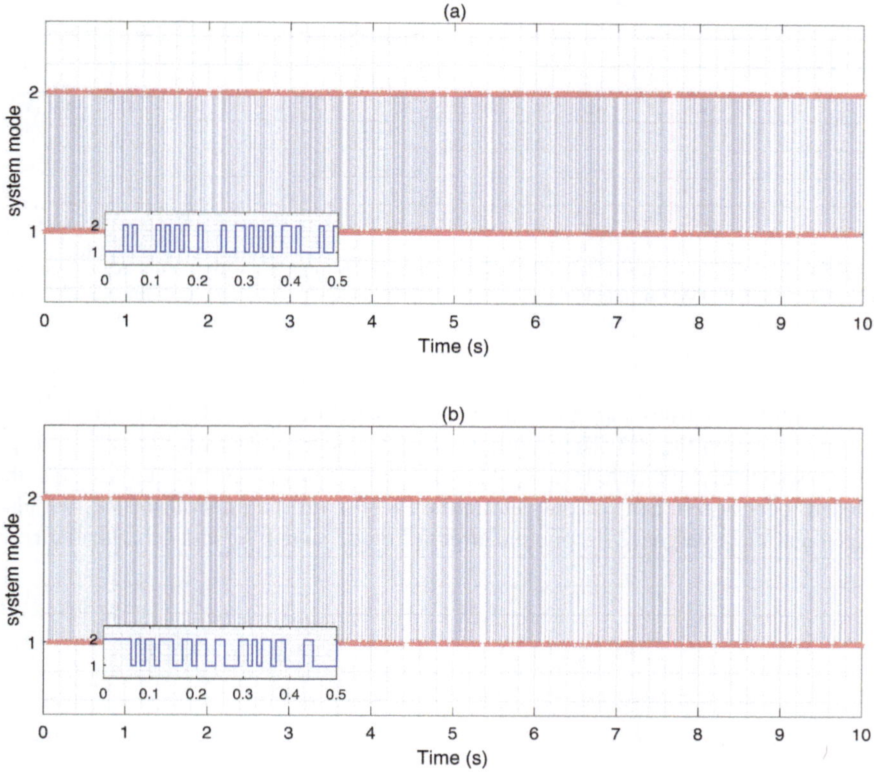

Fig. 3.3 Two possible mode transition sequences of aircraft engine system: **a** mode transition sequence $r_t^{(1)}$; **b** mode transition sequence $r_t^{(2)}$

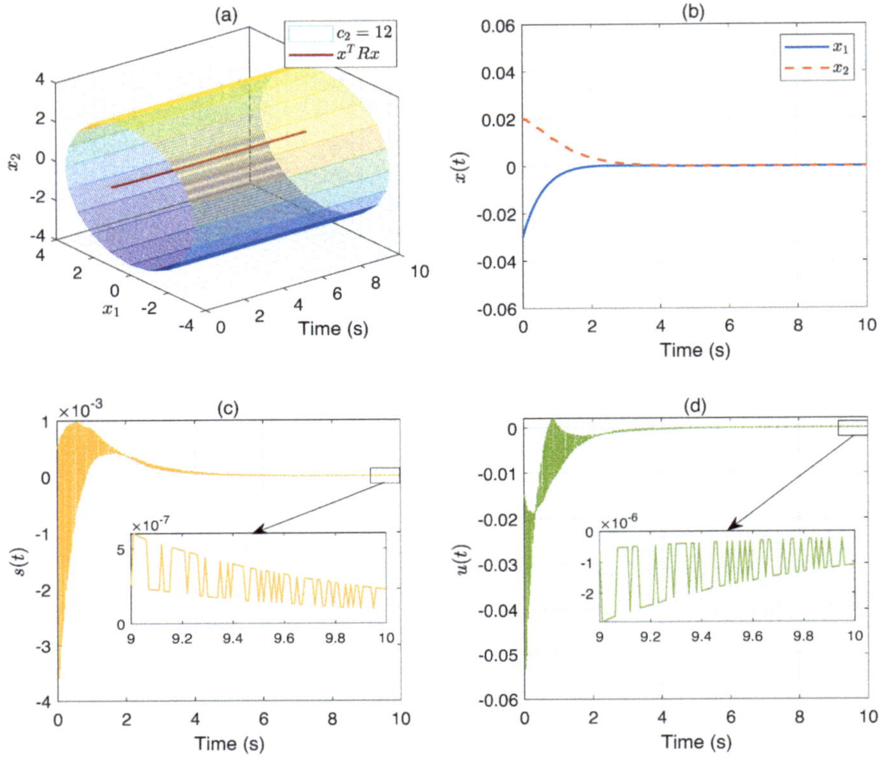

Fig. 3.4 Simulation results of aircraft engine system under mode transition sequence $r_t^{(1)}$: **a** evolution of weighted system state variable $x^{\mathrm{T}}(t)Rx(t)$; **b** state responses $x(t)$; **c** sliding variable $s(t)$; **d** control signal $u(t)$

we operate twice trials and randomly generate two different mode sequences $r_t^{(1)}$ and $r_t^{(2)}$, as shown in Fig. 3.3. The simulation results under mode sequences $r_t^{(1)}$ and $r_t^{(2)}$ are, respectively, exhibited in Figs. 3.4 and 3.5, from which we can see that the closed-loop system can not only attain the SFTB over the interval [0, 10], but also has a quick convergence despite the effect of mode switching and actuator faults.

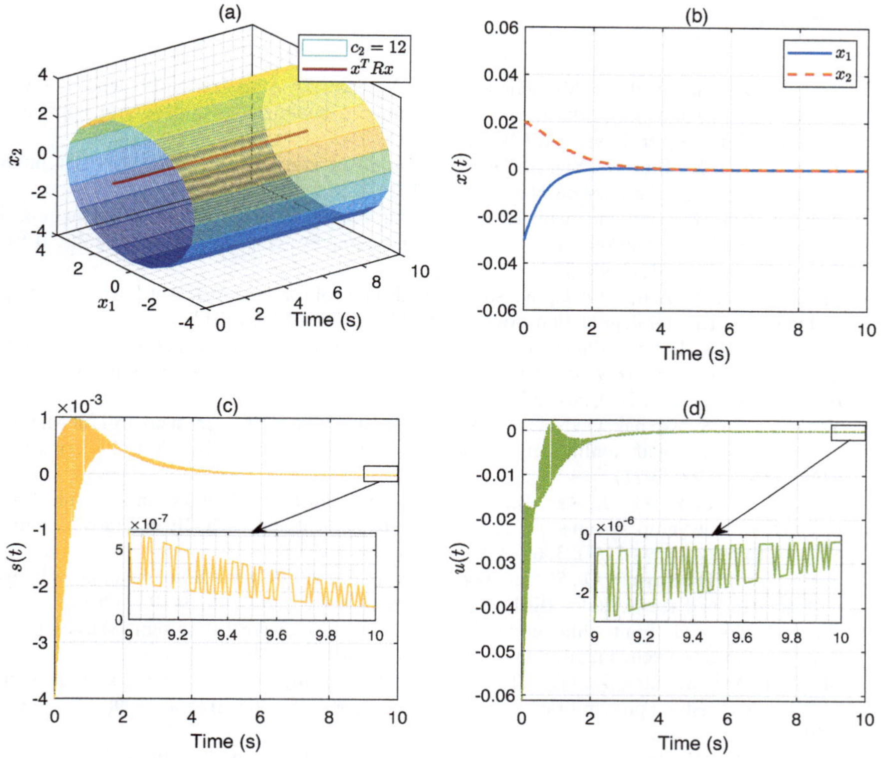

Fig. 3.5 Simulation results of controlled systems under mode transition sequence $r_t^{(2)}$: **a** evolution of weighted system state variable $x^{\mathrm{T}}(t)Rx(t)$; **b** state responses $x(t)$; **c** sliding variable $s(t)$; **d** control signal $u(t)$

3.7 Conclusion

In this chapter, we have investigated the SFTB problem via SMC methods for a class of MJSs subject to actuator faults, exogenous disturbance and parameter uncertainties. It has been manifested that the sliding mode reachability with $T^* < T$ for any given T can be attained by a scalar selection criterion. Moreover, by using the "partition–integration" technique, the sufficient conditions achieving SFTB during both reaching phase and sliding motion phase have been obtained, respectively.

References

1. Aravinth, N., Sakthivel, R., Mohammadzadeh, A., Saat, S.: Stabilization of delayed semi-Markov jump neural networks with actuator faults: a quantized hybrid control approach. Nonlinear Anal.: Hybrid Syst. **54**, 101509 (2024)
2. Yang, H., Wu, X., Han, H., Yin, S.: Observer-based self-triggered FTC for Markovian jump networked systems. Automatica **163**, 111602 (2024)
3. Aravind, R.V., Balasubramaniam, P.: Membership-function-dependent design of quantized fuzzy sampled-data controller for semi-Markovian jump systems with actuator faults. IEEE Trans. Fuzzy Syst. **31**(1), 40–52 (2023)
4. Chen, B., Niu, Y., Zou, Y.: Adaptive sliding mode control for stochastic Markovian jumping systems with actuator degradation. Automatica **49**(6), 1748–1754 (2013)
5. Li, M., Chen, X., Liu, M., Zhang, Y., Zhang, H.: Asynchronous adaptive fault-tolerant sliding-mode control for T-S fuzzy singular Markovian jump systems with uncertain transition rates. IEEE Trans. Cybern. **52**(1), 544–555 (2022)
6. Yang, H., Yin, S., Kaynak, O.: Neural network-based adaptive fault-tolerant control for Markovian jump systems with nonlinearity and actuator faults. IEEE Trans. Syst. Man Cybern. Syst. **51**(6), 3687–3698 (2021)
7. Fenga, X., Kao, Y., Xie, J., Kao, B.: Event-triggered fault-tolerant sliding mode control for discrete Markov jump systems via guaranteed cost control approach. IEEE Trans. Circuits Syst. II: Express Briefs **71**(7), 3363–3367 (2024)
8. Yin, S., Yang, H., Kaynak, O.: Sliding mode observer-based FTC for Markovian jump systems with actuator and sensor faults. IEEE Trans. Autom. Control **82**(4), 1933–1939 (2017)
9. Chen, H., Zong, G.: Finite-time output feedback control for discrete-time semi-Markov jump systems with mismatched modes. Inter. J. Syst. Sci. **55**(5), 941–953 (2024)
10. Shen, M., Yan, S., Tang, Z., Gu, Z.: Finite-time H_∞ filtering of Markov jump systems with incomplete transition probabilities: a probability approach. IET Signal Process. **9**(7), 572–578 (2015)
11. Song, J., Niu, Y., Zou, Y.: Finite-time stabilization via sliding mode control. IEEE Trans. Autom. Control **62**(3), 1478–1483 (2017)

Chapter 4
Component-Wise Sliding Mode Control for Dealing with Multiple Actuator Nonlinearities

4.1 Introduction

As discussed in Chap. 3, due to the inevitable physical constraints, actuators always contain a common source of nonlinearities, such as sectors, deadzones, saturation, backlash, and so on. Among them, sectors and deadzones are most common two kinds of nonlinearities occurring in actuators. The actuator deadzone refers to a range of control input values where there is no output response from the system. This situation occurs when the system does not react to changes in input until a certain threshold is exceeded [1, 2]. Sector nonlinearity describes a specific type of nonlinearity found in control systems, especially, in actuators. This phenomenon occurs when the output of an actuator is limited within a defined range or sector of input values [3, 4].

These kinds of nonlinear inputs may deteriorate the performance of control systems and even result in instability [5]. Some interesting works have been recently made on Markov jump systems (MJSs) subject to actuator nonlinearities including, e.g., dissipative control [6], adaptive control [7], sliding mode control (SMC) [8], et al. Among those control methods, SMC is an effective robust control method for a system with parameter uncertainties and external disturbances [9]. Further considering the transient behavior with short working time for MJSs, an asynchronous secure controller has been designed in [10] to ensure the stochastic finite-time boundedness (SFTB) of the closed-loop system under a multi-level probabilistic event-triggered mechanism. Taking the incompletely known transition rates (TRs) into account, both SFTB and input-output stochastic finite-time stabilization (IO-SFTS) performances have been ensured in [11] by asynchronous filtering methods, in which the TRs consist of completely known case and completely unknown case; while in [12], a finite-time fault detection has been investigated for the MJSs with partially known TRs, where all TRs have been assumed to be uncertain.

However, the above works involving in SFTB cannot be extended to the finite-time SMC for solving actuator nonlinearities, especially, under uncertain TRs, since the SMC technique has its particularity in the structure. In this chapter, we deal with

Z. Cao and Y. Niu, *Finite-Time Sliding Mode Control for Markov Jump Systems*,
Studies in Systems, Decision and Control 603,
https://doi.org/10.1007/978-981-96-7651-4_4

the SFTB for a class of MJSs with uncertain TRs under actuator nonlinearities which contain both sector nonlinearities and deadzones in each component. The reachability of the specified mode-dependent sliding surface is ensured so that the arrival time is not longer than the given finite time ($T^* < T$). Then, sufficient conditions for designing component-wise sliding mode controller are achieved to guarantee the SFTB of the closed-loop system over the whole finite time including reaching phase and sliding motion phase.

4.2 Problem Description

Consider a class of MJSs on a complete probability space $(\Omega, \mathscr{F}, \Pr)$ as in system (1.6) ($f_i(x(t)) \equiv 0$) with actuator nonlinearities, that is, for mode $r_t = i \in \mathbb{S}$,

$$\dot{x}(t) = (A_i + \Delta A_i(t))x(t) + B_i\Phi(u(t)) + D_i w(t), \tag{4.1}$$

where $x(t) \in \mathbb{R}^n$ is the state, $u(t) \in \mathbb{R}^m$ is the control signal, $w(t) \in \mathbb{R}^q$ is the external disturbance, and $\Delta A_i(t)$ is the parameter uncertainty. $\Delta A_i(t)$ and $w(t)$ satisfy Assumptions 1.2 and 1.4, respectively. As shown in Fig. 4.1, the uncertain TR and actuator nonlinearity issues are focused in this chapter.

4.2.1 Uncertain TR Issue

The TRs of the jumping process $\{r_t\}_{t\geq 0}$ are assumed to be not exactly known with transition probabilities (TPs) generated by (1.7), including the case that the TR λ_{ij} is completely unknown and the case that the TR λ_{ij} has uncertainties. For instance, for system (4.1) with four operation modes, the TR matrix Λ might be described as

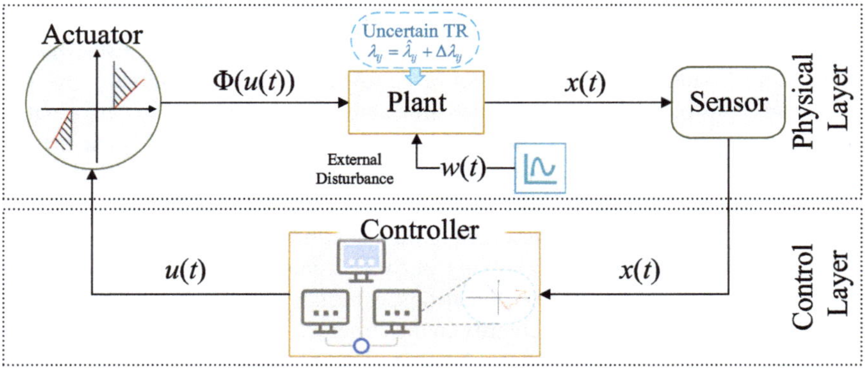

Fig. 4.1 Block diagram of MJSs subject to actuator nonlinearities

follows:

$$\Lambda = \begin{bmatrix} \tilde{\lambda}_{11} & \tilde{\lambda}_{12} & \tilde{\lambda}_{13} & \tilde{\lambda}_{14} \\ ? & \tilde{\lambda}_{22} & ? & \tilde{\lambda}_{24} \\ ? & \tilde{\lambda}_{32} & ? & ? \\ ? & ? & ? & ? \end{bmatrix}, \tag{4.2}$$

where "$\tilde{\lambda}_{ij}$" and "?" represent the uncertain and unknown TRs. $\tilde{\lambda}_{ij} = \hat{\lambda}_{ij} + \Delta\lambda_{ij}$, $\hat{\lambda}_{ij}$ and $\Delta\lambda_{ij} \in \left[-\alpha_{ij}, \alpha_{ij}\right] (\alpha_{ij} \geq 0)$ represent the estimate value and error of the uncertain TR $\tilde{\lambda}_{ij}$, respectively, where $\hat{\lambda}_{ij}$ and α_{ij} are known, and they satisfy $\hat{\lambda}_{ii} = -\sum_{i \neq j} \hat{\lambda}_{ij}$, $\Delta\lambda_{ii} = -\sum_{i \neq j} \Delta\lambda_{ij}$, $|\hat{\lambda}_{ij}| > |\alpha_{ij}|$.

For the elements in TR matrix (4.2), the ith row is composed of the following two sets:

$$\mathbb{S}_k^i \triangleq \{j : \text{the estimate value of } \lambda_{ij} \text{ is known for } j \in \mathbb{S}\},$$
$$\mathbb{S}_{uk}^i \triangleq \{j : \text{the estimate value of } \lambda_{ij} \text{ is unknown for } j \in \mathbb{S}\}.$$

Obviously, $\mathbb{S} = \mathbb{S}_k^i + \mathbb{S}_{uk}^i$. If $\mathbb{S}_k^i \neq \emptyset$, it can be described as $\mathbb{S}_k^i \triangleq \{\mathcal{K}_1^i, \mathcal{K}_2^i, \ldots, \mathcal{K}_{m_i}^i\}$, where $\mathcal{K}_l^i$ denotes the index of the lth element in the ith row of matrix Λ.

4.2.2 Sector Nonlinearities and Deadzones

In system (4.1), $\Phi(u(t))$ represents a nonlinear function on control signal $u(t)$ with $\Phi(u(t)) = [\phi_1(u_1(t))\ \phi_2(u_2(t))\ \cdots\ \phi_m(u_m(t))]^{\mathrm{T}}$ satisfying

$$\phi_l(u_l(t)) = 0, -u_l^- \leq u_l(t) \leq u_l^+, l = 1, 2, \ldots, m \tag{4.3}$$

and

$$(u_l(t) - u_l^+)\phi_l(u_l(t)) \geq k_l^+(u_l(t) - u_l^+), u_l(t) > u_l^+, \tag{4.4}$$
$$(u_l(t) + u_l^-)\phi_l(u_l(t)) \geq k_l^-(u_l(t) + u_l^-), u_l(t) < -u_l^-, \tag{4.5}$$

where u_l^+, u_l^-, k_l^+ and k_l^- are positive constant scalars.

Remark 4.1 It is seen from (4.3)–(4.5) and Fig. 4.2 that for the case $u_l^+ = u_l^- = 0$, the deadzones will be disappear and the $\phi_l(u_l(t))$ will reduce to the special sector nonlinear function. Here, it is assumed that $u_l^- \neq u_l^+$. Hence, the system considered in this chapter is a more general form.

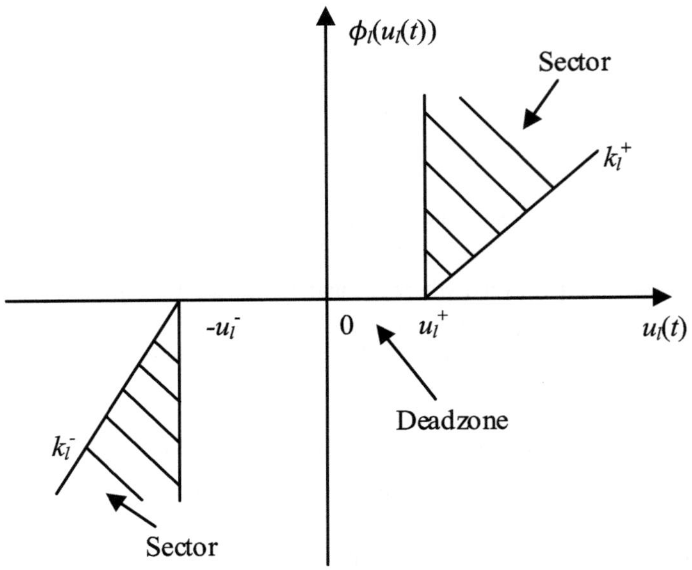

Fig. 4.2 Nonlinear function $\phi_l(u_l(t))$ for the lth input $u_l(t)$

4.3 Component-Wise Controller and Reachability Analysis

It should be pointed out that all modes of MJSs are not independent, but have connections via Markov process. That means that the sliding function corresponding to each mode depends on each other. As such, in this section, a set of mode-dependent sliding functions is constructed as follows:

$$S(x(t), i) = G_i x(t), \quad i \in \mathbb{S}, \tag{4.6}$$

with $S(x(t), i) = [s_1(x(t), i)\ s_2(x(t), i)\ \cdots\ s_m(x(t), i)]^{\mathrm{T}}$ and $G_i = B_i^T X_i \in \mathbb{R}^{m \times n}$, where the matrices X_i will be designed later such that $G_i B_i$ is nonsingular for each $i \in \mathbb{S}$. It is noted that, due to the assumption that B_i is full column rank, the nonsingularity of $B_i^T X_i B_i$ can be ensured for $X_i > 0$ for each $i \in \mathbb{S}$.

Remark 4.2 It is seen from (4.6) that the sliding function $S(x(t), i)(i \in \mathbb{S})$ depends on a set of matrices $X_i (i \in \mathbb{S})$. Apparently, X_i is a key matrix for the reachability of specified sliding surface, whose arbitrary selection come at a cost of trial and error. In order to remove this obstacle, X_i will be obtained via the designed in Theorem 4.4 later. Furthermore, X_i will be solutions of some coupled conditions concerning the TRs of modes. This implies that the effect of Markov switching is reflected in $S(x(t), i)$ through the set of matrices $X_i (i \in \mathbb{S})$. Furthermore, in the resultant SMC systems with Markov switching, the connections among these sliding surfaces corresponding to each mode are also established via the set of matrices $X_i (i \in \mathbb{S})$.

For convenience, some abbreviations are introduced firstly, $\tilde{\lambda}_k^i = \sum_{j\in\mathbb{S}_k^i} \tilde{\lambda}_{ij}$, $\hat{\lambda}_k^i = \sum_{j\in\mathbb{S}_k^i} \hat{\lambda}_{ij}$, $\bar{\lambda}_k^i = \sum_{j\in\mathbb{S}_k^i} (\hat{\lambda}_{ij} + \alpha_{ij})$, $\underline{\lambda}_k^i = \sum_{j\in\mathbb{S}_k^i} (\hat{\lambda}_{ij} - \alpha_{ij})$.

For a given finite-time interval $[0, T]$, we will design a suitable sliding mode controller such that the states trajectories can be driven onto the specified sliding surfaces $S(x(t), i) = 0$ before a finite instant T^* with $T^* < T$ and maintain there during the interval $[T^*, T]$. Considering the actuator nonlinearities occurring in each component and uncertain TRs, the controller will be designed to be component-based and switched. To this end, we give the following theorem:

Theorem 4.1 *For a specified finite (possibly short) interval* $[0, T]$, *with the sliding function* (4.6), *the following SMC law* (4.7)–(4.12) *is designed for the MJS* (4.1), *under which the state trajectories can be driven onto the sliding surface* $S(x(t), i) = 0$ *during the interval* $[0, T^*]$ *with* $T^* < T$ *and remain there in the subsequent time*

$$u_l(t) = \begin{cases} -(\varrho_i + \xi\rho_i(t))\frac{s_l(x(t),i)}{\|S(x(t),i)\|} - u_l^-, & s_l(x(t), i) > 0, \\ 0, & s_l(x(t), i) = 0, \\ -(\varrho_i + \xi\rho_i(t))\frac{s_l(x(t),i)}{\|S(x(t),i)\|} + u_l^+, & s_l(x(t), i) < 0, \end{cases} \tag{4.7}$$
$$l = 1, 2, \ldots, m,$$

where

$$\rho_i(t) \triangleq \gamma_i\|x(t)\| + \mu_i\delta + \frac{\eta_i}{\|S(x(t), i)\|}, \tag{4.8}$$

$$\gamma_i \triangleq \|(G_iB_i)^{-1}G_iA_i\| + \|(G_iB_i)^{-1}G_iM_i\|\|N_i\|, \tag{4.9}$$

$$\mu_i \triangleq \|(G_iB_i)^{-1}G_iD_i\|, \tag{4.10}$$

$$\eta_i \triangleq \begin{cases} \frac{1}{2}\|\sum_{j\in\mathbb{S}_k^i} \bar{\lambda}_{ij} S^{\mathrm{T}}(x(t), j)(G_jB_j)^{-1}S(x(t), j)\| \\ \quad -\frac{1}{2}\underline{\lambda}_k^i\|S^{\mathrm{T}}(x(t), j)(G_jB_j)^{-1}S(x(t), j)\|, i \in \mathbb{S}_k^i, j \in \mathbb{S}_{uk}^i, \\ \frac{1}{2}\|\sum_{j\in\mathbb{S}_k^i} \bar{\lambda}_{ij} S^{\mathrm{T}}(x(t), j)(G_jB_j)^{-1}S(x(t), j)\| \\ \quad -\frac{1}{2}(\lambda_d^i + \underline{\lambda}_k^i)\|S^{\mathrm{T}}(x(t), j)(G_jB_j)^{-1}S(x(t), j)\| \\ \quad \frac{1}{2}\lambda_d^i\|S^{\mathrm{T}}(x(t), i)(G_iB_i)^{-1}S(x(t), i)\|, i \in \mathbb{S}_{uk}^i, j \in \mathbb{S}_{uk}^i, j \neq i, \end{cases} \tag{4.11}$$

$$\varrho_i \geq \frac{\lambda_{\max}\left((G_iB_i)^{-1}\right)}{k(T-\kappa)}\|G_ix(0)\|, \tag{4.12}$$

$\varrho_i > 0$ *is a adjustable parameter with* $0 < \kappa < T$, *and* $\xi > 1/k$ *with* $k = \min\{k_l^+, k_l^-\}$.

Proof It is easily seen from (4.7) that $u_l(t) > u_l^+$, $\phi_l(u_l(t)) > 0$ for $s_l(x(t), i) < 0$ and $u_l(t) < -u_l^-$, $\phi_l(u_l(t)) < 0$ for $s_l(x(t), i) > 0$. Thus, we obtain from (4.4), (4.5) and (4.7) that

$$(u_l(t) - u_l^+)\phi_l(u_l(t)) = -\left(\varrho_i + \xi\rho_i(t)\right)\frac{s_l(x(t), i)}{\|S(x(t), i)\|}\phi_l(u_l(t))$$

$$
\begin{aligned}
&\geq k\big(u_l(t) - u_l^+\big)^2 \\
&\geq k\big(\varrho_i + \xi\rho_i(t)\big)^2 \frac{s_l^2(x(t), i)}{\|S(x(t), i)\|^2}, \qquad (4.13)
\end{aligned}
$$

$$
\begin{aligned}
(u_l(t) + u_l^-)\phi_l(u_l(t)) = &- \left(\big(\varrho_i + \xi\rho_i(t)\big)\frac{s_l(x(t), i)}{\|S(x(t), i)\|}\right)\phi_l(u_l(t)) \\
&\geq k\big(u_l(t) + u_l^-\big)^2 \\
&\geq k\big(\varrho_i + \xi\rho_i(t)\big)^2 \frac{s_l^2(x(t), i)}{\|S(x(t), i)\|^2}. \qquad (4.14)
\end{aligned}
$$

As a result, we have

$$
s_l(x(t), i)\phi_l(u_l(t)) \leq -k\big(\varrho_i + \xi\rho_i(t)\big)\frac{s_l^2(x(t), i)}{\|S(x(t), i)\|}. \qquad (4.15)
$$

For $-u_l^- < u_l(t) < u_l^+$, we have $\phi_l(u_l(t)) = 0$ with $s_l(x(t), i) = 0$. Thus, one can obtain $s_l(x(t), i)\phi_l(u_l(t)) = 0$. Consequentially, for $i \in \mathbb{S}$ and $l = 1, 2, \ldots, m$, we have

$$
s_l(x(t), i)\phi_l(u_l(t)) \leq -k\big(\varrho_i + \xi\rho_i(t)\big)\frac{s_l^2(x(t), i)}{\|S(x(t), i)\|}. \qquad (4.16)
$$

For all $l = 1, 2, \ldots, m$, it yields from (4.16)

$$
\begin{aligned}
S^{\mathrm{T}}(x(t), i)\Phi(u(t)) &= \sum_{l=1}^{m} s_l(x(t), i)\phi_l(u_l(t)) \\
&\leq -k\left(\big(\varrho_i + \xi\rho_i(t)\big)\sum_{l=1}^{m}\frac{s_l^2(x(t), i)}{\|S(x(t), i)\|}\right) \\
&\leq -k\big(\varrho_i + \xi\rho_i(t)\big)\|S(x(t), i)\|. \qquad (4.17)
\end{aligned}
$$

Choose the following Lyapunov function:

$$
V_1(x(t), i) = S^{\mathrm{T}}(x(t), i)(G_iB_i)^{-1}S(x(t), i). \qquad (4.18)
$$

Then, by Definition 1.4, we obtain the infinitesimal operator as

$$
\begin{aligned}
\mathscr{L}V_1(x(t), i) = &\, 2S^{\mathrm{T}}(x(t), i)(G_iB_i)^{-1}G_i(A_i + \Delta A_i)x(t) + 2S^{\mathrm{T}}(x(t), i)\Phi(u(t)) \\
&+ 2S^{\mathrm{T}}(x(t), i)(G_iB_i)^{-1}G_iD_iw(t) \\
&+ \sum_{j\in\mathbb{S}}\Big[\lambda_{ij}S^{\mathrm{T}}(x(t), j)(G_jB_j)^{-1}S(x(t), j)\Big]. \qquad (4.19)
\end{aligned}
$$

It is easily seen that the term $\sum_{j\in\mathbb{S}}\big[\lambda_{ij}S^{\mathrm{T}}(x(t), j)(G_jB_j)^{-1}S(x(t), j)\big]$, cannot be calculated directly due to some unavailable TRs λ_{ij}. In order to obtain a feasible condition, we shall consider two possible cases: $i \in \mathbb{S}_k^i$ and $i \in \mathbb{S}_{uk}^i$, respectively, that is, the diagonal element is known or unknown.

Case 1: $i \in \mathbb{S}_k^i$, that is, the diagonal element is known.

In this case, noting that

$$\tilde{\lambda}_k^i < 0, \quad -\sum_{j \in \mathbb{S}_{uk}^i} \frac{\lambda_{ij}}{\tilde{\lambda}_k^i} = 1, \tag{4.20}$$

we have

$$\begin{aligned}
&\sum_{j \in \mathbb{S}} \lambda_{ij} S^{\mathrm{T}}(x(t), j)(G_j B_j)^{-1} S(x(t), j) \\
&= \sum_{j \in \mathbb{S}_k^i} \tilde{\lambda}_{ij} (S^{\mathrm{T}}(x(t), j)(G_j B_j)^{-1} S(x(t), j) \\
&\quad - \tilde{\lambda}_k^i \Big(- \sum_{j \in \mathbb{S}_{uk}^i} \frac{\lambda_{ij}}{\tilde{\lambda}_k^i} S^{\mathrm{T}}(x(t), j)(G_j B_j)^{-1} S(x(t), j) \Big).
\end{aligned} \tag{4.21}$$

Then, substituting (4.21) into (4.19) yields

$$\begin{aligned}
\mathscr{L} V_1(x(t), i) = &- \sum_{j \in \mathbb{S}_{uk}^i} \frac{\lambda_{ij}}{\tilde{\lambda}_k^i} \Big\{ 2S^{\mathrm{T}}(x(t), i)(G_i B_i)^{-1} G_i (A_i + M_i F_i(t) N_i) x(t) \\
&+ 2S^{\mathrm{T}}(x(t), i)\Phi(u(t)) + 2S^{\mathrm{T}}(x(t), i)(G_i B_i)^{-1} G_i D_i w(t) \\
&+ \sum_{j \in \mathbb{S}_k^i} \Big[\tilde{\lambda}_{ij} S^{\mathrm{T}}(x(t), j)(G_j B_j)^{-1} S(x(t), j) \Big] \\
&- \tilde{\lambda}_k^i S^{\mathrm{T}}(x(t), j)(G_j B_j)^{-1} S(x(t), j) \Big\}.
\end{aligned} \tag{4.22}$$

By means of expression (4.17) and the designed SMC law (4.7)–(4.12), we obtain

$$\begin{aligned}
\mathscr{L} V_1(x(t), i) \le &- \sum_{j \in \mathbb{S}_{uk}^i} \frac{\lambda_{ij}}{\tilde{\lambda}_k^i} 2 \Big\{ \|S(x(t), i)\| \Big(\|(G_i B_i)^{-1} G_i A_i\| \|x(t)\| \\
&+ \|(G_i B_i)^{-1} G_i M_i\| \|N_i\| \|x(t)\| + \|(G_i B_i)^{-1} G_i D_i\| \|w(t)\| \Big) \\
&+ \frac{1}{2} \Big\| \sum_{j \in \mathbb{S}_k^i} \tilde{\lambda}_{ij} S^{\mathrm{T}}(x(t), j)(G_j B_j)^{-1} S(x(t), j) \Big\| \\
&- \frac{1}{2} \tilde{\lambda}_k^i \|S^{\mathrm{T}}(x(t), j)(G_j B_j)^{-1} S(x(t), j)\| + 2S^{\mathrm{T}}(x(t), i)\Phi(u(t)) \Big\} \\
\le &- 2(k\xi - 1)\rho_i(t)\|S(x(t), i)\| - 2k\varrho_i \|S(x(t), i)\|.
\end{aligned} \tag{4.23}$$

Further, due to $\xi > 1/k$, we have

$$\mathscr{L}V_1(x(t), i) \leq -2k\varrho_i \|S(x(t), i)\| < 0. \tag{4.24}$$

Case 2: $i \in \mathbb{S}^i_{uk}$, that is, the diagonal element is unknown.

In this case, noting that

$$\tilde{\lambda}^i_k > 0, \quad -\sum_{j \in \mathbb{S}^i_{uk}, j \neq i} \frac{\lambda_{ij}}{\lambda_{ii} + \tilde{\lambda}^i_k} = 1, \tag{4.25}$$

we have

$$\begin{aligned}
&\sum_{j \in \mathbb{S}} \lambda_{ij} S^{\mathrm{T}}(x(t), j)(G_j B_j)^{-1} S(x(t), j) \\
&= \sum_{j \in \mathbb{S}^i_k} \tilde{\lambda}_{ij} S^{\mathrm{T}}(x(t), j)(G_j B_j)^{-1} S(x(t), j) + \lambda_{ii} S^{\mathrm{T}}(x(t), i)(G_i B_i)^{-1} S(x(t), i) \\
&\quad - (\lambda_{ii} + \tilde{\lambda}^i_k)\Big(- \sum_{j \in \mathbb{S}^i_{uk}, j \neq i} \frac{\lambda_{ij}}{\lambda_{ii} + \tilde{\lambda}^i_k} S^{\mathrm{T}}(x(t), j)(G_j B_j)^{-1} S(x(t), j)\Big).
\end{aligned} \tag{4.26}$$

Then, substituting (4.26) into (4.19) yields

$$\begin{aligned}
\mathscr{L}V_1(x(t), i) = &- \sum_{j \in \mathbb{S}^i_{uk}, j \neq i} \frac{\lambda_{ij}}{\lambda_{ii} + \tilde{\lambda}^i_k} \Big\{ 2S^{\mathrm{T}}(x(t), i)(G_i B_i)^{-1} G_i (A_i + M_i F_i(t) N_i) x(t) \\
&+ 2S^{\mathrm{T}}(x(t), i)\Phi(u(t)) + 2S^{\mathrm{T}}(x(t), i)(G_i B_i)^{-1} G_i D_i w(t) \\
&+ \sum_{j \in \mathbb{S}^i_k} \tilde{\lambda}_{ij} S^{\mathrm{T}}(x(t), j)(G_j B_j)^{-1} S(x(t), j) \\
&+ \lambda_{ii} S^{\mathrm{T}}(x(t), i)(G_i B_i)^{-1} S(x(t), i) \\
&- (\lambda_{ii} + \tilde{\lambda}^i_k) S^{\mathrm{T}}(x(t), j)(G_j B_j)^{-1} S(x(t), j) \Big\}
\end{aligned} \tag{4.27}$$

In this case, a lower bound λ^i_d is given for λ_{ii}. Obviously, $\lambda^i_d \leq \lambda_{ii} \leq -\tilde{\lambda}^i_k$. Thus, similar to **Case 1**, we can obtain

$$\begin{aligned}
\mathscr{L}V_1(x(t), i) \leq &- \sum_{j \in \mathbb{S}^i_{uk}, j \neq i} \frac{\lambda_{ij}}{\lambda_{ii} + \tilde{\lambda}^i_k} 2 \Big\{ \|S(x(t), i)\| \Big(\|(G_i B_i)^{-1} G_i A_i\| \|x(t)\| \\
&+ \|(G_i B_i)^{-1} G_i M_i\| \|N_i\| \|x(t)\| + \|(G_i B_i)^{-1} G_i D_i\| \|w(t)\| \Big) \\
&+ \frac{1}{2} \Big\| \sum_{j \in \mathbb{S}^i_k} \tilde{\lambda}_{ij} S^{\mathrm{T}}(x(t), j)(G_j B_j)^{-1} S(x(t), i)] \Big\|
\end{aligned}$$

$$
\begin{aligned}
&-\frac{1}{2}\lambda_d^i\|S^{\mathrm{T}}(x(t),i)(G_iB_i)^{-1}S(x(t),i)\| - S^{\mathrm{T}}(x(t),i)\Phi(u(t)) \\
&-\frac{1}{2}(\lambda_d^i+\tilde{\lambda}_k^i)\|S^{\mathrm{T}}(x(t),j)(G_jB_j)^{-1}S(x(t),j)\|\Big\} \\
&\le 2\rho_i(t)\|S(x(t),i)\| - 2k(\varrho_i+\xi\rho_i(t))\|S(x(t),i)\| \\
&\le -2k\varrho_i\|S(x(t),i)\| < 0.
\end{aligned} \tag{4.28}
$$

By combing this result with the one in **Case 1**, it can be concluded that the state trajectories will be driven onto the specified switching surfaces $S(x(t),i)=0$ by the SMC law (4.7)–(4.12).

The following will further give the reaching instant T^*. We obtain from (4.18)

$$
V_1(x(t),i) \le \lambda_{\max}\left((G_iB_i)^{-1}\right)\|S(x(t),i)\|^2. \tag{4.29}
$$

Then, we have from (4.29) and $\mathscr{L}V_1(x(t),i) \le -2k\varrho_i\|S(x(t),i)\|$:

$$
\mathscr{L}V_1(x(t),i) \le -\frac{2k\varrho_i}{\sqrt{\lambda_{\max}\left((G_iB_i)^{-1}\right)}}\sqrt{V_1(x(t),i)}. \tag{4.30}
$$

Integrating (4.30) from 0 to t, one has

$$
\mathbb{E}\{\sqrt{V_1(x(t),i)}\} - \sqrt{V_1(x(0),i)} \le -\frac{k\varrho_i}{\sqrt{\lambda_{\max}\left((G_iB_i)^{-1}\right)}}t, \tag{4.31}
$$

from which there must exists an instant T^*

$$
T^* = \frac{\sqrt{\lambda_{\max}[(G_iB_i)^{-1}]}}{k\varrho_i}\sqrt{V_1(x(0),i)}, \tag{4.32}
$$

such that $\mathbb{E}\{V_1(x(t),i)\}=0$ for $t \ge T^*$, consequently, we have $S(x(t),i)=0$ (in the mean square).

Note that from (4.29) and $S(x(0),i)=G_ix(0)$, one has

$$
T^* = \frac{\lambda_{\max}\left((G_iB_i)^{-1}\right)}{k\varrho_i}\|G_ix(0)\|. \tag{4.33}
$$

By choosing a larger ϱ_i satisfying the following condition:

$$
\varrho_i \ge \frac{\lambda_{\max}\left((G_iB_i)^{-1}\right)}{k(T-\kappa)}\|G_ix(0)\| > \frac{\lambda_{\max}\left((G_iB_i)^{-1}\right)}{kT}\|G_ix(0)\|
$$

with a scalar κ satisfying $0<\kappa<T$, we can ensure $T^*<T$. Thus, one obtains $T^*<T$, which means that the state trajectories of system (4.1) will be driven onto

the specified sliding surface $S(x(t), i) = 0$ in finite time T^* with $T^* < T$. This accomplishes the proof. □

Remark 4.3 The SMC law (4.7)–(4.12) includes a set of specified matrices X_i such that the connections among sliding surfaces are reflected in the proposed SMC law just via the matrices X_i. Moreover, the proposed SMC law still depends on the TRs λ_{ij} that reflect the effect of Markov switching from one mode to another. Hence, the present SMC method can effectively deal with the effect of Markov switching, and obtain the desired dynamic performance of SMC systems.

4.4 SFTB Analysis

Next, the SFTB of the controlled system during the reaching phase and the sliding motion phase will be investigated, respectively. If SFTB during the reaching phase and the sliding motion phase can be guaranteed simultaneously, it can be guaranteed that SFTB over the whole finite-time interval $[0, T]$.

4.4.1 Reaching Phase Analysis

For convenience, some abbreviations are introduced as follows:

$$\bar{\lambda}_{ij} = \hat{\lambda}_{ij} + \alpha_{ij}, \quad X_k^i = \sum_{j \in \mathbb{S}_k^i, j \neq i} \tilde{\lambda}_{ij} X_j, \quad \bar{X}_k^i = \sum_{j \in \mathbb{S}_k^i, j \neq i} (\hat{\lambda}_{ij} + \alpha_{ij}) X_j.$$

In the following theorem, the sufficient condition is derived to guarantee SFTB over the reaching phase within $[0, T^*]$.

Theorem 4.2 *For the MJS* (4.1) *with the component-wise controller* (4.7)*, considering the given the parameters* (c_1, c_2, T, R, δ) *and a feasible scalar* φ_i*, if there exist positive constants* σ_i, c^**, and matrices* $X_i > 0$ *and* $W > 0$ $(\forall i \in \mathbb{S})$*, satisfying the following matrix inequalities:*

$$i \in \mathbb{S}_k^i, \begin{bmatrix} \Gamma_{1i} & X_i D_i \\ * & -\varphi_i I \end{bmatrix} < 0, \tag{4.34}$$

$$i \in \mathbb{S}_{uk}^i, \begin{bmatrix} \Gamma_{2i} & X_i D_i \\ * & -\varphi_i I \end{bmatrix} < 0, \tag{4.35}$$

$$c_1 < c^* < c_2, \tag{4.36}$$

$$\frac{\bar{\varpi}_X c_1 + \varphi_i \delta^2 T}{\underline{\varpi}_X} < e^{-\varphi_i T} c^*, \tag{4.37}$$

where

$$\begin{aligned}\Gamma_{1i} &\triangleq \mathrm{He}(X_i A_i) + W + \sigma_i^{-1} X_i M_i M_i^T X_i + \sigma_i N_i^{\mathrm{T}} N_i \\ &\quad - \varphi_i X_i + \bar{X}_k^i + \bar{\lambda}_{ii} X_i - \underline{\lambda}_k^i X_j,\ j \in \mathbb{S}_{uk}^i, \\ \Gamma_{2i} &\triangleq \mathrm{He}(X_i A_i) + W + \sigma_i^{-1} X_i M_i M_i^T X_i + \sigma_i N_i^{\mathrm{T}} N_i \\ &\quad - \varphi_i X_i + \bar{X}_k^i - \lambda_d^i X_i - (\underline{\lambda}_k^i + \lambda_d^i) X_j,\ j \in \mathbb{S}_{uk}^i, \\ \bar{\varpi}_X &\triangleq \max_{i\in\mathbb{S}} \left\{\lambda_{\max}\left(R^{-\frac{1}{2}} X_i R^{-\frac{1}{2}}\right)\right\},\ \underline{\varpi}_X \triangleq \min_{i\in\mathbb{S}} \left\{\lambda_{\min}\left(R^{-\frac{1}{2}} X_i R^{-\frac{1}{2}}\right)\right\},\end{aligned}$$

the closed-loop system is SFTB with respect to $\aleph_{[0,T^*]}$.

Proof For the following Lyapunov functional candidate:

$$V_2(x(t), i) = x^{\mathrm{T}}(t) X_i x(t) + \int_0^t x^{\mathrm{T}}(\tau) W x(\tau) \mathrm{d}\tau, \tag{4.38}$$

we obtain its infinitesimal operator as

$$\begin{aligned}\mathscr{L}V_2(x(t), i) &= \dot{x}^{\mathrm{T}}(t) X_i x(t) + x^{\mathrm{T}}(t) X_i \dot{x}(t) + x^{\mathrm{T}}(t) W x(t) + x^{\mathrm{T}}(t) \sum_{j\in\mathbb{S}} (\lambda_{ij} X_j) x(t) \\ &= x^{\mathrm{T}}(t)\left(A_i^T X_i + X_i A_i + \sum_{j\in\mathbb{S}} (\lambda_{ij} X_j) + W\right) x(t) + 2x^{\mathrm{T}}(t) X_i B_i \Phi(u(t)) \\ &\quad + w^{\mathrm{T}}(t) D_i^T X_i x(t) + x^{\mathrm{T}}(t) X_i D_i w(t) + 2x^{\mathrm{T}}(t) X_i \Delta A_i(t) x(t).\end{aligned}$$

It follows from (4.17) that $2x^{\mathrm{T}}(t) X_i B_i \Phi(u(t)) = 2S^{\mathrm{T}}(x(t), i)\Phi(u(t)) < 0$, which is substituted into the above equation to derive

$$\begin{aligned}\mathscr{L}V_2(x(t), i) &\leq x^{\mathrm{T}}(t)\left(\mathrm{He}(X_i A_i) + \sum_{j\in\mathbb{S}} (\lambda_{ij} X_j) + W + \sigma_i^{-1} X_i M_i M_i^T X_i + \sigma_i N_i^{\mathrm{T}} N_i\right) x(t) \\ &\quad + w^{\mathrm{T}}(t) D_i^T X_i x(t) + x^{\mathrm{T}}(t) X_i D_i w(t),\end{aligned} \tag{4.39}$$

for $\sigma_i > 0$.

Define an auxiliary function as follows:

$$J_1(x(t), i) = \mathscr{L}V_2(x(t), i) - \varphi_i V_2(x(t), i) - \varphi_i w^{\mathrm{T}}(t) w(t). \tag{4.40}$$

It is easily shown that $J_1(x(t), i) + \int_0^t x^{\mathrm{T}}(\tau) W x(\tau) \mathrm{d}\tau < 0$ can be ensured by the following inequality:

$$\begin{bmatrix} \Psi_i & X_i D_i \\ * & -\varphi_i I \end{bmatrix} < 0, \tag{4.41}$$

where $\Psi_i \triangleq \mathrm{He}(X_i A_i) + \sum_{j\in\mathbb{S}} (\lambda_{ij} X_j) + W + \sigma_i^{-1} X_i M_i M_i^T X_i + \sigma_i N_i^{\mathrm{T}} N_i - \varphi_i X_i$.

Case 1. the diagonal element is known ($i \in \mathbb{S}_k^i$)

In this cases, due to the following fact

$$\lambda_k^i < 0, \quad -\sum_{j\in\mathbb{S}_{uk}^i} \frac{\lambda_{ij}}{\tilde{\lambda}_k^i} = 1, \tag{4.42}$$

we have

$$\begin{aligned}\sum_{j\in\mathbb{S}} \lambda_{ij} X_j &= \sum_{j\in\mathbb{S}_k^i} \tilde{\lambda}_{ij} X_j + \sum_{j\in\mathbb{S}_{uk}^i} \lambda_{ij} X_j = X_k^i + \tilde{\lambda}_{ii} X_i - \tilde{\lambda}_k^i \Big(- \sum_{j\in\mathbb{S}_{uk}^i} \frac{\lambda_{ij}}{\tilde{\lambda}_k^i} X_j \Big) \\ &\le \bar{X}_k^i + \bar{\lambda}_{ii} X_i - \underline{\lambda}_k^i \Big(- \sum_{j\in\mathbb{S}_{uk}^i} \frac{\lambda_{ij}}{\tilde{\lambda}_k^i} X_j \Big). \end{aligned} \tag{4.43}$$

Thus, one has from (4.41):

$$-\sum_{j\in\mathbb{S}_{uk}^i} \frac{\lambda_{ij}}{\tilde{\lambda}_k^i} \begin{bmatrix} \Gamma_{1i} & X_i D_i \\ * & -\varphi_i I \end{bmatrix} < 0, \tag{4.44}$$

which can be ensured by (4.34).

Case 2. the diagonal element is unknown ($i \in \mathbb{S}_{uk}^i$)

In this cases, due to the following fact

$$\lambda_k^i > 0, \quad -\sum_{j\in\mathbb{S}_{uk}^i} \frac{\lambda_{ij}}{\tilde{\lambda}_k^i + \lambda_{ii}} = 1, \tag{4.45}$$

we have

$$\begin{aligned}\sum_{j\in\mathbb{S}} \lambda_{ij} X_j =& \sum_{j\in\mathbb{S}_k^i} \tilde{\lambda}_{ij} X_j + \lambda_{ii} X_i + \sum_{j\in\mathbb{S}_{uk}^i, j\neq i} \lambda_{ij} X_j \\ \le& \bar{X}_k^i - \underline{\lambda}_k^i X_i - (\underline{\lambda}_k^i + \lambda_d^i) \Big(- \sum_{j\in\mathbb{S}_{uk}^i} \frac{\lambda_{ij}}{\tilde{\lambda}_k^i + \lambda_{ii}} X_j \Big), \end{aligned} \tag{4.46}$$

Thus, one has from (4.41)

$$-\sum_{j\in\mathbb{S}_{uk}^i} \frac{\lambda_{ij}}{\tilde{\lambda}_k^i + \lambda_{ii}} \begin{bmatrix} \Gamma_{2i} & X_i D_i \\ * & -\varphi_i I \end{bmatrix} < 0, \tag{4.47}$$

which can be ensured by the expression (4.35).

Moreover, it follows from $J_1(x(t), i) < 0$ that

$$\mathscr{L} V_2(x(t), i) < \varphi_i V_2(x(t), i) + \varphi_i w^{\mathrm{T}}(t) w(t). \tag{4.48}$$

Multiplying both sides of expression (4.48) by $e^{-\varphi_i t}$ and integrating the obtained inequality from 0 to t with $t \in [0, T^*]$, we have

$$\begin{aligned} e^{-\varphi_i t}\mathbb{E}\{V_2(x(t), i)\} &< V_2(x(0), r_0) + \varphi_i \int_0^t e^{-\varphi_i \tau} w^{\mathrm{T}}(\tau)w(\tau)\mathrm{d}\tau \\ &\leq \bar{\varpi}_X c_1 + \varphi_i \delta^2 T. \end{aligned} \tag{4.49}$$

On the other hand, it follows from (4.38) that

$$\begin{aligned} e^{-\varphi_i t}\mathbb{E}\{V_2(x(t), i)\} &\geq e^{-\varphi_i t}\mathbb{E}\{x^{\mathrm{T}}(t)X_i x(t)\} + \int_0^t x^{\mathrm{T}}(\tau)(e^{-\varphi_i t}W)x(\tau)\mathrm{d}\tau \\ &\geq e^{-\varphi_i T}\underline{\varpi}_X \mathbb{E}\{x^{\mathrm{T}}(t)Rx(t)\}. \end{aligned} \tag{4.50}$$

It yields from (4.36), (4.37), (4.49) and (4.50) that

$$\mathbb{E}\{x^{\mathrm{T}}(t)Rx(t)\} < \frac{\bar{\varpi}_X c_1 + \varphi_i \delta^2 T}{e^{-\varphi_i T}\underline{\varpi}_X}. \tag{4.51}$$

Further, it follows from (4.51) that we have $\mathbb{E}\{x^{\mathrm{T}}(t)Rx(t)\} < c^*$ for all $t \in [0, T^*]$. Thus, according to Definition 1.4, the resulted closed-loop system is SFTB with respect to $\aleph_{[0,T^*]}$, which accomplishes this proof. □

4.4.2 Sliding Motion Phase Analysis

Now, it will be analyzed that the sliding mode dynamics are SFTB during the interval $[T^*, T]$. According to SMC theory, when the state trajectories maintain in the sliding surface $S(x(t), i) = 0$ for each $i \in \mathbb{S}$, we have the following equivalent controller:

$$\Phi_{eq}(u(t)) = -(G_i B_i)^{-1} G_i \big(A_i + \Delta A_i(t)\big)x(t) - (G_i B_i)^{-1} G_i D_i w(t). \tag{4.52}$$

By substituting (4.52) into the model (4.1), we get the following sliding mode dynamics:

$$\dot{x}(t) = (I - \tilde{G}_i)\big(A_i + \Delta A_i(t)\big)x(t) + (I - \tilde{G}_i)D_i w(t), \tag{4.53}$$

where $\tilde{G}_i \triangleq B_i(G_i B_i)^{-1}G_i$.

Theorem 4.3 *For the sliding mode dynamics* (4.53)*, considering the given parameters* (c_1, c_2, T, R, δ) *and feasible scalar* φ_i*, if there exist positive constants* ζ_i*,* c^* *and matrices* $X_i > 0$ $(\forall i \in \mathbb{S})$*, satisfying the following matrix inequalities:*

$$i \in \mathbb{S}_k^i, \begin{bmatrix} \Upsilon_{1i} & \Omega_{12i} \\ * & \Omega_{22i} \end{bmatrix} < 0, \tag{4.54}$$

$$i \in \mathbb{S}_{uk}^i, \begin{bmatrix} \Upsilon_{2i} & \Omega_{12i} \\ * & \Omega_{22i} \end{bmatrix} < 0, \tag{4.55}$$

$$c_1 < c^* < c_2, \tag{4.56}$$

$$\frac{\bar{\varpi}_X c^* + \varphi_i \delta^2 T}{\underline{\varpi}_X} < e^{-\varphi_i T} c_2, \tag{4.57}$$

where

$$\begin{aligned}\Upsilon_{1i} \triangleq\ & \mathrm{He}(X_i A_i) + A_i^T X_i A_i - \varphi_i X_i + \bar{X}_k^i + \bar{\lambda}_{ii} X_i - \underline{\lambda}_k^i X_j + \sigma_i^{-1} X_i M_i M_i^T X_i \\ & + \sigma_i N_i^{\mathrm{T}} N_i + A_i^T X_i A_i + A_i^T X_i M_i (\zeta_i I - M_i^T X_i M_i)^{-1} M_i^T X_i A_i + \zeta_i N_i^T N_i \\ & + 2 X_i B_i (B_i^T X_i B_i)^{-1} B_i^T X_i,\ j \in \mathbb{S}_{uk}^i, \\ \Upsilon_{2i} \triangleq\ & \mathrm{He}(X_i A_i) + A_i^T X_i A_i - \varphi_i X_i + \bar{X}_k^i - \underline{\lambda}_d^i X_i - (\underline{\lambda}_k^i + \lambda_d^i) X_j + \sigma_i^{-1} X_i M_i M_i^T X_i \\ & + \sigma_i N_i^{\mathrm{T}} N_i + A_i^T X_i A_i + A_i^T X_i M_i (\zeta_i I - M_i^T X_i M_i)^{-1} M_i^T X_i A_i + \zeta_i N_i^T N_i \\ & + 2 X_i B_i (B_i^T X_i B_i)^{-1} B_i^T X_i,\ j \in \mathbb{S}_{uk}^i,\end{aligned}$$

the sliding mode dynamics (4.53) *are SFTB with respect to* $\aleph_{[T^*, T]}$.

Proof Choose the following Lyapunov functional candidate:

$$V_3(x(t), i) = x^{\mathrm{T}}(t) X_i x(t). \tag{4.58}$$

We obtain its infinitesimal operator as

$$\begin{aligned}\mathscr{L} V_3(x(t), i) =\ & \dot{x}^{\mathrm{T}}(t) X_i x(t) + x^{\mathrm{T}}(t) X_i \dot{x}(t) + x^{\mathrm{T}}(t) \sum_{j \in \mathbb{S}} (\lambda_{ij} X_j) x(t) \\ =\ & x^{\mathrm{T}}(t)(\mathrm{He}(X_i A_i)) x(t) + 2 x^{\mathrm{T}}(t) X_i \Delta A_i(t) x(t) \\ & - 2 x^{\mathrm{T}}(t) X_i \tilde{G}_i \big(A_i + \Delta A_i(t)\big) x(t) + x^{\mathrm{T}}(t) \sum_{j \in \mathbb{S}} (\lambda_{ij} X_j) x(t) \\ & - 2 x^{\mathrm{T}}(t) X_i \tilde{G}_i D_i w(t) + w^{\mathrm{T}}(t) D_i^T X_i x(t) + x^{\mathrm{T}}(t) X_i D_i w(t).\end{aligned} \tag{4.59}$$

For any $X_i > 0$, we have

$$\begin{aligned}& - 2 x^{\mathrm{T}}(t) X_i B_i (B_i^T X_i B_i)^{-1} B_i^T X_i D_i w(t) \\ & \le x^{\mathrm{T}}(t) X_i B_i (B_i^T X_i B_i)^{-1} B_i^T X_i x(t) + w^{\mathrm{T}}(t) D_i^T X_i D_i w(t),\end{aligned} \tag{4.60}$$

and from Lemma 1.2, for any $\zeta_i > 0$ satisfying $\zeta_i I - M_i^T X_i M_i > 0$, one has

$$- 2 x^{\mathrm{T}}(t) X_i B_i (B_i^T X_i B_i)^{-1} B_i^T X_i \big(A_i + \Delta A_i(t)\big) x(t)$$

$$
\begin{aligned}
&\leq x^{\mathrm{T}}(t)\Big(A_i^T X_i A_i + A_i^T X_i M_i(\zeta_i I - M_i^T X_i M_i)^{-1} M_i^T X_i A_i + \zeta_i N_i^T N_i\Big)x(t) \\
&\quad + x^{\mathrm{T}}(t) X_i B_i (B_i^T X_i B_i)^{-1} B_i^T X_i x(t).
\end{aligned}
\tag{4.61}
$$

Define an auxiliary function as follows:

$$
J_2(x(t), i) = \mathscr{L} V_3(x(t), i) - \varphi_i V_3(x(t), i) - \varphi_i w^{\mathrm{T}}(t) w(t). \tag{4.62}
$$

It is easily shown that $J_2(x(t), i) < 0$ can be ensured by the following inequality:

$$
\begin{bmatrix} \hbar_i & X_i D_i \\ * & D_i^T X_i D_i - \varphi_i I \end{bmatrix} < 0, \tag{4.63}
$$

where

$$
\begin{aligned}
\hbar_i \triangleq\ & \mathrm{He}(X_i A_i) + \sum_{j \in \mathbb{S}} (\lambda_{ij} X_j) + \sigma_i^{-1} X_i M_i M_i^T X_i + 2 X_i B_i (B_i^T X_i B_i)^{-1} B_i^T X_i \\
& + \sigma_i N_i^{\mathrm{T}} N_i + A_i^T X_i A_i + \zeta_i N_i^T N_i + A_i^T X_i M_i (\zeta_i I - M_i^T X_i M_i)^{-1} M_i^T X_i A_i - \varphi_i X_i.
\end{aligned}
$$

Case 1. the diagonal element is known ($i \in \mathbb{S}_k^i$)

In this case, according to (4.42)–(4.43), one has

$$
-\sum_{j \in \mathbb{S}_{uk}^i} \frac{\lambda_{ij}}{\tilde{\lambda}_k^i} \begin{bmatrix} \Upsilon_{1i} & X_i D_i \\ * & D_i^T X_i D_i - \varphi_i I \end{bmatrix} < 0, \tag{4.64}
$$

which can be ensured by the condition (4.54).

Case 2. The diagonal element is unknown ($i \in \mathbb{S}_{uk}^i$)

In this case, according to the expression (4.46), we have

$$
-\sum_{j \in \mathbb{S}_{uk}^i, j \neq i} \frac{\lambda_{ij}}{\tilde{\lambda}_k^i + \lambda_{ii}} \begin{bmatrix} \Upsilon_{2i} & X_i D_i \\ * & D_i^T X_i D_i - \varphi_i I \end{bmatrix} < 0, \tag{4.65}
$$

which can be ensured by the condition (4.55).

By means of $J_2(x(t), i) < 0$, we obtain

$$
\mathscr{L} V_3(x(t), i) < \varphi_i V_3(x(t), i) + \varphi_i w^{\mathrm{T}}(t) w(t). \tag{4.66}
$$

According to the result in the Theorem 4.2, one has $\mathbb{E}\{x^{\mathrm{T}}(T^*) R x(T^*)\} < c^*$. Thus, by multiplying both sides of expression (4.66) by $e^{-\varphi_i t}$ and integrating the obtained inequality from T^* to t with $t \in [T^*, T]$, we have

$$
\begin{aligned}
e^{-\varphi_i t}\mathbb{E}\{V_3(x(t), i)\} <&e^{-\varphi_i T^*}\mathbb{E}\{V_3(x(T^*), r_{T^*})\} + \varphi_i \int_{T^*}^{t} e^{-\varphi_i \tau} w^{\mathrm{T}}(\tau)w(\tau)\mathrm{d}\tau \\
<&\bar{\varpi}_X \mathbb{E}\{x^{\mathrm{T}}(T^*)Rx(T^*)\} + \varphi_i \delta^2 T \\
<&\bar{\varpi}_X c^* + \varphi_i \delta^2 T. \qquad (4.67)
\end{aligned}
$$

In addition, it follows from (4.58) that

$$
e^{-\varphi_i t}\mathbb{E}\{V_3(x(t), i)\} \geq e^{-\varphi_i T} \underline{\varpi}_X \mathbb{E}\{x^{\mathrm{T}}(t)Rx(t)\}. \qquad (4.68)
$$

Putting together (4.67) and (4.68), we have

$$
\mathbb{E}\{x^{\mathrm{T}}(t)Rx(t)\} < \frac{\bar{\varpi}_X c^* + \varphi_i \delta^2 T}{e^{-\varphi_i T} \underline{\varpi}_X}. \qquad (4.69)
$$

Further, if there exists a scalar c^* satisfying (4.56), (4.57), it follows from (4.69) that we have $\mathbb{E}\{x^{\mathrm{T}}(t)Rx(t)\} < c_2$ for all $t \in [T^*, T]$, which accomplishes this proof. □

4.4.3 Solving Algorithm for Controller Gains

It is worthy of pointing out that the X_i should be designed to ensure the conditions in Theorems 4.2 and 4.3 to be feasible simultaneously. However, it is noted that some matrix inequalities in Theorems 4.2 and 4.3 are nonlinear, which are difficult to solve. Hence, we shall give a sufficient condition based on linear matrix inequalities (LMIs), which can guarantee that the conditions in Theorems 4.2 and 4.3 hold simultaneously.

Theorem 4.4 *For the given the parameters* (c_1, c_2, T, R, δ) *and a feasible scalar* φ_i*, consider the MJS* (4.1) *with the component-wise controller* (4.7)*. If there exist positive constants* σ_i*,* ζ_i*,* θ_i*,* c^* *and matrices* $X_i > 0$*,* $W > 0$*,* $(\forall i \in \mathbb{S})$ *satisfying the following LMIs:*

$$
\begin{aligned}
& i \in \mathbb{S}_k^i, \begin{bmatrix} \Theta_{1i} & \Xi_{12i} \\ * & \Xi_{22i} \end{bmatrix} < 0, && (4.70) \\
& i \in \mathbb{S}_{uk}^i, \begin{bmatrix} \Theta_{2i} & \Xi_{12i} \\ * & \Xi_{22i} \end{bmatrix} < 0, && (4.71) \\
& c_1 < c^* < c_2, && (4.72) \\
& \theta_i R < X_i < 2R, && (4.73) \\
& \begin{bmatrix} -e^{-\varphi_i T} c^* & \sqrt{2c_1 + \varphi_i \delta} \\ * & -\theta_i \end{bmatrix} < 0, && (4.74) \\
& -e^{-\varphi_i T} c_2 \theta_i + 2c^* + \varphi_i \delta < 0, && (4.75)
\end{aligned}
$$

where

$$\Theta_{1i} \triangleq \begin{bmatrix} \text{He}(X_iA_i)+A_i^TX_iA_i+W-\varphi_iX_i+\bar{X}_k^i+\bar{\lambda}_{ii}X_i-\underline{\lambda}_k^iX_j & X_iD_i \\ * & D_i^TX_iD_i-\varphi_iI \end{bmatrix},\ j \in \mathbb{S}_{uk}^i,$$

$$\Theta_{2i} \triangleq \begin{bmatrix} \text{He}(X_iA_i)+A_i^TX_iA_i+W-\varphi_iX_i+\bar{X}_k^i-\lambda_d^iX_i-(\underline{\lambda}_k^i+\lambda_d^i)X_j & X_iD_i \\ * & D_i^TX_iD_i-\varphi_iI \end{bmatrix},\ j \in \mathbb{S}_{uk}^i,$$

$$\Xi_{12i} \triangleq \begin{bmatrix} X_iM_i & \sigma_iN_i^{\mathrm{T}} & \sqrt{2}X_iB_i & A_i^TX_iM_i & \zeta_iN_i^{\mathrm{T}} \\ 0 & 0 & 0 & 0 & 0 \end{bmatrix},$$

$$\Xi_{22i} \triangleq -\text{diag}\{\sigma_i,\ \sigma_i,\ B_i^TX_iB_i,\ \zeta_iI - M_i^TX_iM_i,\ \zeta_iI\},$$

the closed-loop system is SFTB with respect to $\aleph_{[0,T]}$.

Proof Taking the conditions (4.34)–(4.37) in Theorem 4.2 and conditions (4.54)-(4.57) in Theorem 4.3 simultaneously, it is easily shown that (4.34), (4.54) can be ensured by (4.70); and (4.35), (4.55) can be ensured by (4.71). Then, we have the following inequalities:

$$c_1 < c^* < c_2, \tag{4.76}$$

$$\frac{\bar{\varpi}_X c_1 + \varphi_i\delta}{\underline{\varpi}_X} < e^{-\varphi_iT}c^*, \tag{4.77}$$

$$\frac{\bar{\varpi}_X c^* + \varphi_i\delta}{\underline{\varpi}_X} < e^{-\varphi_iT}c_2, \tag{4.78}$$

On the other hand, by the condition (4.73), one has $\theta_i < \underline{\varpi}_X$, $\bar{\varpi}_X < 2$. Thus, it is easily shown that the inequalities (4.77)–(4.78) can be ensured by (4.74)–(4.75). This accomplishes the proof. □

4.5 Simulation

In this section, a numerical example is firstly given, and then, a wheeled mobile manipulator is provided to demonstrate the effectiveness of the proposed finite-time SMC method subject to uncertain TRs and actuator nonlinearities.

4.5.1 Numerical Example

Consider the MJSs (4.1) with four operation modes as

$$A_1 = \begin{bmatrix} -1.2 & 0.8 \\ 0.5 & -2.4 \end{bmatrix},\ A_2 = \begin{bmatrix} -1.0 & 0.7 \\ 0.6 & -1.5 \end{bmatrix},\ A_3 = \begin{bmatrix} -1.2 & 1.5 \\ 1.8 & -0.6 \end{bmatrix},\ A_4 = \begin{bmatrix} -1.4 & 1.0 \\ 0.6 & -2.4 \end{bmatrix},$$

$$B_1 = \begin{bmatrix} 1.4 \\ 2.5 \end{bmatrix},\ B_2 = \begin{bmatrix} 1.5 \\ 1.4 \end{bmatrix},\ B_3 = \begin{bmatrix} 0.5 \\ 1.4 \end{bmatrix},\ B_4 = \begin{bmatrix} 1.8 \\ 2.1 \end{bmatrix},\ D_1 = \begin{bmatrix} 0.03 \\ 0.01 \end{bmatrix},\ D_2 = \begin{bmatrix} 0.02 \\ 0.03 \end{bmatrix},$$

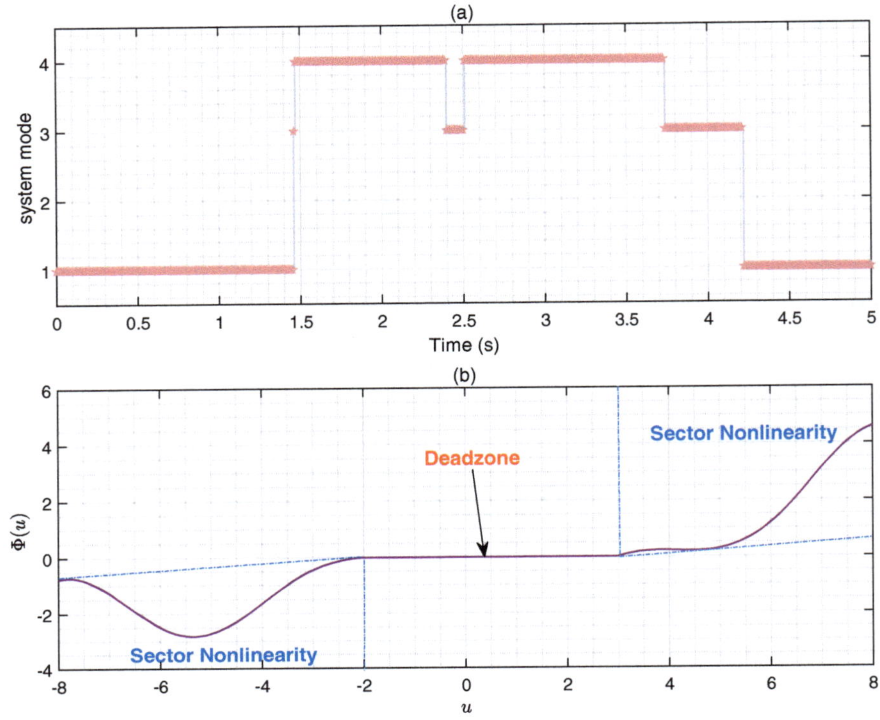

Fig. 4.3 A possible mode transition sequence and actuator nonlinearity: **a** mode transition sequence r_t; **b** nonlinear function of actuator $\Phi(u(t))$

$$D_3 = \begin{bmatrix} 0.01 \\ 0.02 \end{bmatrix}, D_4 = \begin{bmatrix} 0.02 \\ 0.01 \end{bmatrix}, M_1 = \begin{bmatrix} 0.01 \\ 0.02 \end{bmatrix}, M_2 = \begin{bmatrix} 0.02 \\ 0.01 \end{bmatrix}, M_3 = \begin{bmatrix} 0.01 \\ 0.01 \end{bmatrix},$$

$$M_4 = \begin{bmatrix} 0.01 \\ 0.02 \end{bmatrix}, N_1 = \begin{bmatrix} 0.3 & 0.2 \end{bmatrix}, N_2 = \begin{bmatrix} 0.2 & 0.2 \end{bmatrix}, N_3 = \begin{bmatrix} 0.2 & 0.1 \end{bmatrix}, N_4 = \begin{bmatrix} 0.1 & 0.3 \end{bmatrix},$$

$F_1 = 0.3\sin(t)$, $F_2 = 0.4\sin(t)$, $F_3 = 0.3\cos(t)$, $F_4 = 0.4\cos(t)$, $w(t) = 0.5\sin(t)$, and the nonlinear function of the actuator is

$$\Phi(u(t)) = \begin{cases} (0.4 + 0.4\sin(u(t)) + 0.1e^{0.2|\sin(u(t))|})(u(t) - 3), & u(t) > 3, \\ 0, & -2 \le u(t) \le 3, \\ (0.4 + 0.4\sin(u(t)) + 0.1e^{0.2|\sin(u(t))|})(u(t) + 2), & u(t) < -2. \end{cases}$$

The following form of the TRs matrix is considered in this simulation:

$$\Lambda = \begin{bmatrix} -1.5 + \Delta\lambda_{11} & 0.2 + \Delta\lambda_{12} & ? & ? \\ ? & ? & 0.1 + \Delta\lambda_{23} & 0.3 + \Delta\lambda_{24} \\ 0.4 + \Delta\lambda_{31} & ? & -2.5 + \Delta\lambda_{33} & ? \\ 0.1 + \Delta\lambda_{41} & 0.3 + \Delta\lambda_{42} & 0.5 + \Delta\lambda_{43} & -0.9 + \Delta\lambda_{44} \end{bmatrix},$$

where $\Delta\lambda_{11}$, $\Delta\lambda_{23}$, $\Delta\lambda_{33}$ and $\Delta\lambda_{41} \in [-0.05\ 0.05]$; $\Delta\lambda_{12}$, $\Delta\lambda_{24}$, $\Delta\lambda_{31}$ and $\Delta\lambda_{42} \in [-0.01\ 0.01]$; $\Delta\lambda_{43} \in [-0.02\ 0.02]$; $\Delta\lambda_{44} \in [-0.03\ 0.03]$; the lower bound $\lambda_d^2 = -1.5$ of λ_{22}.

Set $c_1 = 2$, $c_2 = 5$, $T = 5$, $R = \text{diag}\{7, 7\}$. To this end, by solving the LMIs (4.70)–(4.75) in Theorem 4.4 with the given parameters $\varphi_i = 3, i = 1, 2, 3, 4$, the sliding function (4.6) and SMC law (4.7) can be designed as

$$S(x(t), 1) = [13.2349 \quad 20.5524]x(t), \quad S(x(t), 2) = [16.2668 \quad 17.2393]x(t),$$
$$S(x(t), 3) = [10.9895 \quad 15.3457]x(t), \quad S(x(t), 4) = [8.7268 \quad 16.1451]x(t),$$

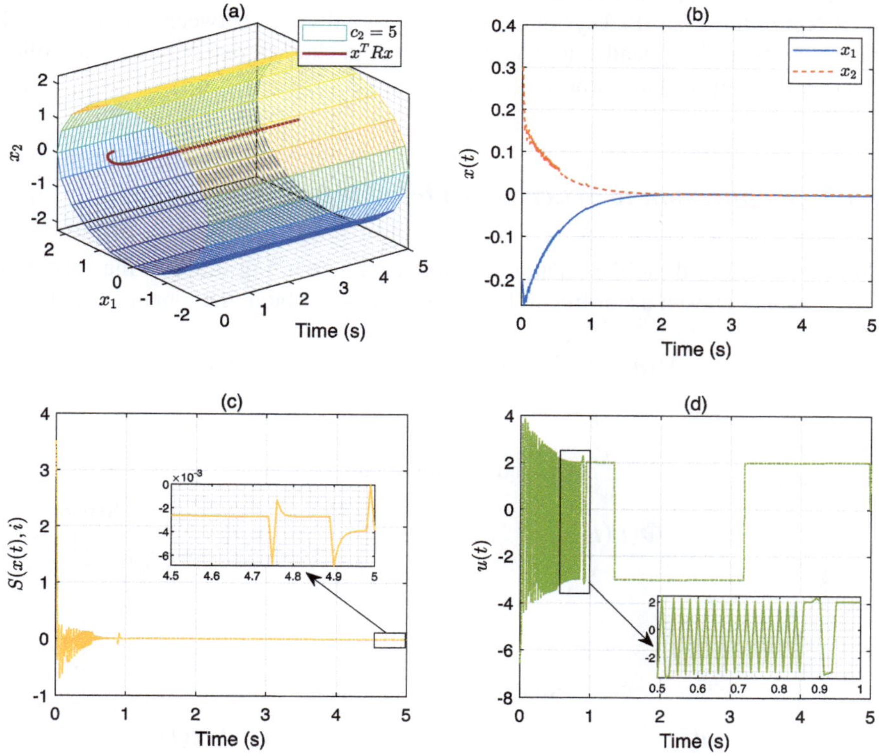

Fig. 4.4 Simulation results of controlled system: **a** evolution of weighted system state variable $x^{\mathrm{T}}(t)Rx(t)$; **b** state responses $x(t)$; **c** sliding variable $S(x(t), i)$; **d** control signal $u(t)$

$$u(t)=\begin{cases}-(\varrho_i+\xi(\gamma_i\|x(t)\|+\mu_i\delta+\frac{\eta_i}{\|S(x(t),i)\|}))\frac{S(x(t),i)}{\|S(x(t),i)\|}-2, & S(x(t),i)>0,\\ 0, & S(x(t),i)=0,\\ -(\varrho_i+\xi(\gamma_i\|x(t)\|+\mu_i\delta+\frac{\eta_i}{\|S(x(t),i)\|}))\frac{S(x(t),i)}{\|S(x(t),i)\|}+3, & S(x(t),i)<0,\end{cases}$$

with $k=0.1$, $\xi=11>1/k$, $\delta=5$, $\gamma_1=0.5627$, $\gamma_2=0.3251$, $\gamma_3=0.6014$, $\gamma_4=0.6099$, $\mu_1=0.0086$, $\mu_2=0.0174$, $\mu_3=0.0154$, $\mu_4=0.0068$, $\eta_1=0.0334$, $\eta_2=0.0297$, $\eta_3=0.0647$ and $\eta_4=0.0051$.

A possible mode variation r_t is given as in Fig. 4.3a, under which the simulation is undertaken with the initial state $x(0)=[-0.2\ \ 0.3]^{\mathrm{T}}$. Figure 4.3b shows that there exist the deadzones and sector nonlinearities happened in the actuator. We replace $S(x(t),i)/\|S(x(t),i)\|$ with $S(x(t),i)/(\|S(x(t),i)\|+0.1)$ to prevent the control signal from chattering. Thus, from Fig. 4.4d, the chattering can be alleviated effectively in the latter phase of control signal $u(t)$, although there are intense in the initial phase of control signal $u(t)$. It is seen from Figs. 4.4a–c that the resultant closed-loop systems are SFTB under the effect of the proposed SMC law (4.7) despite the uncertain TRs and actuator nonlinearity. Besides, it is from Fig. 4.4d that after attaining the SFTB of the controlled system, the control signal varies between $u(t)\in[-3,2]$ due to the bounds of deadzone, which exhibits the characteristics of the controller (4.7) for dealing with actuator nonlinearity.

4.5.2 Application to Wheeled Mobile Manipulator

Consider a wheeled mobile manipulator model shown in Fig. 4.5, and the variables in Fig. 4.5 are chosen according to [13]. Then, we can obtain the dynamic model:

$$M(q)\ddot{q}+V(q,\dot{q})\dot{q}+G(q)+d(t)=B(q)\tau$$

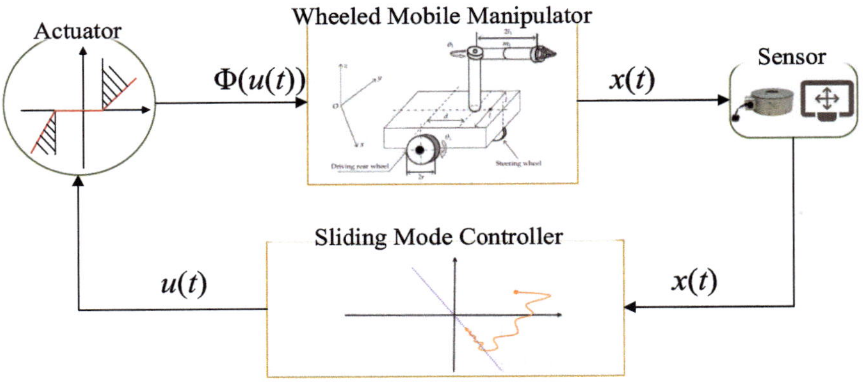

Fig. 4.5 System diagram of wheeled mobile manipulator

where $q = [q_v^{\mathrm{T}}, q_a^{\mathrm{T}}]^{\mathrm{T}} \in \mathbb{R}^n$ with $q_v = [x, y, \vartheta]^{\mathrm{T}} \in \mathbb{R}_v^n$ denotes the generalized coordinates for the mobile platform, $q_a \in \mathbb{R}_a^n$ denotes the coordinates of the robotic manipulator joints, $n = n_v + n_a$. $M(q)$ denotes the symmetric positive definite inertia matrix, $V(q, \dot{q})$ denotes the Centripetal and Coriolis torques, $G(q)$ denotes the gravitational torque vector, $B(q)$ denotes the input transformation matrix, τ denotes the control inputs, $d(t)$ is the bounded external disturbance from the environments. The detail description of this model refers to [13, 14]. Under the nonholonomic constraints and hybrids joints, the linearized dynamic with Markov switching is obtained as follows:

$$A_1 = \begin{bmatrix} 0 & 0 & 1 & 0 \\ 0 & 0 & 0 & 1 \\ 0.0040 & 0.0012 & 0.0653 & -0.0728 \\ -0.0047 & -0.0010 & -0.0717 & 0.0647 \end{bmatrix}, B_1 = \begin{bmatrix} 0 & 0 \\ 0 & 0 \\ 0.03 & 0.54 \\ -0.03 & -0.20 \end{bmatrix},$$

$$A_2 = \begin{bmatrix} 0 & 0 & 1 & 0 \\ 0 & 0 & 0 & 1 \\ 0.0040 & 0.0012 & 0.0653 & -0.0728 \\ -0.0047 & -0.0010 & -0.0717 & 0.0647 \end{bmatrix}, B_2 = \begin{bmatrix} 0 & 0 \\ 0 & 0 \\ 0.03 & 0.54 \\ -0.03 & -0.33 \end{bmatrix},$$

$$A_3 = \begin{bmatrix} 0 & 0 & 1 & 0 \\ 0 & 0 & 0 & 1 \\ 0.0057 & 0.0014 & 0.0725 & -0.0764 \\ -0.0064 & -0.0011 & -0.0790 & 0.0676 \end{bmatrix}, B_3 = \begin{bmatrix} 0 & 0 \\ 0 & 0 \\ 0.35 & 0.82 \\ -0.35 & -0.49 \end{bmatrix},$$

$$A_4 = \begin{bmatrix} 0 & 0 & 1 & 0 \\ 0 & 0 & 0 & 1 \\ 0.0057 & 0.0014 & 0.0725 & -0.0764 \\ -0.0064 & -0.0011 & -0.0790 & 0.0676 \end{bmatrix}, B_4 = \begin{bmatrix} 0 & 0 \\ 0 & 0 \\ 0.35 & 0.82 \\ -0.35 & -0.33 \end{bmatrix},$$

$$A_5 = \begin{bmatrix} 0 & 0 & 1 & 0 \\ 0 & 0 & 0 & 1 \\ 0.0042 & 0.0016 & 0.0628 & -0.0686 \\ -0.0048 & -0.0013 & -0.0691 & 0.0606 \end{bmatrix}, B_5 = \begin{bmatrix} 0 & 0 \\ 0 & 0 \\ -0.22 & 0.75 \\ 0.22 & 0.58 \end{bmatrix},$$

$$A_6 = \begin{bmatrix} 0 & 0 & 1 & 0 \\ 0 & 0 & 0 & 1 \\ 0.0042 & 0.0016 & 0.0628 & -0.0686 \\ -0.0048 & -0.0013 & -0.0691 & 0.0606 \end{bmatrix}, B_6 = \begin{bmatrix} 0 & 0 \\ 0 & 0 \\ -0.22 & 0.75 \\ 0.22 & 0.33 \end{bmatrix},$$

$$A_7 = \begin{bmatrix} 0 & 0 & 1 & 0 \\ 0 & 0 & 0 & 1 \\ 0.0055 & 0.0010 & 0.0753 & -0.0809 \\ -0.0062 & -0.0008 & -0.0819 & 0.0719 \end{bmatrix}, B_7 = \begin{bmatrix} 0 & 0 \\ 0 & 0 \\ 0.68 & 0.08 \\ -0.68 & -0.75 \end{bmatrix},$$

$$A_8 = \begin{bmatrix} 0 & 0 & 1 & 0 \\ 0 & 0 & 0 & 1 \\ 0.0055 & 0.0010 & 0.0753 & -0.0809 \\ -0.0062 & -0.0008 & -0.0819 & 0.0719 \end{bmatrix}, B_8 = \begin{bmatrix} 0 & 0 \\ 0 & 0 \\ 0.68 & 0.08 \\ -0.68 & -0.33 \end{bmatrix},$$

$$D_1 = \begin{bmatrix} 0.004 \\ 0.002 \\ 0.001 \\ -0.002 \end{bmatrix}, D_2 = \begin{bmatrix} 0.003 \\ 0.001 \\ 0.001 \\ 0.003 \end{bmatrix}, D_3 = \begin{bmatrix} -0.002 \\ 0.002 \\ -0.003 \\ 0.002 \end{bmatrix}, D_4 = \begin{bmatrix} 0.005 \\ 0.002 \\ 0.002 \\ -0.001 \end{bmatrix},$$

$$D_5 = \begin{bmatrix} 0.002 \\ 0.001 \\ 0.001 \\ -0.0025 \end{bmatrix}, D_6 = \begin{bmatrix} -0.003 \\ 0.002 \\ 0.0015 \\ 0.003 \end{bmatrix}, D_7 = \begin{bmatrix} 0.002 \\ -0.001 \\ -0.001 \\ 0.002 \end{bmatrix}, D_8 = \begin{bmatrix} 0.003 \\ -0.002 \\ 0.0025 \\ 0.001 \end{bmatrix},$$

$$M_1 = \begin{bmatrix} 0.4 \\ 0.2 \\ 0.1 \\ -0.2 \end{bmatrix}, M_2 = \begin{bmatrix} 0.3 \\ 0.1 \\ 0.1 \\ 0.3 \end{bmatrix}, M_3 = \begin{bmatrix} -0.2 \\ 0.2 \\ -0.3 \\ 0.2 \end{bmatrix}, M_4 = \begin{bmatrix} 0.5 \\ 0.2 \\ 0.2 \\ -0.1 \end{bmatrix},$$

$$M_5 = \begin{bmatrix} 0.2 \\ 0.1 \\ 0.1 \\ -0.25 \end{bmatrix}, M_6 = \begin{bmatrix} -0.3 \\ 0.2 \\ 0.15 \\ 0.3 \end{bmatrix}, M_7 = \begin{bmatrix} 0.2 \\ -0.1 \\ -0.1 \\ 0.2 \end{bmatrix}, M_8 = \begin{bmatrix} 0.3 \\ -0.2 \\ 0.25 \\ 0.1 \end{bmatrix},$$

$N_1 = \begin{bmatrix} 0.2 & 0.1 & -0.4 & 0.1 \end{bmatrix}$, $N_2 = \begin{bmatrix} 0.1 & -0.2 & 0.3 & 0.5 \end{bmatrix}$, $N_3 = \begin{bmatrix} 0.2 & 0.1 & -0.2 & 0.1 \end{bmatrix}$, $N_4 = \begin{bmatrix} -0.2 & 0.3 & 0.2 & 0.3 \end{bmatrix}$, $N_5 = \begin{bmatrix} 0.1 & -0.3 & -0.3 & 0.1 \end{bmatrix}$, $N_6 = \begin{bmatrix} 0.4 & 0.2 & 0.1 & -0.5 \end{bmatrix}$, $N_7 = \begin{bmatrix} -0.2 & 0.5 & 0.2 & 0.3 \end{bmatrix}$, $N_8 = \begin{bmatrix} -0.5 & 0.2 & 0.7 & 0.3 \end{bmatrix}$, $F_5 = F_1 = 0.3\sin(t)$, $F_6 = F_2 = 0.4\sin(t)$, $F_7 = F_3 = 0.3\cos(t)$, $F_8 = F_4 = 0.4\cos(t)$, $w(t) = 0.3\cos(t)$, and the nonlinear function for each component $l = 1, 2$

$$\phi_l(u_l(t)) = \begin{cases} (1 - 0.03\cos(u_l(t)))(u_l(t) - 0.5), & u_l(t) > 0.5, \\ 0, & -0.5 \le u_l(t) \le 0.5, \\ (1 - 0.02\sin(u_l(t)))(u_l(t) + 0.5), & u_l(t) < -0.5. \end{cases}$$

The following form of the transition rates matrix is considered in the simulation:

$$\Lambda = \begin{bmatrix} -0.72+\Delta\lambda_{11} & 0.22+\Delta\lambda_{12} & ? & ? & 0.14+\Delta\lambda_{15} & 0 & 0 & 0 \\ ? & ? & 0.42+\Delta\lambda_{23} & 0.1+\Delta\lambda_{24} & 0 & 0.1+\Delta\lambda_{26} & 0 & 0 \\ 0.26+\Delta\lambda_{31} & ? & -0.78+\Delta\lambda_{33} & ? & 0 & 0 & 0.1+\Delta\lambda_{37} & 0 \\ 0.16+\Delta\lambda_{41} & 0.26+\Delta\lambda_{42} & 0.26+\Delta\lambda_{43} & -0.78+\Delta\lambda_{44} & 0 & 0 & 0 & 0.1+\Delta\lambda_{48} \\ 0 & 0 & 0 & 0 & -0.78+\Delta\lambda_{55} & 0.26+\Delta\lambda_{56} & 0.26+\Delta\lambda_{57} & 0.26+\Delta\lambda_{58} \\ 0 & 0 & 0 & 0 & 0.26+\Delta\lambda_{65} & -0.78+\Delta\lambda_{66} & 0.26+\Delta\lambda_{67} & 0.26+\Delta\lambda_{68} \\ 0 & 0 & 0 & 0 & 0.26+\Delta\lambda_{75} & 0.26+\Delta\lambda_{76} & -0.78+\Delta\lambda_{77} & 0.26+\Delta\lambda_{78} \\ 0 & 0 & 0 & 0 & 0.26+\Delta\lambda_{85} & 0.26+\Delta\lambda_{86} & 0.26+\Delta\lambda_{87} & -0.78+\Delta\lambda_{88} \end{bmatrix}$$

Here, given the estimate error $\Delta\lambda_{ij} \in [-\alpha_{ij}, \alpha_{ij}]$, $\alpha_{ij} = 0.01$ and the lower bound $\lambda_d^2 = -1$ of λ_{22}. "?" represent the completely unknown elements.

Now, set $c_1 = 3, c_2 = 10, T = 3, R = \text{diag}\{10, 10\}$. By solving the LMIs (4.70)–(4.75) in Theorem 4.4 with the given parameters $\varphi_i = 5, i = 1, 2, 3, 4$, we can obtain the controller parameters as

$$X_1 = \begin{bmatrix} 8.4286 & 0.6749 & -0.3544 & -0.8558 \\ 0.6749 & 11.4314 & -1.7524 & -0.9876 \\ -0.3544 & -1.7524 & 17.8942 & -1.8466 \\ -0.8558 & -0.9876 & -1.8466 & 14.5800 \end{bmatrix}, X_2 = \begin{bmatrix} 10.1371 & -0.6383 & 4.0449 & 0.9580 \\ -0.6383 & 9.5590 & -0.3550 & 3.7197 \\ 4.0449 & -0.3550 & 16.5079 & 0.2653 \\ 0.9580 & 3.7197 & 0.2653 & 17.4652 \end{bmatrix},$$

$$X_3 = \begin{bmatrix} 11.3444 & 1.6425 & -1.8012 & 0.2236 \\ 1.6425 & 12.7437 & -1.7684 & 0.0923 \\ -1.8012 & -1.7684 & 15.9485 & -1.9001 \\ 0.2236 & 0.0923 & -1.9001 & 15.2662 \end{bmatrix}, X_4 = \begin{bmatrix} 11.3322 & -1.5750 & -0.6639 & -0.7686 \\ -1.5750 & 13.8172 & 1.1171 & 1.8445 \\ -0.6639 & 1.1171 & 14.7399 & 0.5207 \\ -0.7686 & 1.8445 & 0.5207 & 16.0226 \end{bmatrix},$$

$$X_5 = \begin{bmatrix} 10.2309 & 0.4638 & -0.3326 & 0.4749 \\ 0.4638 & 13.5186 & 2.4128 & -1.3244 \\ -0.3326 & 2.4128 & 16.1466 & -0.3402 \\ 0.4749 & -1.3244 & -0.3402 & 13.9947 \end{bmatrix}, X_6 = \begin{bmatrix} 13.5135 & 0.3790 & 0.2221 & -2.5118 \\ 0.3790 & 9.4622 & -0.2817 & -0.7855 \\ 0.2221 & -0.2817 & 13.9166 & -1.2259 \\ -2.5118 & -0.7855 & -1.2259 & 16.0473 \end{bmatrix},$$

$$X_7 = \begin{bmatrix} 11.8593 & -1.2855 & -0.1355 & -0.2321 \\ -1.2855 & 15.1966 & 0.9690 & 1.8411 \\ -0.1355 & 0.9690 & 14.3212 & 0.7752 \\ -0.2321 & 1.8411 & 0.7752 & 15.6616 \end{bmatrix}, X_8 = \begin{bmatrix} 11.2344 & -1.3468 & -1.3251 & -0.2181 \\ -1.3468 & 12.2654 & 1.0560 & 0.3869 \\ -1.3251 & 1.0560 & 17.1269 & 1.1401 \\ -0.2181 & 0.3869 & 1.1401 & 15.1409 \end{bmatrix},$$

and $k = 0.2$, $\xi = 6 > 1/k$, $\delta = 0.3$, $\varrho_1 = 780.1816$, $\varrho_2 = 2725.7$, $\varrho_3 = 40.0168$, $\varrho_4 = 9.1267$, $\varrho_5 = 1.1569$, $\varrho_6 = 1.4504$, $\varrho_7 = 2.9637$, $\varrho_8 = 9.8162$, $\gamma_1 = 13.0715$, $\gamma_2 = 47.3497$, $\gamma_3 = 1.0695$, $\gamma_4 = 0.2030$, $\gamma_5 = 0.6865$, $\gamma_6 = 1.4870$, $\gamma_7 = 0.3740$, $\gamma_8 = 1.6970$, $\mu_1 = 0.1143$, $\mu_2 = 0.3788$, $\mu_3 = 0.0034$, $\mu_4 = 0.0022$, $\mu_5 = 0.0087$, $\mu_6 = 0.0097$, $\mu_7 = 0.0019$, $\mu_8 = 0.0132$, $\eta_1 = 17.3237$, $\eta_2 = 350.1999$, $\eta_3 = 186.3046$, $\eta_4 = 98.2697$, $\eta_5 = 0.3010$, $\eta_6 = 0.2913$, $\eta_7 = 0.1781$ and $\eta_8 = 0.6999$.

Given a possible mode variations r_t as in Fig. 4.6a, the simulation is undertaken with the initial state $x(0) = [-0.2\,0.3\,-0.1\,0.2]^{\mathrm{T}}$. The simulation results are accordingly obtained as in Fig. 4.7. Figure 4.6b shows that there exist deadzones and sector

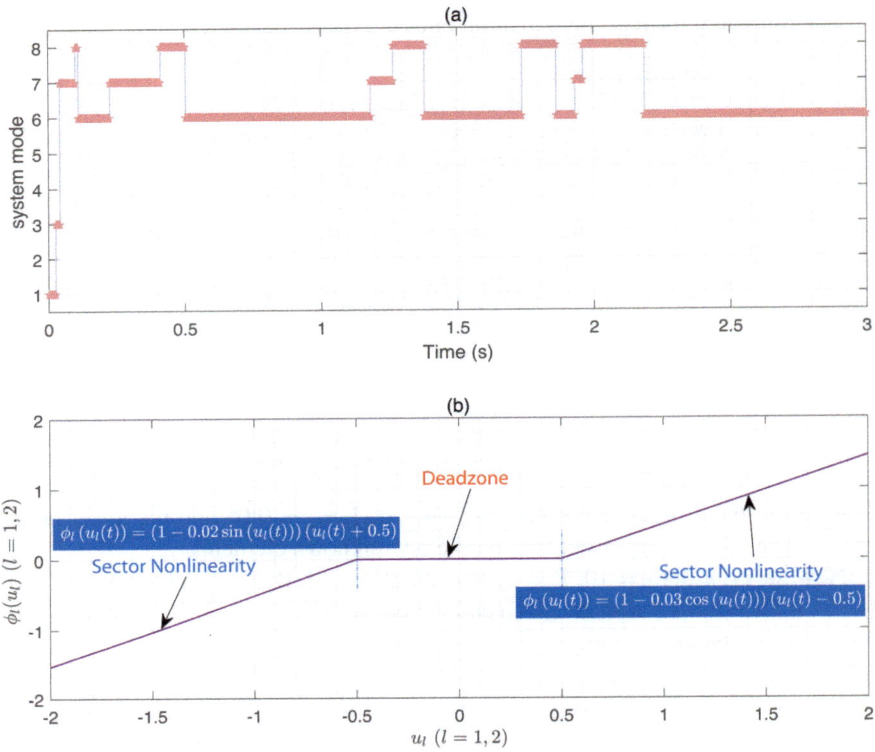

Fig. 4.6 A possible mode transition sequence and actuator nonlinearity of wheeled mobile manipulator: **a** mode transition sequence r_t; **b** nonlinear function of actuator $\phi_l(u_l(t))$

nonlinearities happened in actuators. It is seen from Fig. 4.7a that the open-loop system doesn't attain SFTB, while the proposed SMC law (4.7) can reduce the effect of actuator nonlinearity and uncertain TRs to ensure the SFTB of the resultant closed-loop system. Figure 4.7b–d exhibits state responses, sliding variables and control signals, respectively. Specifically, it can be seen from Fig. 4.7b that during the given time interval [0, 3], although the closed-loop system is not stable, the weighting state does not exceed the given level $c_2 = 10$, that is, the transient performance can be ensured; there exist some sudden changes in Fig. 4.7c, which results from the mode jumping; and after attaining the SFTB of the controlled system, the control signal varies between $u(t) \in [-0.5, 0.5]$ as in Fig. 4.7d according to the component-wise controller (4.7), which can effectively handel actuator nonlinearities.

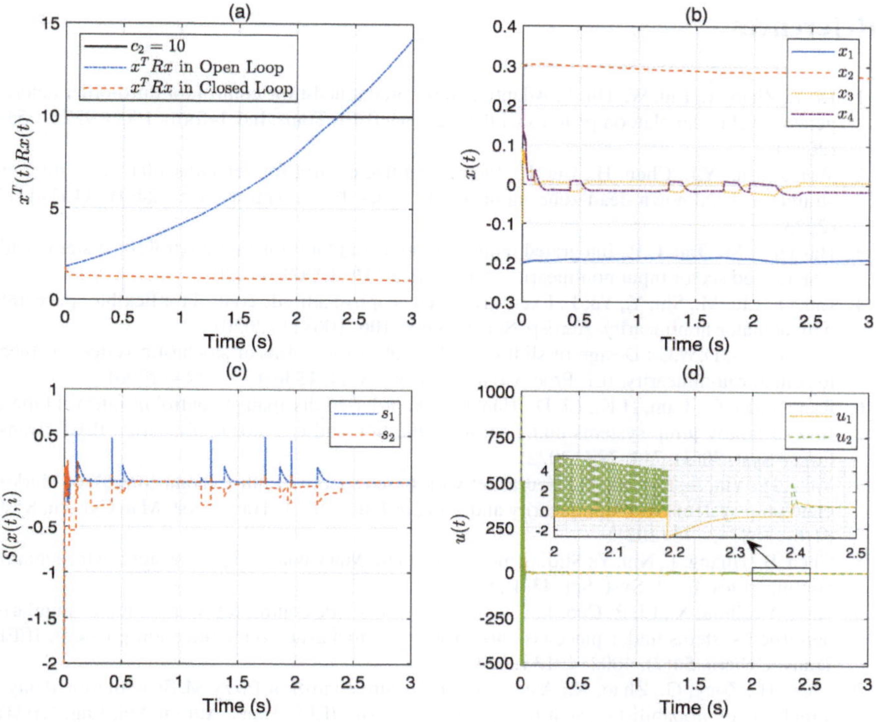

Fig. 4.7 Simulation results of wheeled mobile manipulator: **a** evolution of weighted system state variable $x^{\mathrm{T}}(t)Rx(t)$; **b** state responses $x(t)$; **c** sliding variable $S(x(t), i)$; **d** control signal $u(t)$

4.6 Conclusion

In this chapter, the problem of actuator nonlinearities including deadzones and sector nonlinearities have been addressed for a class of MJSs, which occurs in each component. Accordingly, a component-wise sliding mode controller has been designed to ensure the reachability of the specified mode-dependent sliding surface strictly before the given time T. Then, sufficient conditions have been attained such that the closed-loop MJSs are SFTB over both reaching phase and sliding motion phase. In addition, the designed gain matrices have established the connections among the specified sliding surfaces corresponding to each mode.

References

1. Li, Y., Zhao, Y., Liu, W., Hu, J.: Adaptive fuzzy predefined-time control for third-order heterogeneous vehicular platoon systems with dead zone. IEEE Trans. Ind. Inform. **19**(9), 9525–9534 (2022)
2. Wei, J., Liu, Y.J., Chen, H., Liu, L.: Fuzzy adaptive control for vehicular platoons with constraints and unknown dead-zone input. IEEE Trans. Intell. Transp. Syst. **24**(4), 4403–4412 (2023)
3. Hashemi, M., Tan, C.P.: Integrated fault estimation and fault tolerant control for systems with generalized sector input nonlinearity. Automatica **119**, 109098 (2020)
4. Liu, Q., Liu, M., Shi, Y., Yu, J.: Event-triggered adaptive attitude control for flexible spacecraft with actuator nonlinearity. Aerosp. Sci. Technol. **106**, 106111 (2020)
5. Niu, Y., Ho, D.W.C.: Design of sliding mode control for nonlinear stochastic systems subject to actuator nonlinearity. IEE Proc.-Control Theory Appl. **153**(6) 737–744 (2006)
6. Ran, G., Li, C., Lam, H.K., Li, D., Han, C.: Event-based dissipative control of interval type-2 fuzzy Markov jump systems under sensor saturation and actuator nonlinearity. IEEE Trans. Fuzzy Syst. **30**(3), 714–727 (2022)
7. Yang, H., Yin, S., Kaynak, O.: Neural network-based adaptive fault-tolerant control for Markovian jump systems with nonlinearity and actuator faults. IEEE Trans. Syst. Man Cybern. Syst. **51**(6), 3687–3698 (2021)
8. Chen, B., Huang, J., Niu, Y.: Sliding mode control for Markovian jumping systems with actuator nonlinearities. Int. J. Syst. Sci. **43**(4), 656–664 (2012)
9. Tang, Y., Zhou, X., Li, P., Cao, J., Cheng, J.: Sliding mode control for Markov jump singularly perturbed systems under piecewise homogeneous stochastic communication protocol. IEEE Trans. Cybern. **54**(7), 4002–4013 (2024)
10. Chen, H., Zong, G., Zhao, X.: Asynchronous secure control of fuzzy MJSs with time delays: a multi-level probabilistic event-triggered mechanism. IEEE Trans. Autom. Sci. Eng. (2024). https://doi.org/10.1109/TASE.2024.3421530
11. Ren, H., Zong, G., Karimi, H.R.: Asynchronous finite-time filtering of Markov jump nonlinear systems and its applications. IEEE Trans. Syst. Man Cybern. Syst. **51**(3), 1725–1734 (2021)
12. Sakthivel, R., Nithya, V., Suveetha, V.T.: Resilient finite-time fault detection dissipative-based filter for semi-Markovian jump systems with incomplete measurements. Int. J. Adapt. Control Signal Process. **36**(5), 1216–1230 (2022)
13. Ding, L., Gao, H., Xia, K.: Adaptive sliding mode control of mobile manipulators with Markovian switching joints. J. Appl. Math. **2012**, 1–24 (2012)
14. Zheng, B., Yang, G.: Sliding mode control for Markov jump linear uncertain systems with partly unknown transition rates. Int. J. Syst. Sci. **45**(10), 1999–2011 (2014)

Chapter 5
Non-Fragile Sliding Mode Control for Addressing Controller Gain Perturbations

5.1 Introduction

In many practical situations, the gains of the controller may occur random perturbations due to the parameter drift, accuracy problem, external disturbances, and other factors. Accordingly, it is essential to synthesize a resilient controller insensitive to gain perturbations. Besides, the uncertainties can occur in a random way in many device (e.g., actuator, controller, and sensor). The random characteristics in the system model can be represented as randomly occurring uncertainties (ROUs) [1, 2], and the random characteristics in the controller can be regard as randomly occurring gain perturbations (ROGPs). Recently, some remarkable works on ROGPs problems have been investigated for Markov jump systems (MJSs) in [3–5], where the state feedback controllers have been designed to realize the stochastic stability of the considered MJSs. However, under the consideration of the stochastic finite-time boundedness (SFTB), how to ensure the control performance of MJSs subject to ROUs and ROGPs, especially, by using sliding mode control (SMC) method, is meaningful and challenging. To this end, the following issues had to be dealt with:

(1) Covering ROUs and ROGPs modeled by two independent random variables, the closed-loop system under investigation is quite comprehensive. Due mainly to occurring in different components for ROUs and ROGPs, the dealing methods should be distinguished. As such, how to handle the negative effects of ROUs and ROGPs is the first difficulty.
(2) Based on the SMC theory, the control process for MJSs is bifurcated into two distinct stages: the reaching phase and the sliding motion phase. It is discernible that performance analysis during the reaching phase is influenced by ROGPs, whereas the analysis in the sliding motion phase remains unaffected. The second challenge lies in devising a viable SMC law that guarantees the desired performance and accomplishes SFTB throughout both stages within a specified finite-time interval.

Z. Cao and Y. Niu, *Finite-Time Sliding Mode Control for Markov Jump Systems*,
Studies in Systems, Decision and Control 603,
https://doi.org/10.1007/978-981-96-7651-4_5

In order to solve above issues, a mode-dependent sliding variable is firstly constructed, based on which a non-fragile SMC approach is developed to attain the reachability for the specified sliding surface despite the ROUs and ROGPs. *Especially, different from the traditional congruence-transformation-based approach for solving controller gains in Chaps.* 2 *and* 3, *we introduce sufficiently small scalars in this chapter to decouple the Lyapunov matrices and controller gains.* Then, the solvable conditions are obtained to ensure the SFTB of the resultant closed-loop system.

5.2 Preliminaries

Consider a class of MJSs on a complete probability space $(\Omega, \mathscr{F}, \Pr)$ as in system (1.6) $(f_i(x(t)) \equiv 0)$ with ROUs. The transition probabilities (TPs) are given as in (1.7) with the assumptions that all transition rates (TRs) are completely available. Accordingly, for mode $r_t = i \in \mathbb{S}$, the considered MJS with ROUs is expressed as

$$\dot{x}(t) = (A_i + \rho(t)\Delta A_i(t))x(t) + B_i(u(t) + f_i(x(t))) + D_i w(t), \tag{5.1}$$

where $x(t) \in \mathbb{R}^n$ is the state; $u(t) \in \mathbb{R}^m$ is the control signal; the parameter uncertainty $\Delta A_i(t)$, matched nonlinear perturbation $f_i(x(t))$ and unmatched external disturbance $w(t)$ satisfy Assumptions 1.2–1.4, respectively; the stochastic variable $\rho(t)$ characterizes ROUs obeying Bernoulli distribution with $\Pr\{\rho(t) = 1\} = \tilde{\rho}$, $\Pr\{\rho(t) = 0\} = 1 - \tilde{\rho}$, and known constant $\tilde{\rho} \in [0, 1]$. As shown in Fig. 5.1, the uncertain TR and actuator nonlinearity issues are focused in this chapter.

The objective of this chapter is to design the non-fragile SMC scheme for ensuring the SFTB of the MJSs (5.1) during the given interval $[0, T]$, namely, (1) to construct a sliding mode controller for reducing the negative effects of ROUs; and (2) to

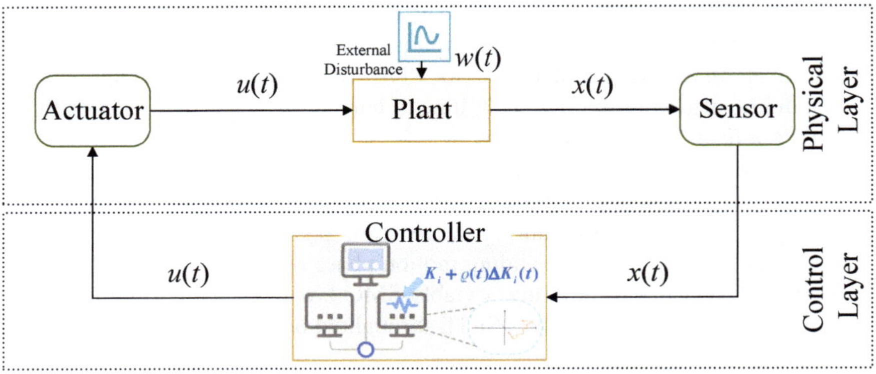

Fig. 5.1 Block diagram of MJSs subject to controller gain perturbations

derive solvable conditions on controller gains for covering the effects of ROGPs and guaranteeing the SFTB of the MJSs during the given interval $[0, T]$.

5.3 ROGP Issues in Controller

Firstly, we construct the following sliding function dependent on mode $i \in \mathbb{S}$:

$$s(x(t), i) = G_i x(t), \tag{5.2}$$

where $G_i (i \in \mathbb{S})$ are the designing matrices to ensure that $G_i B_i$ is nonsingular. Since B_i is of full column rank, the nonsingularity of $G_i B_i$ can be ensued by choosing $G_i = B_i^{\mathrm{T}} P_i$ with $P_i > 0$.

Next, the desirable SMC law is constructed as

$$u_{\text{ideal}}(t) = K_i x(t) - \eta_i(t)\text{sgn}(s(x(t), i)). \tag{5.3}$$

Considering that the gain perturbation phenomena might be occurred in the controller, the output of the controller (i.e., the input of actuator/plant) can be expressed as

$$u(t) = (K_i + \varrho(t)\Delta K_i(t))x(t) - \eta_i(t)\text{sgn}(s(x(t), i)), \tag{5.4}$$

where matrices $K_i \in \mathbb{R}^{m \times n}$ $(i \in \mathbb{S})$ are designed later, and the matrices $\Delta K_i(t)$ represent the unknown controller gain perturbations with $\Delta K_i(t) = E_i J_i(t) H_i$ and the matrices E_i, H_i being known and $J_i(t)$ being unknown time-varying matrices satisfying $J_i^{\mathrm{T}}(t) J_i(t) \leq I$. The random variable $\varrho(t) \in \mathbb{R}$ describes the phenomenon of ROGPs, which obeys Bernoulli distribution with $\Pr\{\varrho(t) = 1\} = \tilde{\varrho}$, $\Pr\{\varrho(t) = 0\} = 1 - \tilde{\varrho}$, and known constant $\tilde{\varrho} \in [0, 1]$. Here, the random variables $\rho(t)$ and $\varrho(t)$ are supposed to be mutually uncorrelated.

In SMC law (5.4), the robust term $\eta_i(t)$ is designed as

$$\eta_i(t) = \varpi_i + \varpi_{1i} + \varpi_{2i}(t) + \varpi_{3i}(t), \tag{5.5}$$

where

$$\varpi_i \geq \frac{\lambda_{\max}[(B_i^{\mathrm{T}} P_i B_i)^{-1}] \| B_i^{\mathrm{T}} P_i x(0) \|}{T - \varphi}, \tag{5.6}$$

$$\varpi_{1i} \triangleq \delta \| (B_i^{\mathrm{T}} P_i B_i)^{-1} B_i^{\mathrm{T}} P_i D_i \|, \tag{5.7}$$

$$\begin{aligned} \varpi_{2i}(t) \triangleq {} & [\| (B_i^{\mathrm{T}} P_i B_i)^{-1} B_i^{\mathrm{T}} P_i \| (\| A_i \| + \tilde{\rho} \| M_i \| \| N_i \|) \\ & + \| K_i \| + \tilde{\varrho} \| E_i \| \| H_i \| + \epsilon_i)] \, \| x(t) \| , \end{aligned} \tag{5.8}$$

$$\varpi_{3i}(t) \triangleq \frac{\| x(t) \|^2}{2 \, \| s(x(t), i) \|} \left\| \sum_{j \in \mathbb{S}} \lambda_{ij} P_j B_j (B_j^{\mathrm{T}} P_j B_j)^{-1} B_j^{\mathrm{T}} P_j \right\|, \tag{5.9}$$

where φ is a given small constant satisfying $0 < \varphi < T$.

5.4 Non-fragile Performance Analysis

5.4.1 Reachability Analysis

In this subsection, we will show that the specified sliding surface $s(x(t), i) = 0$ can be reached in the interval $[0, T^*]$ with $T^* < T$:

Theorem 5.1 *Considering the MJSs* (5.1) *under ROUs and ROGPs, the SMC law* (5.3) *can force the system state onto the sliding surface* $s(x(t), i) = 0$ *before the instant* T^* *with* $T^* < T$.

Proof Select the mode-dependent Lyapunov function as:

$$V_1(x(t), i) = s^{\mathrm{T}}(x(t), i)(B_i^{\mathrm{T}} P_i B_i)^{-1} s(x(t), i), \tag{5.10}$$

whose infinitesimal operator is given as

$$\begin{aligned}\mathscr{L}V_1(x(t), i) = & \, 2s^{\mathrm{T}}(x(t), i)(B_i^{\mathrm{T}} P_i B_i)^{-1} B_i^{\mathrm{T}} P_i (A_i + \rho(t)\Delta A_i)x(t) \\ & + 2s^{\mathrm{T}}(x(t), i)(B_i^{\mathrm{T}} P_i B_i)^{-1} B_i^{\mathrm{T}} P_i D_i w(t) \\ & + \sum_{j\in\mathbb{S}} \left(\lambda_{ij} s^{\mathrm{T}}(x(t), j)(B_j^{\mathrm{T}} P_j B_j)^{-1} s(x(t), j)\right) \\ & + 2s^{\mathrm{T}}(x(t), i)(K_i x(t) + \varrho(t)\Delta K_i(t) \\ & - \eta_i(t)\mathrm{sgn}(s(x(t), i)) + 2s^{\mathrm{T}}(x(t), i) f_i(x(t)).\end{aligned} \tag{5.11}$$

Noting $\|s(x(t), i)\| \le |s(x(t), i)|$, we obtain from (5.4) and (5.11):

$$\begin{aligned}&\mathscr{L}V_1(x(t), i) \\ &\le 2\|s(x(t), i)\|\Big(\|(B_i^{\mathrm{T}} P_i B_i)^{-1} B_i^{\mathrm{T}} P_i\|(\|A_i\| + \tilde{\rho}\|M_i\|\|N_i\|)\|x(t)\| \\ &\quad + \|(B_i^{\mathrm{T}} P_i B_i)^{-1} B_i^{\mathrm{T}} P_i D_i\|\|w(t)\| + (\|K_i\| + \tilde{\varrho}\|E_i\|\|H_i\| + 2\epsilon_i)\|x(t)\|\Big) \\ &\quad + \|x(t)\|\left\|\sum_{j\in\mathbb{S}} \lambda_{ij} P_j B_j (B_j^{\mathrm{T}} P_j B_j)^{-1} B_j^{\mathrm{T}} P_j\right\|\|x(t)\| \\ &\quad - 2\|s(x(t), i)\|(\varpi_i + \varpi_{3i}(t))\|x(t)\| - 2\|s(x(t), i)\|\eta_i(t).\end{aligned} \tag{5.12}$$

Then, combining (5.9) and (5.12) yields

$$
\begin{aligned}
& - 2\varpi_{3i}(t)\|s(x(t),i)\|\|x(t)\| + \|x(t)\|\left\|\sum_{j\in\mathbb{S}}\lambda_{ij}P_jB_j(B_j^{\mathrm{T}}P_jB_j)^{-1}B_j^{\mathrm{T}}P_j\right\|\|x(t)\| \\
= & -2\frac{\|x(t)\|}{\|s(x(t),i)\|}\left\|\sum_{j\in\mathbb{S}}\lambda_{ij}P_jB_j(B_j^{\mathrm{T}}P_jB_j)^{-1}B_j^{\mathrm{T}}P_j\right\|\|s(x(t),i)\|\|x(t)\| \\
& + \|x(t)\|\left\|\sum_{j\in\mathbb{S}}\lambda_{ij}P_jB_j(B_j^{\mathrm{T}}P_jB_j)^{-1}B_j^{\mathrm{T}}P_j\right\|\|x(t)\| < 0.
\end{aligned}
\tag{5.13}
$$

From (5.12) and (5.13), one has

$$
\mathscr{L}V_1(x(t),i) \leq -2\varpi_i\|s(x(t),i)\| < 0. \tag{5.14}
$$

Accordingly, we obtain that the state trajectories will be driven onto the specified sliding surfaces $s(x(t),i)=0$ by the SMC law (5.4)–(5.9).

The following will further give the reaching instant T^*. Similar to the derivation between (4.29)–(4.33), it is easy to see that if there exists an instant T^* satisfying

$$
T^* = \frac{\lambda_{\max}((B_i^{\mathrm{T}}P_iB_i)^{-1})}{\varpi_i}\|B_i^{\mathrm{T}}P_ix(0)\|, \tag{5.15}
$$

we have $\mathbb{E}\{\sqrt{V_1(x(t),i)}\}=0$ with $t \geq T^*$. Further, if follows from (5.10) that we have $s(x(t),i)=0$ in mean square. Thus, together with (5.6) and (5.15), we have $T^* \leq T-\varphi < T$, that is, the state trajectories of the uncertain MJS (5.1) will arrive at the specified sliding surface $s(x(t),i)=0$ before the instant T^* with $T^* < T$. This completes the proof. □

From Theorem 5.1, the designed controller (5.3) can ensure the reachability of the specified sliding surface $s(x(t),i)=0$ before $T^* < T$. Compared with the controller (2.2) in Chap. 2, the robust term of designed controller (5.3) includes $\tilde{\rho}\|(B_i^{\mathrm{T}}P_iB_i)^{-1}B_i^{\mathrm{T}}P_i\|\|M_i\|\|N_i\|\|x(t)\|$ and $\tilde{\varrho}\|(B_i^{\mathrm{T}}P_iB_i)^{-1}B_i^{\mathrm{T}}P_i\|\|E_i\|\|H_i\|$, which increases the robustness for the ROUs and ROGPs. Owing to the capability of insensitivity to ROGPs, the controller (5.3) is called as non-fragile controller.

5.4.2 SFTB During Reaching Phase

In the sequel, we further analyze the SFTB of the MJSs (5.1) over $[0, T^*]$ under the designed non-fragile SMC law (5.3). Applying the controller (5.4) under ROGPs into the system (5.1), we obtain the closed-loop system as

$$
\begin{aligned}
\dot{x}(t) = & (\breve{A}_i + \rho(t)\Delta A_i(t) + \varrho(t)B_i\Delta K_i(t))x(t) + D_iw(t) \\
& + B_i(f_i(x(t)) - \eta_i(t)\mathrm{sgn}(s(x(t),i)))
\end{aligned}
\tag{5.16}
$$

with $\breve{A}_i \triangleq A_i + B_i K_i$.

The following theorem gives the SFTB conditions during the reaching phase within $[0, T^*]$.

Theorem 5.2 *The closed-loop MJS* (5.16) *is SFTB with respect to* $\aleph_{[0,T^*]}$, *if there exist scalars* $\psi_i > 0$, $c^* > 0$, *matrices* $P_i > 0$, K_i $(\forall i \in \mathbb{S})$ *satisfying the following conditions:*

$$\begin{bmatrix} \Upsilon_i & P_i D_i \\ * & -\psi_i I \end{bmatrix} < 0, \tag{5.17}$$

$$c_1 < c^* < c_2, \tag{5.18}$$

$$\frac{\bar{\lambda}(P_i)c_1 + \delta^2 T \psi_i}{\underline{\lambda}(P_i)} < e^{-\psi_i T} c^* \tag{5.19}$$

with

$$\begin{aligned}
\Upsilon_i \triangleq\ & P_i A_i + A_i^{\mathrm{T}} P_i + P_i B_i K_i + (B_i K_i)^{\mathrm{T}} P_i - \psi_i P_i + \varepsilon_{1i} \tilde{\rho} P_i M_i M_i^{\mathrm{T}} P_i \\
& + \varepsilon_{1i}^{-1} \tilde{\rho} N_i^{\mathrm{T}} N_i + \varepsilon_{2i} \tilde{\varrho} P_i B_i E_i E_i^{\mathrm{T}} B_i^{\mathrm{T}} P_i + \varepsilon_{2i}^{-1} \tilde{\varrho} H_i^{\mathrm{T}} H_i + \sum_{j=1}^{N} \lambda_{ij} P_j, \\
\bar{\lambda}(P_i) \triangleq\ & \max_{i \in \mathbb{S}} \{\lambda_{\max}(R^{-\frac{1}{2}} P_i R^{-\frac{1}{2}})\},\ \underline{\lambda}(P_i) \triangleq \min_{i \in \mathbb{S}} \{\lambda_{\min}(R^{-\frac{1}{2}} P_i R^{-\frac{1}{2}})\}.
\end{aligned}$$

Proof Choosing the mode-dependent Lyapunov function as follows:

$$V_2(x(t), i) = x^{\mathrm{T}}(t) P_i x(t), \tag{5.20}$$

one has

$$\begin{aligned}
\mathscr{L} V_2(x(t), i) =\ & x^{\mathrm{T}}(t) \left(P_i A_i + A_i^{\mathrm{T}} P_i + P_i B_i K_i + (B_i K_i)^{\mathrm{T}} P_i + \sum_{j=1}^{N} \lambda_{ij} P_j \right) x(t) \\
& + 2\tilde{\rho} x^{\mathrm{T}}(t) P_i \Delta A_i(t) x(t) + 2\tilde{\varrho} x^{\mathrm{T}}(t) P_i B_i \Delta K_i x(t) \\
& + w^{\mathrm{T}}(t) D_i^{\mathrm{T}} P_i^{\mathrm{T}} x(t) + x^{\mathrm{T}}(t) P_i D_i w(t) + 2x^{\mathrm{T}}(t) P_i B_i f_i(x(t)) \\
& - 2x^{\mathrm{T}}(t) P_i B_i \eta_i(t) \mathrm{sgn}(s(x(t), i)).
\end{aligned} \tag{5.21}$$

For $\varepsilon_{1i} > 0$ and $\varepsilon_{2i} > 0$, one obtains

$$2x^{\mathrm{T}}(t) P_i \Delta A_i(t) x(t) \le \varepsilon_{1i} x^{\mathrm{T}}(t) P_i M_i M_i^{\mathrm{T}} P_i x(t) + \varepsilon_{1i}^{-1} x^{\mathrm{T}}(t) N_i^{\mathrm{T}} N_i x(t), \tag{5.22}$$

$$2x^{\mathrm{T}}(t) P_i B_i \Delta K_i x(t) \le \varepsilon_{2i} x^{\mathrm{T}}(t) P_i B_i E_i E_i^{\mathrm{T}} B_i^{\mathrm{T}} P_i x(t) + \varepsilon_{2i}^{-1} x^{\mathrm{T}}(t) H_i^{\mathrm{T}} H_i x(t). \tag{5.23}$$

Noting the fact, $\|s(x(t), i))\| \le |s(x(t), i))|$, one has

$$
\begin{aligned}
&2x^{\mathrm{T}}(t)P_iB_i(f_i(x(t)) - \eta_i(t)\mathrm{sgn}(s(x(t), i))) \\
\leq\ &2\epsilon_i\|x(t)\|\|s(x(t), i))\| - 2\eta_i(t)|s(x(t), i))| < 0.
\end{aligned} \tag{5.24}
$$

Then, we have from (5.21)–(5.24)

$$
\begin{aligned}
\mathscr{L}V_2(x(t), i) \leq x^{\mathrm{T}}(t)\Bigg(&P_iA_i + A_i^{\mathrm{T}}P_i + P_iB_iK_i + (B_iK_i)^{\mathrm{T}}P_i + \sum_{j=1}^{N}\lambda_{ij}P_j \\
&+ \varepsilon_{1i}\tilde{\rho}P_iM_iM_i^{\mathrm{T}}P_i + \varepsilon_{1i}^{-1}\tilde{\rho}N_i^{\mathrm{T}}N_i \\
&+ \varepsilon_{2i}\tilde{\varrho}P_iB_iE_iE_i^{\mathrm{T}}B_i^{\mathrm{T}}P_i + \varepsilon_{2i}^{-1}\tilde{\varrho}H_i^{\mathrm{T}}H_i\Bigg)x(t) \\
&+ w^{\mathrm{T}}(t)D_i^{\mathrm{T}}P_i^{\mathrm{T}}x(t) + x^{\mathrm{T}}(t)P_iD_iw(t).
\end{aligned} \tag{5.25}
$$

For some scalars $\psi_i > 0$, define the following auxiliary functions:

$$
\Pi(x(t), i) = \mathscr{L}V_2(x(t), i) - \psi_iV_2(x(t), i) - \psi_iw^{\mathrm{T}}(t)w(t). \tag{5.26}
$$

Obviously, $\Pi(x(t), i) < 0$ can be guaranteed by the condition (5.17). Hence, we have

$$
\mathscr{L}V_2(x(t), i) < \psi_iV_2(x(t), i) + \psi_iw^{\mathrm{T}}(t)w(t). \tag{5.27}
$$

Integrating (5.27) from 0 to t with $t \in [0, T^*]$, we obtain

$$
\begin{aligned}
e^{-\psi_it}\mathbb{E}\{V_2(x(t), i)\} <&V_2(x(0), r_0) + \psi_i\int_0^t e^{-\psi_i\tau}w^{\mathrm{T}}(\tau)w(\tau)\mathrm{d}\tau \\
\leq&\bar{\lambda}(P_i)c_1 + \delta^2T\psi_i.
\end{aligned} \tag{5.28}
$$

Besides, it is obtained from (5.20) that

$$
\mathbb{E}\{V_2(x(t), i)\} \geq \underline{\lambda}(P_i)\mathbb{E}\{x^{\mathrm{T}}(t)Rx(t)\}. \tag{5.29}
$$

By means of the expressions (5.18), (5.19), (5.29), and (5.30), we obtain

$$
\mathbb{E}\{x^{\mathrm{T}}(t)Rx(t)\} < \frac{(\bar{\lambda}(P_i)c_1 + \delta^2T\psi_i)e^{\psi_iT}}{\underline{\lambda}(P_i)}. \tag{5.30}
$$

If the condition (5.19) holds, the above inequality is less than $c^*$with $t \in [0, T^*]$, which, according to Definition 1.5, means the SFTB of the closed-loop system (5.16) with respect to $\aleph_{[0,T^*]}$. This completes the proof. □

5.4.3 SFTB During Sliding Motion Phase

Further, we analyze SFTB of sliding motion during the interval $[T^*, T]$. According to the SMC theory, in the sliding surface $s(x(t), i) = 0$ for each $i \in \mathbb{S}$, we can obtain the following equivalent control law from (5.1) and (5.2)

$$u_{eq}(t) = -(B_i^{\mathrm{T}} P_i B_i)^{-1} B_i^{\mathrm{T}} P_i((A_i + \rho(t)\Delta A_i(t))x(t) + D_i w(t)) - f_i(x(t)). \tag{5.31}$$

Substituting (5.31) into (5.1) yields the sliding mode dynamics as:

$$\dot{x}(t) = (\grave{A}_i + \rho(t)\Delta\grave{A}_i(t))x(t) + \grave{D}_i w(t) \tag{5.32}$$

with $\grave{A}_i \triangleq \Sigma_i A_i$, $\Delta\grave{A}_i \triangleq \Sigma_i \Delta A_i(t)$, $\grave{D}_i \triangleq \Sigma_i D_i$, $\Sigma_i \triangleq I - B_i(B_i^{\mathrm{T}} P_i B_i)^{-1} B_i^{\mathrm{T}} P_i$.

The following theorem gives the SFTB conditions during the sliding motion phase within $[T^*, T]$.

Theorem 5.3 *If there exist scalars $\psi_i > 0$, $\xi_i > 0$, and $c^* > 0$ and matrices $P_i > 0$, K_i ($\forall i \in \mathbb{S}$) satisfying:*

$$\begin{bmatrix} \Omega_i & P_i D_i \\ * & D_i^{\mathrm{T}} P_i D_i - \psi_i I \end{bmatrix} < 0, \tag{5.33}$$

$$c_1 < c^* < c_2, \tag{5.34}$$

$$\frac{\bar{\lambda}(P_i)c^* + \delta^2 T \psi_i}{\underline{\lambda}(P_i)} < e^{-\psi_i T} c_2 \tag{5.35}$$

with

$$\begin{aligned}\Omega_i \triangleq\ & P_i A_i + A_i^{\mathrm{T}} P_i - \psi_i P_i + \sum_{j=1}^{N} \lambda_{ij} P_j + 2P_i B_i (B_i^{\mathrm{T}} P_i B_i)^{-1} B_i^{\mathrm{T}} P_i \\ & + A_i^{\mathrm{T}} P_i M_i (\xi_i I - M_i^{\mathrm{T}} P_i M_i)^{-1} M_i^{\mathrm{T}} P_i A_i + A_i^{\mathrm{T}} P_i A_i \\ & + \xi_i \tilde{\rho}^2 N_i^{\mathrm{T}} N_i + \varepsilon_{1i}^{-1} \tilde{\varrho} P_i M_i M_i^{\mathrm{T}} P_i + \varepsilon_{1i} \tilde{\rho} N_i^{\mathrm{T}} N_i,\end{aligned}$$

then, the sliding mode dynamics (5.32) *are SFTB with respect to $\aleph_{[T^*,T]}$.*

Proof For the Lyapunov function (5.20), its weak infinitesimal operator is given as

$$\begin{aligned}\mathscr{L}V_2(x(t), i) =\ & x^{\mathrm{T}}(t)\left(\mathrm{He}(P_i \grave{A}_i) + \sum_{j=1}^{N} \lambda_{ij} P_j\right)x(t) + 2\tilde{\rho} x^{\mathrm{T}}(t)\Delta\grave{A}_i(t)x(t) \\ & + w^{\mathrm{T}}(t)\grave{D}_i^{\mathrm{T}} P_i x(t) + x^{\mathrm{T}}(t) P_i \grave{D}_i w(t) \\ =\ & x^{\mathrm{T}}(t)\left(\mathrm{He}(P_i A_i) + \sum_{j=1}^{N} \lambda_{ij} P_j\right)x(t) + 2x^{\mathrm{T}}(t) P_i \tilde{\rho} \Delta A_i x(t)\end{aligned}$$

$$
\begin{aligned}
&- 2x^{\mathrm{T}}(t)P_iB_i(B_i^{\mathrm{T}}P_iB_i)^{-1}B_i^{\mathrm{T}}P_i\big(A_i + \tilde{\rho}\Delta A_i(t)\big)x(t) \\
&- 2x^{\mathrm{T}}(t)P_iB_i(B_i^{\mathrm{T}}P_iB_i)^{-1}B_i^{\mathrm{T}}P_iD_iw(t) + w^{\mathrm{T}}(t)D_i^{\mathrm{T}}P_i^{\mathrm{T}}x(t) \\
&+ x^{\mathrm{T}}(t)P_iD_iw(t).
\end{aligned} \tag{5.36}
$$

According to Lemma 1.2, for any $\xi_i > 0$ satisfying $\xi_i I - M_i^{\mathrm{T}}P_iM_i > 0$, we have

$$
\begin{aligned}
&- 2x^{\mathrm{T}}(t)P_iB_i(B_i^{\mathrm{T}}P_iB_i)^{-1}B_i^{\mathrm{T}}P_i\big(A_i + \tilde{\rho}\Delta A_i(t)\big)x(t) \\
&\le x^{\mathrm{T}}(t)P_iB_i(B_i^{\mathrm{T}}P_iB_i)^{-1}B_i^{\mathrm{T}}P_ix(t) + x^{\mathrm{T}}(t)\big(A_i + \tilde{\rho}\Delta A_i(t)\big)^{\mathrm{T}}P_i\big(A_i + \tilde{\rho}\Delta A_i(t)\big)x(t) \\
&\le x^{\mathrm{T}}(t)\Big(A_i^{\mathrm{T}}P_iA_i + A_i^{\mathrm{T}}P_iM_i(\xi_iI - M_i^{\mathrm{T}}P_iM_i)^{-1}M_i^{\mathrm{T}}P_iA_i + u_i\tilde{\rho}^2N_i^{\mathrm{T}}N_i\Big)x(t) \\
&\quad + x^{\mathrm{T}}(t)P_iB_i(B_i^{\mathrm{T}}P_iB_i)^{-1}B_i^{\mathrm{T}}P_ix(t),
\end{aligned} \tag{5.37}
$$

which yields

$$
\begin{aligned}
&- 2x^{\mathrm{T}}(t)P_iB_i(B_i^{\mathrm{T}}P_iB_i)^{-1}B_i^{\mathrm{T}}P_iD_iw(t) \\
&\le x^{\mathrm{T}}(t)P_iB_i(B_i^{\mathrm{T}}P_iB_i)^{-1}B_i^{\mathrm{T}}P_ix(t) + w^{\mathrm{T}}(t)D_i^{\mathrm{T}}P_iD_iw(t).
\end{aligned} \tag{5.38}
$$

Moreover, for the auxiliary function as (5.26), if the condition (5.33) holds, we have $\Pi(x(t), i) < 0$. Thus, one has

$$
\mathscr{L}V_2(x(t), i) < \psi_i V_2(x(t), i) + \psi_i w^{\mathrm{T}}(t)w(t). \tag{5.39}
$$

From Theorem 5.2, one has $\mathbb{E}\{x^{\mathrm{T}}(T^*)Rx(T^*)\} < c^*$. Similarly, we have

$$
\begin{aligned}
e^{-\psi_i\tau}\mathbb{E}\{V_2(x(t), i)\} &< e^{-\psi_i\tau}\mathbb{E}\{V_2(x(T^*), r_{T^*})\} + \psi_i\int_{T^*}^{t} e^{-\psi_i\tau}w^{\mathrm{T}}(\tau)w(\tau)\mathrm{d}\tau \\
&< \bar{\lambda}(P_i)c^* + \delta^2T\psi_i.
\end{aligned} \tag{5.40}
$$

Besides, one has

$$
\mathbb{E}\{V_2(x(t), i)\} \ge \underline{\lambda}(P_i)\mathbb{E}\{x^{\mathrm{T}}(t)Rx(t)\}, \tag{5.41}
$$

which, together with (5.40), implies that

$$
\mathbb{E}\{x^{\mathrm{T}}(t)Rx(t)\} < \frac{(\bar{\lambda}(P_i)c^* + \delta^2T\psi_i)e^{\psi_it}}{\underline{\lambda}(P_i)}. \tag{5.42}
$$

Obviously, if there exists a scalar c^* satisfying (5.34) and (5.35), it can be obtained from the condition (5.42) that $\mathbb{E}\{x^{\mathrm{T}}(t)Rx(t)\} < c_2$ for $\forall t \in [T^*, T]$. Then, the SFTB for the sliding mode dynamics (5.32) is ensured with respect to $\aleph_{[T^*,T]}$. This completes the proof. □

Remark 5.1 From the derivation processes of Theorems 5.2 and 5.3, it can be summarized that during the reaching phase, the SFTB of closed-loop system (5.16) governed by the designed SMC law (5.3) has been achieved and during the sliding motion phase, the SFTB of sliding mode dynamics (5.32) under equivalent control law (5.31) has been guaranteed. Discriminatively, the closed-loop system (5.16) involves the effects of ROUs and ROGPs, while the sliding mode dynamics (5.32) do not. This phenomenon results from the SMC characteristic and thus it needs to derive the feasible conditions for solving controller gains that are immune to ROUs and ROGPs. Accordingly, over the whole finite-time SMC framework, the ROUs and ROGPs are addressed in two steps: 1) designing robust term of sliding mode controller to weaken the effects of ROUs and ROGPs; and 2) providing the solvable conditions to obtain the non-fragile controller gains.

5.4.4 Quadratic-Decomposition-Based Solving Algorithm

Note that the sufficient conditions obtained in the reaching stage and the sliding motion stage should be solved concurrently according to the designed sliding matrices P_i and controller gains K_i $(i \in \mathbb{S})$. However, these sufficient conditions are nonlinear due to the existence of coupling term $P_i B_i K_i$. To this end, we shall firstly give the follow result in Lemma 5.1.

Lemma 5.1 *For a set of sufficiently small scalar $\nu_i > 0$, the following inequalities* (5.43) *hold for $i \in \mathbb{S}$:*

$$\Theta_i + P_i B_i K_i + (B_i K_i)^{\mathrm{T}} P_i < 0, \tag{5.43}$$

if the following matrix inequalities hold for $i \in \mathbb{S}$:

$$\begin{bmatrix} \Theta_i - 2P_i + \nu_i I & P_i + \nu_i (B_i K_i)^{\mathrm{T}} \\ * & -\nu_i I \end{bmatrix} < 0. \tag{5.44}$$

where Θ_i is a set of mode-dependent matrices with appropriate dimension.

Proof For a set of sufficiently small scalar ν_i, inequalities (5.43) holds if and only if the following inequality holds:

$$\Theta_i + P_i B_i K_i + (B_i K_i)^{\mathrm{T}} P_i + \nu_i (B_i K_i)(B_i K_i)^{\mathrm{T}} < 0, \tag{5.45}$$

which is equivalent to

$$\begin{aligned} &\Theta_i + P_i B_i K_i + (B_i K_i)^{\mathrm{T}} P_i + \nu_i (B_i K_i)(B_i K_i)^{\mathrm{T}} \\ &+ \nu_i^{-1} P_i P_i - \nu_i^{-1} P_i P_i < 0. \end{aligned} \tag{5.46}$$

The inequality (5.46) can be rewritten as follows:

$$\Theta_i + (P_i + \nu_i (B_i K_i))^{\mathrm{T}} \nu_i^{-1} I (P_i + \nu_i (B_i K_i)) - \nu_i^{-1} P_i P_i < 0. \tag{5.47}$$

From Lemma 1.2, we have $-\nu_i^{-1} P_i P_i < -2P_i + \nu_i I$. Thus, (5.46) can be ensured by the following inequalities:

$$\Theta_i + (P_i + \nu_i(B_i K_i))^{\mathrm{T}} \nu_i^{-1} I (P_i + \nu_i(B_i K_i)) - 2P_i + \nu_i I < 0, \tag{5.48}$$

which is equivalent to (5.44) by Schur Complement (Lemma 1.3). This completes the proof. □

Remark 5.2 The decoupled method proposed in Lemma 5.1 is different from the traditional congruence-transformation-based handling method that multiplies the inverse of the Lyapunov matrix on both sides of the inequality and then substitutes the variables. The traditional method sometimes may cause a coupled problem, which can be avoided by the proposed method in Lemma 5.1.

Theorem 5.4 *Given the positive scalars ψ_i and parameters (c_1, c_2, T, R, δ), if for any $i \in \mathbb{S}$, there exist scalars $\xi_i > 0$, $\nu_i > 0$, $\vartheta_i > 0$ and matrices $P_i > 0$ and $\bar{K}_i$ $(\forall i \in \mathbb{S})$ satisfying:*

$$\begin{bmatrix} \Gamma_{11i} & \Gamma_{12i} \\ * & \Gamma_{22i} \end{bmatrix} < 0, \tag{5.49}$$

$$c_1 < c^* < c_2, \tag{5.50}$$

$$\vartheta_i R < P_i < 2R, \tag{5.51}$$

$$\begin{bmatrix} -e^{-\psi_i T} c^* & \sqrt{2c_1 + \delta^2 T \psi_i} \\ * & -\vartheta_i \end{bmatrix} < 0, \tag{5.52}$$

$$2c^* + \delta^2 T \psi_i - e^{-\psi_i T} c_2 \vartheta_i < 0 \tag{5.53}$$

with

$$\Gamma_{11i} \triangleq \begin{bmatrix} \bar{\Gamma}_{11i} & P_i D_i \\ * & D_i^{\mathrm{T}} P_i D_i - \psi_i I \end{bmatrix},$$

$$\Gamma_{12i} \triangleq \left[\begin{matrix} \sqrt{\tilde{\rho}} P_i M_i & \sqrt{\tilde{\rho}} \varepsilon_{1i} N_i^{\mathrm{T}} & \sqrt{\tilde{\varrho}} P_i E_i B_i & \sqrt{\tilde{\varrho}} \varepsilon_{2i} H_i^{\mathrm{T}} & A_i^{\mathrm{T}} P_i E_i & \xi_i \tilde{\rho} N_i^{\mathrm{T}} \\ 0 & 0 & 0 & 0 & 0 & 0 \end{matrix}\right.$$
$$\left.\begin{matrix} \sqrt{2} P_i B_i & P_i + (B_i \bar{K}_i)^{\mathrm{T}} \\ 0 & 0 \end{matrix}\right],$$

$$\Gamma_{22i} \triangleq -\mathrm{diag}\{\varepsilon_{1i} I,\ \varepsilon_{1i} I, \varepsilon_{2i} I,\ \varepsilon_{2i} I,\ \xi_i I - E_i^{\mathrm{T}} P_i E_i,\ \xi_i I,\ B_i^{\mathrm{T}} P_i B_i,\ \nu_i I\},$$

$$\bar{\Gamma}_{11i} \triangleq P_i A_i + A_i^{\mathrm{T}} P_i + A_i^{\mathrm{T}} P_i A_i - (\psi_i + 2) P_i + \nu_i I + \sum_{j=1}^{N} \lambda_{ij} P_j,$$

then, the closed-loop MJS (5.16) *is SFTB with respect to $\aleph_{[0,T]}$ during the interval $[0, T]$. Besides, we have the control gains $K_i = \nu_i^{-1} \bar{K}_i$.*

Proof Considering the conditions (5.17)–(5.19) and (5.33)–(5.35) simultaneously, we have

$$\begin{bmatrix} \Phi_i & P_i D_i \\ * & D_i^{\mathrm{T}} P_i D_i - \psi_i I \end{bmatrix} < 0, \tag{5.54}$$

$$c_1 < c^* < c_2, \tag{5.55}$$

$$\bar{\lambda}(P_i)c_1 + \delta^2 T \psi_i < e^{-\psi_i T} \underline{\lambda}(P_i)c^*, \tag{5.56}$$

$$\bar{\lambda}(P_i)c^* + \delta^2 T \psi_i < e^{-\psi_i T} \underline{\lambda}(P_i)c_2 \tag{5.57}$$

with

$$\begin{aligned} \Phi_i \triangleq\ & P_i A_i + A_i^{\mathrm{T}} P_i + P_i B_i K_i + (B_i K_i)^{\mathrm{T}} P_i - \psi_i P_i + \sum_{j=1}^{N} \lambda_{ij} P_j \\ & + \varepsilon_{1i} \tilde{\rho} P_i M_i M_i^{\mathrm{T}} P_i + \varepsilon_{1i}^{-1} \tilde{\rho} N_i^{\mathrm{T}} N_i + \varepsilon_{2i}^{-1} \tilde{\varrho} H_i^{\mathrm{T}} H_i \\ & + \varepsilon_{2i} \tilde{\varrho} P_i B_i E_i E_i^{\mathrm{T}} B_i^{\mathrm{T}} P_i + 2 P_i B_i (B_i^{\mathrm{T}} P_i B_i)^{-1} B_i^{\mathrm{T}} P_i \\ & + A_i^{\mathrm{T}} P_i E_i (\xi_i I - E_i^{\mathrm{T}} P_i E_i)^{-1} E_i^{\mathrm{T}} P_i A_i + A_i^{\mathrm{T}} P_i A_i + \xi_i \tilde{\rho}^2 N_i^{\mathrm{T}} N_i. \end{aligned}$$

This implies that the conditions (5.54)–(5.57) can ensure the SFTB with respect to $\aleph_{[0,T]}$ in the finite-time interval $[0, T]$.

If the condition (5.49) holds, one has

$$\begin{bmatrix} \Xi_i & P_i D_i & P_i + (B_i \bar{K}_i)^{\mathrm{T}} \\ * & D_i^{\mathrm{T}} P_i D_i - \psi_i I & 0 \\ * & * & -\psi_i I \end{bmatrix} < 0 \tag{5.58}$$

with $\bar{K}_i \triangleq \nu_i K_i$,

$$\begin{aligned} \Xi_i \triangleq\ & P_i A_i + A_i^{\mathrm{T}} P_i - 2P_i + \nu_i I - \psi_i P_i + \sum_{j=1}^{N} \lambda_{ij} P_j \\ & + \varepsilon_{1i} \tilde{\rho} P_i M_i M_i^{\mathrm{T}} P_i + \varepsilon_{1i}^{-1} \tilde{\rho} N_i^{\mathrm{T}} N_i + \varepsilon_{2i} \tilde{\varrho} P_i B_i E_i E_i^{\mathrm{T}} B_i^{\mathrm{T}} P_i \\ & + \varepsilon_{2i}^{-1} \tilde{\varrho} H_i^{\mathrm{T}} H_i + 2 P_i B_i (B_i^{\mathrm{T}} P_i B_i)^{-1} B_i^{\mathrm{T}} P_i \\ & + A_i^{\mathrm{T}} P_i E_i (\xi_i I - E_i^{\mathrm{T}} P_i E_i)^{-1} E_i^{\mathrm{T}} P_i A_i + A_i^{\mathrm{T}} P_i A_i + \xi_i \tilde{\rho}^2 N_i^{\mathrm{T}} N_i. \end{aligned}$$

By Lemma 5.1, for sufficiently small scalars $\nu_i > 0$, we further obtain that the inequality (5.54) holds.

In addition, from the condition (5.51), it shows $\vartheta_i < \underline{\lambda}(P_i)$, $\bar{\lambda}(P_i) < 2$. Thus, the inequalities (5.56), (5.57) can be guaranteed by (5.51) and (5.52). This completes the proof. □

In the above sections, by designing the non-fragile SMC law (5.4), the finite-time SMC problem has been addressed for the uncertain MJSs subject to ROUs in system model and ROGPs in controller. The given interval $[0, T]$ is divided into the reaching stage and sliding motion stage and their corresponding SFTB are analyzed with the

Table 5.1 Modes of the parameters $\mathcal{M}_i$ and $\mathcal{J}_i$

Mode i	Parameter $\mathcal{M}_i$ (kg)	Parameter $\mathcal{J}_i$ (N·m)
1	0.24	4
2	0.46	7
3	0.61	9

aid of the "partition–integration" technique. Thus, the SFTB can be guaranteed over the whole finite-time interval $[0, T]$.

Remark 5.3 This chapter provides the systematic investigation on the non-fragile SMC issue for MJSs subject to ROUs and ROGPs over a given finite-time interval. Different from the existing works, the designed control scheme exhibit the following novelties: (1) the system model (5.1) is more practical and embraces several well-known phenomena (i.e. stochastic parameter uncertainties, nonlinearities, and disturbances) in an unified framework; and (2) under the considered SFTB performance, a new non-fragile SMC scheme is developed to decrease the effects from ROUs and ROGPs.

5.5 Simulation

Consider a single-link robot arm (as in [6]), in which the dynamic equality of the robot arm is modeled as

$$\ddot{\theta}(t) = -\frac{g\mathcal{L}}{\mathcal{J}_i}\left(\frac{1}{2}m + \mathcal{M}_i\right)\sin(\theta(t)) - \frac{\mathcal{D}(t)}{\mathcal{J}_i}\dot{\theta}(t) + \frac{1}{\mathcal{J}_i}u(t) + \frac{\mathcal{L}}{\mathcal{J}_i}w(t),\ i = 1, 2, 3 \tag{5.59}$$

where $\theta(t)$ is the angle position of the arm; $u(t)$ is the control input; $w(t)$ is the external disturbance; $\mathcal{M}_i$ is the mass of the payload; $\mathcal{J}_i$ is the moment of inertia; g is the acceleration of gravity; $\mathcal{L}$ is the length of the arm; and $\mathcal{D}(t)$ is the uncertain coefficient of viscous friction.

For the considered single-link robot arm model, set $m = 0.05$ kg, $\mathcal{L} = 1.5$ m, and $g = 9.8$ m/s^2. $\mathcal{D}(t)$ is assumed to be bounded within 10% of its nominal value $\mathcal{D}_0 = 2$, that is, $\mathcal{D}(t) = \mathcal{D}_0 + \Delta\mathcal{D}$ with $|\Delta\mathcal{D}| \leq 0.2$, and the uncertainty occurs randomly with the probability $\tilde{\rho} = 0.75$. The parameters $\mathcal{M}_i$ and $\mathcal{J}_i$ for three different modes are given in Table 5.1.

The TR matrix is given as

$$\Lambda = \begin{bmatrix} -0.20 & 0.14 & 0.06 \\ 0.25 & -0.53 & 0.28 \\ 0.12 & 0.16 & -0.28 \end{bmatrix},$$

under which, denoting $x_1(t) = \theta(t)$ and $x_2(t) = \dot{\theta}(t)$, a uncertain system model can be formulated for $i = 1, 2, 3$

$$\dot{x}(t) = \begin{bmatrix} 0 & 1 \\ -\frac{g\mathcal{L}}{\mathcal{J}_i}(\frac{1}{2}m + \mathcal{M}_i) & -\frac{\mathcal{D}_0}{\mathcal{J}_i} \end{bmatrix} x(t) + \begin{bmatrix} 0 & 1 \\ 0 & -\frac{\Delta\mathcal{D}}{\mathcal{J}_i} \end{bmatrix} x(t) + \begin{bmatrix} 0 \\ \frac{1}{\mathcal{J}_i} \end{bmatrix} u(t) + \begin{bmatrix} 0 \\ \frac{\mathcal{L}}{\mathcal{J}_i} \end{bmatrix} w(t).$$

Specifically, the system matrices can be calculated as

$$A_1 = \begin{bmatrix} 0 & 1 \\ -0.9739 & -0.5000 \end{bmatrix}, \; B_1 = \begin{bmatrix} 0 \\ 0.2500 \end{bmatrix}, \; D_1 = \begin{bmatrix} 0 \\ 0.3750 \end{bmatrix},$$
$$A_2 = \begin{bmatrix} 0 & 1 \\ -1.0185 & -0.2857 \end{bmatrix}, \; B_2 = \begin{bmatrix} 0 \\ 0.1429 \end{bmatrix}, \; D_2 = \begin{bmatrix} 0 \\ 0.2143 \end{bmatrix},$$
$$A_3 = \begin{bmatrix} 0 & 1 \\ -1.0372 & -0.2222 \end{bmatrix}, \; B_3 = \begin{bmatrix} 0 \\ 0.1111 \end{bmatrix}, \; D_3 = \begin{bmatrix} 0 \\ 0.1667 \end{bmatrix},$$

and according to $|\Delta\mathcal{D}| \leq 0.2$, the uncertainty parameters can be chosen as

$$M_1 = \begin{bmatrix} 0 \\ 1 \end{bmatrix}, \; N_1 = \begin{bmatrix} 0 & 0.2 \end{bmatrix}, \; F_1(t) = 0.2\sin(t),$$
$$M_2 = \begin{bmatrix} 0 \\ 2 \end{bmatrix}, \; N_2 = \begin{bmatrix} 0 & 0.1 \end{bmatrix}, \; F_2(t) = 0.3\sin(t),$$
$$M_3 = \begin{bmatrix} 0 \\ -1 \end{bmatrix}, \; N_3 = \begin{bmatrix} 0 & 0.2 \end{bmatrix}, \; F_3(t) = 0.25\cos(t).$$

The nonlinear functions are denoted as $f_i(x(t)) = 0$ $(i = 1, 2, 3)$. And the external disturbances are assumed as $w(t) = \delta e^{-t}$ with $\delta = 0.3$. In addition, the sliding mode controller suffers from ROGPs with the probability $\tilde{\varrho} = 0.85$ and the gain perturbations parameters as follows:

$$E_1 = 0.60, \; H_1 = \begin{bmatrix} 0.25 & 0.12 \end{bmatrix}, \; J_1 = \sin(6t),$$
$$E_2 = 0.50, \; H_2 = \begin{bmatrix} 0.20 & 0.10 \end{bmatrix}, \; J_2 = e^{-t},$$
$$E_3 = 0.75, \; H_3 = \begin{bmatrix} 0.15 & 0.22 \end{bmatrix}, \; J_3 = \cos(7t).$$

Let $T = 4$, $c_2 = 0.029$, $R = 10^{-3}\text{diag}\{1, 1\}$, and $\psi_i = 0.02$ $(i = 1, 2, 3)$. By solving the conditions (5.49)–(5.53) in Theorem 5.4, we obtain

$$P_1 = \begin{bmatrix} 0.0014 & 0.0004 \\ 0.0004 & 0.0016 \end{bmatrix}, \; P_2 = \begin{bmatrix} 0.0017 & 0.0003 \\ 0.0003 & 0.0017 \end{bmatrix}, \; P_3 = \begin{bmatrix} 0.0012 & 0.0002 \\ 0.0002 & 0.0013 \end{bmatrix},$$
$$K_1 = \begin{bmatrix} -2.7700 & -11.5582 \end{bmatrix}, \; K_2 = \begin{bmatrix} -2.9329 & -15.9452 \end{bmatrix},$$
$$K_3 = \begin{bmatrix} -6.2594 & -35.4096 \end{bmatrix}.$$

The simulations are undergone by using the discretization approach with initial state $x(0) = [1.4 \quad -3]^{\mathrm{T}}$ and stepsize $\Delta t = 0.01\mathrm{s}$, whose results are shown in Figs. 5.2 and 5.3. There are some notable variations in the control signal $u(t)$ as shown in Fig. 5.2a, which is just the feature of the SMC method. Due to the effects of ROGPs, there are some differences between the actual control signal $u(t)$ and ideal control signal $u_{\text{ideal}}(t)$, which is exhibited in Fig. 5.2b with $\Delta u(t) \triangleq u(t) - u_{\text{ideal}}(t)$. Under one possible mode transition sequence r_t in Fig. 5.3b, the evolution $x^{\mathrm{T}}(t)Rx(t)$ of the closed-loop system in Fig. 5.3a never across the given constraint c_2 in the settled finite-time interval $[0, 3]$s. It is observed from Fig. 5.3d that according to the proposed non-fragile SMC law, the specified sliding surface $s(x(t), i) = 0$ is reached with $T^* < T$. Hence, the proposed non-fragile SMC approach is feasible and effective to ensure the SFTB of the uncertain MJSs in spite of containing ROUs and ROGPs.

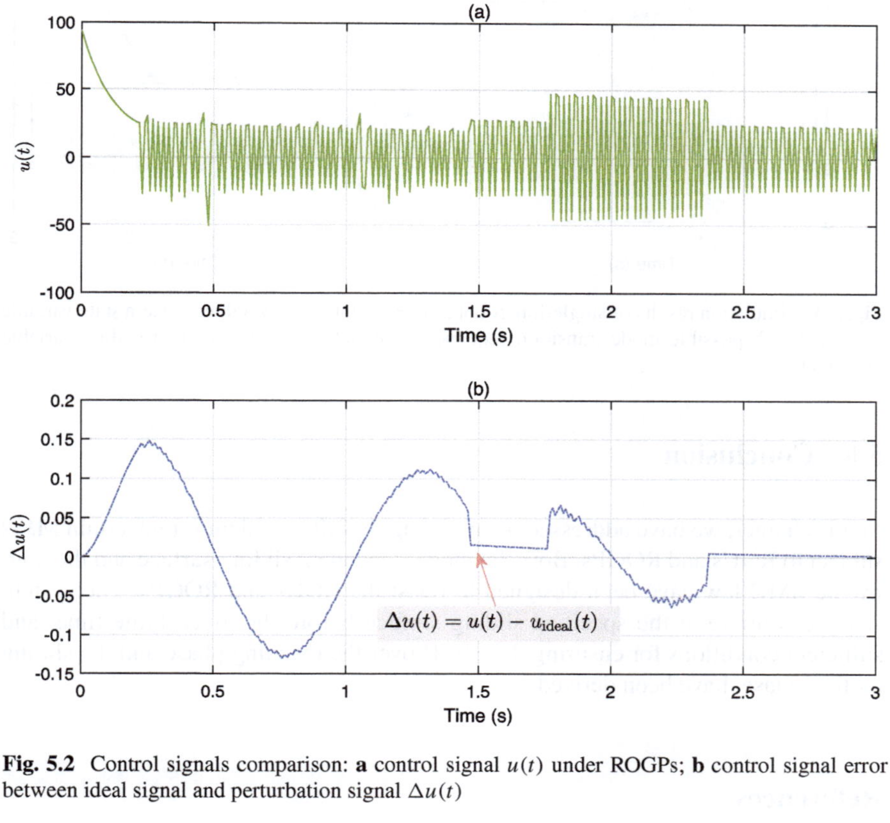

Fig. 5.2 Control signals comparison: **a** control signal $u(t)$ under ROGPs; **b** control signal error between ideal signal and perturbation signal $\Delta u(t)$

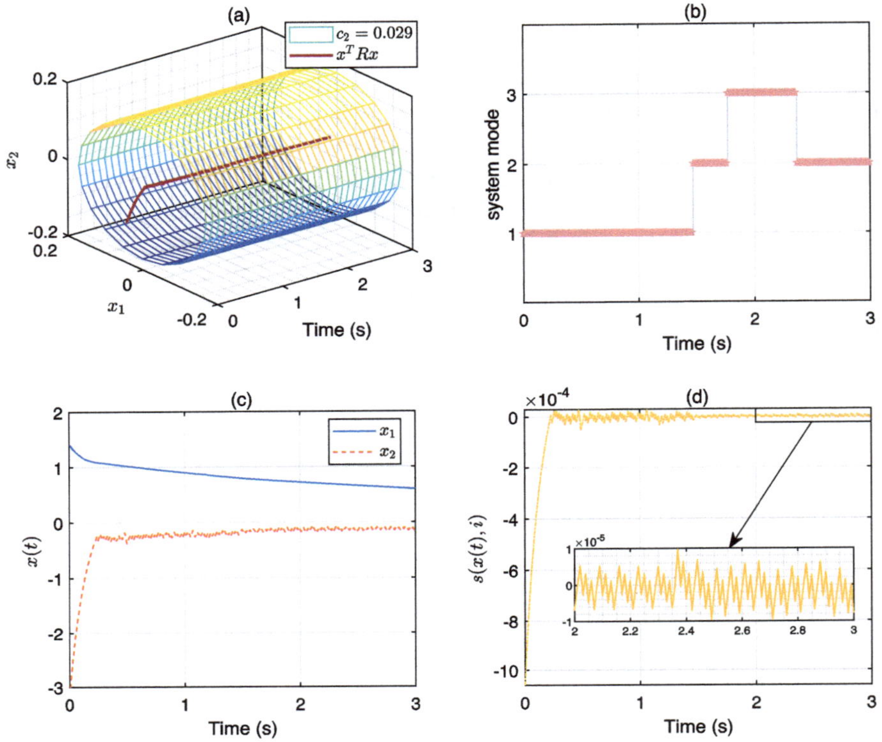

Fig. 5.3 Simulation results of single-link robot arm: **a** evolution of weighted system state variable $x^{\mathrm{T}}(t)Rx(t)$; **b** possible mode transition sequence r_t; **c** state responses $x(t)$; **d** sliding variable $s(x(t), i)$

5.6 Conclusion

In this chapter, we have addressed the non-fragile SMC problem for uncertain MJSs subject to ROUs and ROGPs. Both the mode-dependent sliding surface and the non-fragile SMC law have been designed to resist the ROUs and ROGPs. The system states can arrive at the specified sliding surface before the given finite time, and sufficient conditions for ensuring the SFTB over the reaching phase and the sliding motion phase, have been derived.

References

1. Song, J., Niu, Y., Wang, S.: Robust finite-time dissipative control subject to randomly occurring uncertainties and stochastic fading measurements. J. Frankl. Inst. **354**(9), 3706–3723 (2017)

2. Sakthivel, R., Joby, M., Anthoni, S.M.: Resilient dissipative based controller for stochastic systems with randomly occurring gain fluctuations. Inf. Sci. **418**, 447–462 (2017)
3. Li, X., Nguang, S.K., She, K., Cheng, J., Zhong, S.: Resilient controller synthesis for Markovian jump systems with probabilistic faults and gain fluctuations under stochastic sampling operational mechanism. Appl. Math. Comput. **392**, 125623 (2021)
4. Liu, Z., Su, L., Chen, X., Shi, K., Shen, H.: Extended dissipative fault-tolerant control for fuzzy Markov jump nonlinear systems with randomly occurring gain variations. Int. J. Fuzzy Syst. **24**(3), 1708–1718 (2022)
5. Dai, M., Xia, J., Xia, H., Shen, H.: Event-triggered passive synchronization for Markov jump neural networks subject to randomly occurring gain variations. Neurocomputing **331**, 403–411 (2019)
6. Wu, H.-N., Cai, K.-Y.: Mode-independent robust stabilization for uncertain Markovian jump nonlinear systems via fuzzy control. IEEE Trans. Syst. Man Cybern. B Cybern. **36**(3), 509–519 (2005)

Chapter 6
Attack-Resistant Sliding Mode Control Subject to Random Injection Attacks

6.1 Introduction

The cyber-physical systems (CPSs) merge the cyber-world of computing and communications with the physical world and have become ubiquitous in the control of complex dynamical systems, such as safety-critical aerospace systems, transportation systems, large-scale manufacturing systems and health-care systems (see [1–3] and the references therein). Recently, the security of CPSs has received considerable attention due to its vulnerability to adversarial attacks. The existing methods, like IT security encryption techniques, are inadequate to ensure the security of CPSs against the attacks from physical devices or interactions between physical and cyber layers. Hence, it is crucial to investigate the CPSs security from the systems' perspective [4].

In the existing literature, cyber-attacks can be divided into three types: the denial-of-services (DoS) attacks [5, 6], replay attacks [7, 8], and deception attacks [9, 10]. Deception attacks can be also called as false data injection attacks, in which the desired data is injected into the control input signals transmitted via the network layer. Due to the unknown attack modes designed by adversaries, it is difficult to detect the injected false data. Hence, some adaptive control architectures have been developed to mitigate the effect of the false data injection attacks in [10, 11]. It should be pointed out that the above injection attacks are launched in a deterministic way, while the adversaries may send false data in a probabilistic way to inject the cyber-attacks. To this end, the randomly occurring injection attacks (ROIAs) will be introduced in this chapter to describe this particularly stochastic phenomena.

It is noticed that the aforementioned works just focused on the *asymptotic behavior* of the controlled system over an infinite working time interval. Actually, in many engineering applications, there may be only a finite or short working time with considering the prescribed constraints on *transient dynamics*, which has motivated the recent increasing researches on the finite-time control problem [12, 13]. Some interesting results have been also obtained on sliding mode control (SMC) method in, e.g., [14–16], under the *secure signal transmission*. However, since the adversarial attacks

Z. Cao and Y. Niu, *Finite-Time Sliding Mode Control for Markov Jump Systems*,
Studies in Systems, Decision and Control 603,
https://doi.org/10.1007/978-981-96-7651-4_6

may be inevitable over a common shared communication network, it is necessary and interesting to further investigate the above problem subject to cyber attacks.

Motivated by the above discussion, this chapter will consider the design of SMC for ensuring the stochastic finite-time boundedness (SFTB) of Markov jump cyber-physical systems, for which only a finite-time/short-time behavior will be involved via an insecure communication network. Differing from the false data injection attacks as in [10, 11], the ROIAs will be considered in this chapter, which means that the adversaries may send false data in a probabilistic way.

The main contributions of this chapter are highlighted as follows:

(1) The *randomly occurring injection attacks* is introduced to describe a stochastic cyber-attack case happened in controller-to-actuator (C/A) channel. An attack-tolerant SMC law is proposed to guarantee the reachability of an explicit sliding domain during the interval $[0, T^*]$ with $T^* < T$ under ROIAs and randomly occurring uncertainties (ROUs).
(2) By introducing a sufficiently small mode-independent scalar into some coupled Lyapunov inequalities, the sufficient conditions for the SFTB of the closed-loop systems are derived, which provides an effective co-designing scheme for sliding gains and controller gains.

6.2 Randomly Injection Attack Model

Consider a class of Markov jump systems (MJSs) on a complete probability space $(\Omega, \mathscr{F}, \Pr)$ as in system (1.6) $(f_i(x(t)) \equiv 0)$ with ROUs and cyber attacks. The transition probabilities (TPs) are generated by (1.7) with the assumptions that all transition rates (TRs) are completely available. Accordingly, for mode $r_t = i \in \mathbb{S}$, the considered MJS with ROUs and cyber attacks is expressed as

$$\dot{x}(t) = (A_i + \alpha(t)\Delta A_i(t))x(t) + B_i u^A(t) + D_i w(t), \tag{6.1}$$

where $x(t) \in \mathbb{R}^n$ is the state; $u^A(t) \in \mathbb{R}^m$ is the control signal inputting to the actuator after suffering from attacks; $w(t)$ is the external disturbance satisfying Assumption 1.4; and $\Delta A_i(t)$ is the parameter uncertainty satisfying Assumption 1.2 with stochastic variable $\alpha(t) \in \mathbb{R}$ characterizing the phenomena of ROUs. The stochastic variable $\alpha(t)$ obeys the exponential distribution of switching with $\Pr\{\alpha(t) = 1\} = \mathbb{E}\{\alpha(t)\} = \bar{\alpha}, \Pr\{\alpha(t) = 0\} = 1 - \mathbb{E}\{\alpha(t)\} = 1 - \bar{\alpha}$ with known constants $\bar{\alpha} \in [0, 1]$.

As shown in Fig. 6.1, there may inevitably happen attacks on control signals transmitted via an insecure communication network. Hence, in this chapter, we consider the following ROIA model:

$$u^A(t) = u(t) + \varphi(t)H_i(t)\Psi(x(t), t, i), \tag{6.2}$$

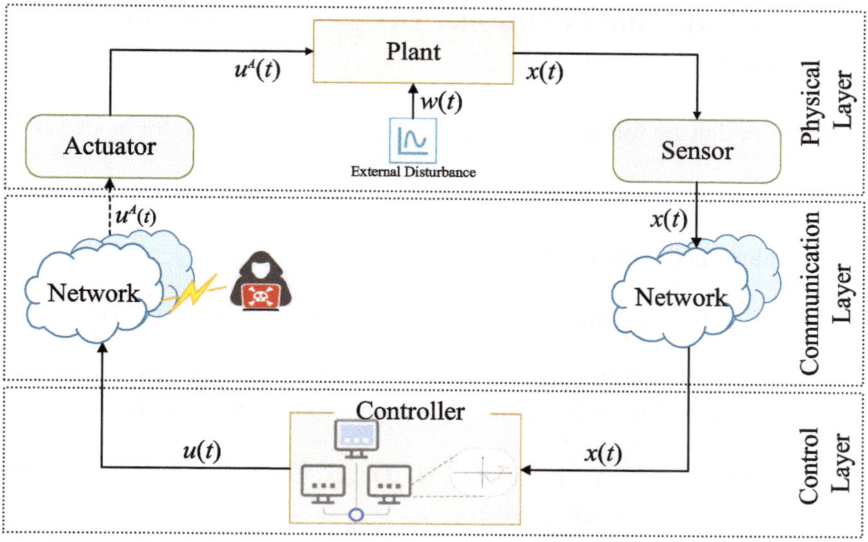

Fig. 6.1 Block diagram of MJSs under injection attacks

where $H_i(t)$ is an unknown weighting matrix able to representing the attacks' pattern, and the nonlinear function $\Psi(x(t), t, i)$ denotes that the system information used by the adversaries. It is assumed that $\|H_i(t)\| \leq \varsigma_i$, $\|\Psi(x(t), t, i)\| \leq \psi(x(t), t, i)$, where both the constant $\varsigma_i > 0$ and the function $\psi(x(t), t, i) > 0$ are known. $\varphi(t)$ obeys the exponential distribution of switching with $\Pr\{\varphi(t) = 1\} = \mathbb{E}\{\varphi(t)\} = \bar{\varphi}$, $\Pr\{\varphi(t) = 0\} = 1 - \mathbb{E}\{\varphi(t)\} = 1 - \bar{\varphi}$ with known constants $\bar{\varphi} \in [0, 1]$.

Remark 6.1 The ROIA (6.2) includes two key features: (1) the adversaries may send false data in a probabilistic way into the control input signals, which describes a particular stochastic phenomena. When $\bar{\varphi} = 1$, the case considered in this chapter will reduce to the certain attack case as in most existing works; (2) the attacks in (6.2) are mode-dependent, which will be more general and also more difficultly handled than the common injection attacks for all mode.

Remark 6.2 It should be pointed out that although the attackers in practical applications may often gain access to network layers and manipulate the exchanged data among sensors, controllers, and actuators, the injected information is usually subject to physical constraints including channel capacity, amplitude, etc. [17]. Therefore, it is reasonably assumed that $H_i(t)$ and $\Psi(x(t), t, i)$ are bounded. In this work, the bound constant and bound function are assumed to be known. For the unknown case, some estimation methods, e.g., parameter adaptive method [18], neural network approximation [11], may be effectively utilized.

The objective of this chapter is to design a sliding mode controller such that the SFTB of the resultant closed-loop MJSs can be ensured despite ROUs and ROIAs.

6.3 Attack-Resistant Controller Design and Reachability Analysis

Firstly, we design the following mode-dependent sliding function (for mode $i \in \mathbb{S}$):

$$s(x(t), i) = B_i^{\mathrm{T}} X_i x(t), \tag{6.3}$$

and the sliding mode controller as:

$$u(t) = K_i x(t) - \rho_i(t)\mathrm{sgn}(s(x(t), i)) - \bar{\varphi}\varsigma_i \psi(x(t), t, i)\mathrm{sgn}(s(x(t), i)), \tag{6.4}$$

where matrices $X_i \in \mathbb{R}^{n\times n}$ and $K_i \in \mathbb{R}^{m\times n}$ will be designed later such that $B_i^{\mathrm{T}} X_i B_i$ is nonsingular for each $i \in \mathbb{S}$, and the robust term $\rho_i(t)$ is constructed as

$$\rho_i(t) \triangleq \eta_{1i} + \eta_{2i} + (\gamma_{1i} + \gamma_{2i})\|x(t)\|, \tag{6.5}$$

$$\eta_{1i} \geq \frac{\lambda_{\max}((B_i^{\mathrm{T}} X_i B_i)^{-1})}{T - \kappa}\|B_i^{\mathrm{T}} X_i x(0)\|, \tag{6.6}$$

$$\eta_{2i} \triangleq \|(B_i^{\mathrm{T}} X_i B_i)^{-1} B_i^{\mathrm{T}} X_i D_i\|\delta, \tag{6.7}$$

$$\gamma_{1i} \triangleq \|(B_i^{\mathrm{T}} X_i B_i)^{-1} B_i^{\mathrm{T}} X_i\| \left(\|A_i\| + \bar{\alpha}\|M_i\|\|N_i\|\right) + \|K_i\|, \tag{6.8}$$

$$\gamma_{2i} \triangleq \frac{1}{2}\|B_i^{\mathrm{T}} X_i\|^{-1}\left\| \sum_{j\in\mathbb{S}} \lambda_{ij} X_j B_j (B_j^{\mathrm{T}} X_j B_j)^{-1} B_j^{\mathrm{T}} X_j \right\| \tag{6.9}$$

with κ being a positive scalar and satisfying $0 < \kappa < T$. In the above design, the nonsingularity of $B_i^{\mathrm{T}} X_i B_i$ can be ensured for $X_i > 0$ ($i \in \mathbb{S}$) due to the full column rank of matrix B_i.

As another step in SMC design, the finite-time reachability of the specified sliding surfaces $s(x(t), i) = 0$ before a finite instant T^* with $T^* < T$ will be further investigated in the sequel.

Theorem 6.1 *For a given finite interval* $[0, T]$ *and the specified the sliding function* (6.3), *the designed SMC law* (6.4)–(6.9) *can drive the state trajectories of the MJS* (6.16) *onto the sliding surface* $s(x(t), i) = 0$ *during the interval* $[0, T^*]$ *with* $T^* < T$ *and remain there in the subsequent time.*

Proof Choose the Lyapunov function:

$$U(x(t), i) = s^{\mathrm{T}}(x(t), i)(B_i^{\mathrm{T}} X_i B_i)^{-1} s(x(t), i). \tag{6.10}$$

Then, by Definition 1.4, we obtain the infinitesimal operator as

$$\begin{aligned}
&\mathscr{L}U(x(t), i)\\
&= 2s^{\mathrm{T}}(x(t), i)(B_i^{\mathrm{T}} X_i B_i)^{-1} B_i^{\mathrm{T}} X_i (A_i + \bar{\alpha}\Delta A_i)x(t)\\
&\quad + 2s^{\mathrm{T}}(x(t), i)(B_i^{\mathrm{T}} X_i B_i)^{-1} B_i^{\mathrm{T}} X_i D_i w(t)
\end{aligned}$$

$$
\begin{aligned}
&+\sum_{j\in\mathbb{S}}\lambda_{ij}s^{\mathrm{T}}(x(t),j)(B_j^{\mathrm{T}}X_jB_j)^{-1}s(x(t),j)\\
&+2s^{\mathrm{T}}(x(t),i)\Big(K_ix(t)-\rho_i(t)\mathrm{sgn}(s(x(t),i))\\
&-\bar{\varphi}\varsigma_i\psi(x(t),t,i)\mathrm{sgn}(s(x(t),i))\Big)\\
&+2s^{\mathrm{T}}(x(t),i)\bar{\varphi}H_i(t)\Psi(x(t),t,i).
\end{aligned}
\tag{6.11}
$$

Note the fact $\|s(x(t),i)\|\le|s(x(t),i)|$. One has from (6.11)

$$
\begin{aligned}
&\mathscr{L}U(x(t),i)\\
\le&\,2\|s(x(t),i)\|\|(B_i^{\mathrm{T}}X_iB_i)^{-1}B_i^{\mathrm{T}}X_i\|(\|A_i\|+\bar{\alpha}\|M_i\|\|N_i\|)\|x(t)\|\\
&+2\|s(x(t),i)\|\|(B_i^{\mathrm{T}}X_iB_i)^{-1}B_i^{\mathrm{T}}X_iD_i\|\|w(t)\|\\
&+2\|s(x(t),i)\|\bar{\varphi}\varsigma_i\psi(x(t),t,i)\\
&+2\|x(t)\|\Big\|\sum_{j\in\mathbb{S}}\lambda_{ij}X_jB_j(B_j^{\mathrm{T}}X_jB_j)^{-1}B_j^{\mathrm{T}}X_j\Big\|\|x(t)\|\\
&+2\|s(x(t),i)\|\|K_i\|\|x(t)\|-2\,s(x(t),i)|\rho_i(t)\|x(t)\|)\\
&-2\,s(x(t),i)|\bar{\varphi}\varsigma_i\psi(x(t),t,i)\\
\le&\,\|x(t)\|\Big\|\sum_{j\in\mathbb{S}}\lambda_{ij}X_jB_j(B_j^{\mathrm{T}}X_jB_j)^{-1}B_j^{\mathrm{T}}X_j\Big\|\|x(t)\|\\
&-2\|s(x(t),i)\|(\eta_{1i}+\gamma_{2i}\|x(t)\|),
\end{aligned}
\tag{6.12}
$$

where the expressions (6.6)–(6.9) are utilized.

Next, we will prove that

$$
\begin{aligned}
&\|x(t)\|\Big\|\sum_{j\in\mathbb{S}}\lambda_{ij}X_jB_j(B_j^{\mathrm{T}}X_jB_j)^{-1}B_j^{\mathrm{T}}X_j\Big\|\|x(t)\|\\
&-\gamma_{2i}\|s(x(t),i)\|\|x(t)\|<0,
\end{aligned}
\tag{6.13}
$$

which can ensure $\mathscr{L}U(x(t),i)<0$ from (6.12). Actually, we have

$$
\begin{aligned}
&\|x(t)\|\Big\|\sum_{j\in\mathbb{S}}\lambda_{ij}X_jB_j(B_j^{\mathrm{T}}X_jB_j)^{-1}B_j^{\mathrm{T}}X_j\Big\|\|x(t)\|-\|s(x(t),i)\|\gamma_{2i}\|x(t)\|\\
\le&\Big\|\sum_{j\in\mathbb{S}}\lambda_{ij}X_jB_j(B_j^{\mathrm{T}}X_jB_j)^{-1}B_j^{\mathrm{T}}X_j\Big\|\Big(\|x(t)\|-\|s(x(t),i)\|\|B_i^{\mathrm{T}}X_i\|^{-1}\Big)\\
=&\Big\|\sum_{j\in\mathbb{S}}\lambda_{ij}X_jB_j(B_j^{\mathrm{T}}X_jB_j)^{-1}B_j^{\mathrm{T}}X_j\Big\|\left(\|x(t)\|-\|s(x(t),i)\|\frac{\|x(t)\|}{\|B_i^{\mathrm{T}}X_i\|\|x(t)\|}\right)
\end{aligned}
$$

$$= \left\| \sum_{j\in\mathbb{S}} \lambda_{ij} X_j B_j (B_j^{\mathrm{T}} X_j B_j)^{-1} B_j^{\mathrm{T}} X_j \right\| \left(\|x(t)\| - \|s(x(t), i)\| \frac{\|x(t)\|}{\|B_i^{\mathrm{T}} X_i x(t)\|} \right) < 0, \tag{6.14}$$

which, together with the expression (6.12), yields

$$\mathscr{L}U(x(t), i) \leq -2\eta_{1i}\|s(x(t), i)\| < 0.$$

Accordingly, we obtain that the state trajectories will be driven onto the specified sliding surfaces $s(x(t), i) = 0$ by the SMC law (6.4)–(6.9).

The following will further give the reaching instant T^*. Similar to the derivation between (4.29)–(4.33), it is easy to see that if there exists an instant T^* satisfying

$$T^* = \frac{\lambda_{\max}((B_i^{\mathrm{T}} X_i B_i)^{-1})}{\eta_{1i}} \|B_i^{\mathrm{T}} X_i x(0)\| \tag{6.15}$$

such that for $t \geq T^*$, $\mathbb{E}\{U(x(t), i)\} = 0$, consequently, the sliding surface $s(x(t), i) = 0$ is attained (in the mean square).. Thus, from (6.6) and (6.15), one obtains $T^* \leq T - \kappa < T$, which means that the state trajectories of system (6.1) will be driven onto the specified sliding surface $s(x(t), i) = 0$ in a finite time T^* with $T^* < T$ by SMC law (6.4)–(6.9). This completes the proof. □

6.4 SFTB Analysis

For a given time interval $[0, T]$, it is known that there exist two phases for the state trajectories of SMC systems: the reaching phase within $[0, T^*]$ and the sliding motion phase within $[T^*, T]$. Thus, one has to investigate the SFTB of the controlled system during the reaching phase and the sliding motion phase simultaneously.

6.4.1 Reaching Phase Analysis

By substituting SMC law (6.4) into the system (6.1), we obtain the following closed-loop system:

$$\begin{aligned} \dot{x}(t) = &(\bar{A}_i + \alpha(t)\Delta A_i(t))x(t) - B_i\rho_i(t)\mathrm{sgn}(s(x(t), i)) \\ &- B_i\bar{\varphi}\varsigma_i\psi(x(t), t, i)\mathrm{sgn}(s(x(t), i)) \\ &+ \varphi(t)B_iH_i(t)\Psi(x(t), t, i) + D_iw(t), \end{aligned} \tag{6.16}$$

where $\bar{A}_i \triangleq A_i + B_iK_i$.

The following theorem provides the sufficient conditions to guarantee SFTB over the reaching phase $[0, T^*]$.

Theorem 6.2 *Given the parameters* (c_1, c_2, T, R, δ) *and scalar* $\xi_i > 0$*, if there exist positive constants* ζ_i, c^*, *matrices* $X_i > 0$ *and* K_i $(\forall i \in \mathbb{S})$*, satisfying the following matrix inequalities:*

$$\begin{bmatrix} \Gamma_i & X_i D_i \\ * & -\xi_i I \end{bmatrix} < 0, \tag{6.17}$$

$$c_1 < c^* < c_2, \tag{6.18}$$

$$\bar{\sigma}(X_{Ri})c_1 + \delta^2 T \xi_i < e^{-\xi_i T} \underline{\sigma}(X_{Ri})c^*, \tag{6.19}$$

where

$$\Gamma_i \triangleq \mathrm{He}(X_i \bar{A}_i) + \tilde{X}_i - \xi_i X_i + \zeta_i^{-1} \bar{\alpha} X_i M_i M_i^{\mathrm{T}} X_i + \zeta_i \bar{\alpha} N_i^{\mathrm{T}} N_i,$$

$$\bar{\sigma}(X_{Ri}) \triangleq \max_{i \in \mathbb{S}} \left\{ \lambda_{\max}(R^{-\frac{1}{2}} X_i R^{-\frac{1}{2}}) \right\}, \; \tilde{X}_i \triangleq \sum_{j=1}^{N} \lambda_{ij} X_j,$$

$$\underline{\sigma}(X_{Ri}) \triangleq \min_{i \in \mathbb{S}} \left\{ \lambda_{\min}(R^{-\frac{1}{2}} X_i R^{-\frac{1}{2}}) \right\},$$

the closed-loop system (6.16) *is SFTB with respect to* $\aleph_{[0,T^*]}$.

Proof Choose the following Lyapunov functional candidate:

$$V(x(t), i) = x^{\mathrm{T}}(t) X_i x(t). \tag{6.20}$$

Notice that the closed-loop system (6.16) is an MJS involving in stochastic variables $\alpha(t)$ and $\varphi(t)$, whose infinitesimal operator is given as

$$\begin{aligned} \mathscr{L}V(x(t), i) = {} & x^{\mathrm{T}}(t) \left(\bar{A}_i^{\mathrm{T}} X_i^{\mathrm{T}} + X_i \bar{A}_i + \tilde{X}_i \right) x(t) + 2\bar{\alpha} x^{\mathrm{T}}(t) X_i \Delta A_i x(t) \\ & + w^{\mathrm{T}}(t) D_i^{\mathrm{T}} X_i^{\mathrm{T}} x(t) + x^{\mathrm{T}}(t) X_i D_i w(t) \\ & + 2x^{\mathrm{T}}(t) X_i B_i \Big(\bar{\varphi} H_i(t) \Psi(x(t), t, i) \\ & - \big(\rho_i(t) + \bar{\varphi} \varsigma_i \psi(x(t), t, i) \big) \mathrm{sgn}(s(x(t), i)) \Big). \end{aligned} \tag{6.21}$$

Note the fact $\|s(x(t), i))\| \leq |s(x(t), i))|$, $s(x(t), i)) = B_i^{\mathrm{T}} X_i x(t)$. One has

$$\begin{aligned} & 2x^{\mathrm{T}}(t) X_i B_i \Big(\bar{\varphi} H_i(t) \Psi(x(t), t, i) \\ & - \big(\rho_i(t) + \bar{\varphi} \varsigma_i \psi(x(t), t, i) \big) \mathrm{sgn}(s(x(t), i)) \Big) \\ = {} & 2s^{\mathrm{T}}(x(t), i) \bar{\varphi} \Big(H_i(t) \Psi(x(t), t, i) - \varsigma_i \psi(x(t), t, i) \mathrm{sgn}(s(x(t), i)) \Big) \end{aligned}$$

$$
\begin{aligned}
&- 2s^{\mathrm{T}}(x(t), i)\rho_i(t)\mathrm{sgn}(s(x(t), i)) \\
\leq\, & 2\bar{\varphi}\Big(\varsigma_i \psi(x(t), t, i)\|s(x(t), i)\| - \varsigma_i \psi(x(t), t, i)|s(x(t), i)|\Big) \\
&- 2\rho_i(t)|s(x(t), i)| \\
\leq & \,- 2\rho_i(t)|s(x(t), i))| < 0.
\end{aligned} \tag{6.22}
$$

For $\zeta_i > 0$, we have

$$
2x^{\mathrm{T}}(t)X_i\Delta A_i(t)x(t) \leq \zeta_i^{-1}x^{\mathrm{T}}(t)X_iM_iM_i^{\mathrm{T}}X_ix(t) + \zeta_ix^{\mathrm{T}}(t)N_i^{\mathrm{T}}N_ix(t). \tag{6.23}
$$

Then, by means of expressions (6.23) and (6.22), it follows from (6.21) that

$$
\begin{aligned}
&\mathscr{L}V(x(t), i) \\
\leq\, & x^{\mathrm{T}}(t)\left(\mathrm{He}(X_i\bar{A}_i) + \zeta_i^{-1}\bar{\alpha}X_iM_iM_i^{\mathrm{T}}X_i + \zeta_i\bar{\alpha}N_i^{\mathrm{T}}N_i + \tilde{X}_i\right)x(t) \\
&+ w^{\mathrm{T}}(t)D_i^{\mathrm{T}}X_i^{\mathrm{T}}x(t) + x^{\mathrm{T}}(t)X_iD_iw(t).
\end{aligned} \tag{6.24}
$$

Define an auxiliary function as follows:

$$
J(x(t), i) = \mathscr{L}V(x(t), i) - \xi_iV(x(t), i) - \xi_iw^{\mathrm{T}}(t)w(t). \tag{6.25}
$$

Apparently, $J(x(t), i) < 0$ can be ensured by (6.17). Thus, it follows from $J(x(t), i) < 0$ that

$$
\mathscr{L}V(x(t), i) < \xi_iV(x(t), i) + \xi_iw^{\mathrm{T}}(t)w(t). \tag{6.26}
$$

Multiplying both sides of expression (6.26) by $e^{-\xi_it}$ and integrating the obtained inequality from 0 to t with $t \in [0, T^*]$, we have

$$
\begin{aligned}
e^{-\xi_it}\mathbb{E}\{V(x(t), i)\} &< V(x(0), r_0) + \xi_i\int_0^{\mathrm{T}} e^{-\xi_i\tau}w^{\mathrm{T}}(\tau)w(\tau)\mathrm{d}\tau \\
&\leq \bar{\sigma}(X_{Ri})c_1 + \delta^2T\xi_i.
\end{aligned} \tag{6.27}
$$

On the other hand, it follows from (6.20) that

$$
\begin{aligned}
e^{-\xi_it}\mathbb{E}\{V(x(t), i)\} &= e^{-\xi_it}\mathbb{E}\{x^{\mathrm{T}}(t)X_ix(t)\} \\
&\geq e^{-\xi_iT}\underline{\sigma}(X_{Ri})\mathbb{E}\{x^{\mathrm{T}}(t)Rx(t)\}.
\end{aligned} \tag{6.28}
$$

It yields from (6.18)–(6.19), (6.27) and (6.28) that

$$
\mathbb{E}\{x^{\mathrm{T}}(t)Rx(t)\} < \frac{\bar{\sigma}(X_{Ri})c_1 + \delta^2T\xi_i}{e^{-\xi_iT}\underline{\sigma}(X_{Ri})}. \tag{6.29}
$$

It follows from (6.19) and (6.29) that $\mathbb{E}\{x^{\mathrm{T}}(t)Rx(t)\} < c^*$ for all $t \in [0, T^*]$. Thus, according to Definition 1.5, the resulting closed-loop system (6.16) is SFTB with respect to $\aleph_{[0,T^*]}$. This fulfills the proof. □

6.4.2 Sliding Motion Phase Analysis

According to SMC theory, when the state trajectories maintain in the sliding surface $s(x(t), i) = 0$, we have the following equivalent controller for each $i \in \mathbb{S}$

$$\begin{aligned} u_{eq}(t) = & -\varphi(t)H_i(t)\Psi(x(t), t, i) - (B_i^{\mathrm{T}} X_i B_i)^{-1} B_i^{\mathrm{T}} X_i D_i w(t) \\ & - (B_i^{\mathrm{T}} X_i B_i)^{-1} B_i^{\mathrm{T}} X_i (A_i + \alpha(t)\Delta A_i(t))x(t). \end{aligned} \tag{6.30}$$

By substituting (6.30) into the MJS (6.1), we obtain the following sliding mode dynamics:

$$\dot{x}(t) = (I - Y_i)\left(A_i + \alpha(t)\Delta A_i(t))x(t) + D_i w(t)\right), \tag{6.31}$$

where $Y_i \triangleq B_i(B_i^{\mathrm{T}} X_i B_i)^{-1} B_i^{\mathrm{T}} X_i$.

In this subsection, we give the following sufficient conditions to guarantee SFTB over the sliding motion phase $[T^*, T]$.

Theorem 6.3 *Given the parameters* (c_1, c_2, T, R, δ) *and scalar* $\xi_i > 0$, *if there exist positive constants* ζ_i, μ_i, c^* *and matrices* $X_i > 0$ *and* K_i $(\forall i \in \mathbb{S})$, *satisfying the following matrix inequalities:*

$$\begin{bmatrix} \Upsilon_i & X_i D_i \\ * & D_i^{\mathrm{T}} X_i D_i - \xi_i I \end{bmatrix} < 0, \tag{6.32}$$

$$c_1 < c^* < c_2, \tag{6.33}$$

$$\bar{\sigma}(X_i)c^* + \delta^2 T \xi_i < e^{-\xi_i T}\underline{\sigma}(X_i)c_2, \tag{6.34}$$

where

$$\begin{aligned} \Upsilon_i \triangleq & \operatorname{He}(X_i A_i) - \xi_i X_i + \tilde{X}_i + 2X_i B_i(B_i^{\mathrm{T}} X_i B_i)^{-1} B_i^{\mathrm{T}} X_i + \zeta_i^{-1}\bar{\alpha} X_i M_i M_i^{\mathrm{T}} X_i \\ & + A_i^{\mathrm{T}} X_i M_i(\mu_i I - M_i^{\mathrm{T}} X_i M_i)^{-1} M_i^{\mathrm{T}} X_i A_i + A_i^{\mathrm{T}} X_i A_i + \mu_i \bar{\alpha}^2 N_i^{\mathrm{T}} N_i + \zeta_i \bar{\alpha} N_i^{\mathrm{T}} N_i, \end{aligned}$$

the sliding mode dynamics (6.31) *are SFTB with respect to* $\aleph_{[T^*,T]}$.

Proof For the Lyapunov function as in (6.20), we can obtain the infinitesimal operator as

$$\begin{aligned} & \mathscr{L}V(x(t), i) \\ & = x^{\mathrm{T}}(t)(\operatorname{He}(X_i A_i) + \tilde{X}_i)x(t) + 2\bar{\alpha} x^{\mathrm{T}}(t) X_i \Delta A_i(t) x(t) \end{aligned}$$

$$
\begin{aligned}
&- 2x^{\mathrm{T}}(t)X_iY_i(A_i + \bar{\alpha}\Delta A_i(t))x(t) - 2x^{\mathrm{T}}(t)X_iY_iD_iw(t) \\
&+ w^{\mathrm{T}}(t)D_i^{\mathrm{T}}X_ix(t) + x^{\mathrm{T}}(t)X_iD_iw(t).
\end{aligned} \tag{6.35}
$$

For $X_i > 0$, one has

$$
\begin{aligned}
&- 2x^{\mathrm{T}}(t)X_iB_i(B_i^{\mathrm{T}}X_iB_i)^{-1}B_i^{\mathrm{T}}X_iD_iw(t) \\
\leq\, & x^{\mathrm{T}}(t)X_iB_i(B_i^{\mathrm{T}}X_iB_i)^{-1}B_i^{\mathrm{T}}X_i^{\mathrm{T}}x(t) + w^{\mathrm{T}}(t)D_i^{\mathrm{T}}X_iD_iw(t),
\end{aligned} \tag{6.36}
$$

and from Lemma 1.1, for any $\mu_i > 0$ satisfying $\mu_i I - M_i^{\mathrm{T}}X_iM_i > 0$, one has

$$
\begin{aligned}
&- 2x^{\mathrm{T}}(t)X_iB_i(B_i^{\mathrm{T}}X_iB_i)^{-1}B_i^{\mathrm{T}}X_i\big(A_i + \bar{\alpha}\Delta A_i(t)\big)x(t) \\
\leq\, & x^{\mathrm{T}}(t)\big(A_i + \bar{\alpha}\Delta A_i(t)\big)^{\mathrm{T}}X_i\big(A_i + \bar{\alpha}\Delta A_i(t)\big)x(t) \\
&+ x^{\mathrm{T}}(t)X_iB_i(B_i^{\mathrm{T}}X_iB_i)^{-1}B_i^{\mathrm{T}}X_ix(t) \\
\leq\, & x^{\mathrm{T}}(t)\left(A_i^{\mathrm{T}}X_iM_i(\mu_iI - M_i^{\mathrm{T}}X_iM_i)^{-1}M_i^{\mathrm{T}}X_iA_i + A_i^{\mathrm{T}}X_iA_i + \mu_i\bar{\alpha}^2N_i^{\mathrm{T}}N_i\right)x(t) \\
&+ x^{\mathrm{T}}(t)X_iB_i(B_i^{\mathrm{T}}X_iB_i)^{-1}B_i^{\mathrm{T}}X_ix(t).
\end{aligned} \tag{6.37}
$$

Then, for the auxiliary function $J(x(t), i)$ as (6.25), $J(x(t), i) < 0$ can be ensured by (6.32). Thus, we obtain

$$
\mathscr{L}V(x(t), i) < \xi_iV(x(t), i) + \xi_iw^{\mathrm{T}}(t)w(t). \tag{6.38}
$$

According to the result in the Theorem 6.2, one has $\mathbb{E}\{x^{\mathrm{T}}(T^*)Rx(T^*)\} < c^*$. Thus, by multiplying both sides of expression (6.38) by $e^{-\xi_it}$ and integrating the obtained inequality from T^* to t with $t \in [T^*, T]$, we have

$$
\begin{aligned}
& e^{-\xi_it}\mathbb{E}\{V(x(t), i)\} \\
<\, & e^{-\xi_iT^*}\mathbb{E}\{V(x(T^*), r_{T^*})\} + \xi_i\int_{T^*}^{\mathrm{T}} e^{-\xi_i\tau}w^{\mathrm{T}}(\tau)w(\tau)\mathrm{d}\tau \\
<\, & \bar{\sigma}(\hat{P}_i)\mathbb{E}\{x^{\mathrm{T}}(T^*)E^{\mathrm{T}}REx(T^*)\} + \delta^2T\xi_i \\
<\, & \bar{\sigma}(X_{Ri})c^* + \delta^2T\xi_i.
\end{aligned} \tag{6.39}
$$

In addition, it follows from (6.20) that

$$
e^{-\xi_it}\mathbb{E}\{V(x(t), i)\} \geq e^{-\xi_iT}\underline{\sigma}(X_{Ri})\mathbb{E}\{x^{\mathrm{T}}(t)Rx(t)\}. \tag{6.40}
$$

Putting together (6.39) and (6.40), we have

$$
\mathbb{E}\{x^{\mathrm{T}}(t)Rx(t)\} < \frac{\bar{\sigma}(X_{Ri})c^* + \delta^2T\xi_i}{e^{-\xi_iT}\underline{\sigma}(X_{Ri})}. \tag{6.41}
$$

Furthermore, if there exists a scalar c^* satisfying (6.33), (6.34), we have from (6.41) that $\mathbb{E}\{x^{\mathrm{T}}(t)Rx(t)\} < c_2$ for all $t \in [T^*, T]$. Thus, the sliding mode dynamics (6.31) are SFTB with respect to $\aleph_{[T^*,T]}$, which complete the proof. □

6.4.3 Solving Algorithm for Controller Gains

It is noted that the SFTB of the closed-loop system over the whole interval $[0, T]$ is attained only when the conditions in Theorems 6.2 and 6.3 hold simultaneously. That means that the the matrices X_i and K_i should be designed to ensure the conditions in Theorems 6.2 and 6.3 to be feasible simultaneously. In the sequel, we shall give sufficient conditions in Theorem 6.4, which can guarantee that the conditions in Theorems 6.2 and 6.3 hold simultaneously.

Theorem 6.4 *Given the parameters* (c_1, c_2, T, R, δ) *and scalar* ξ_i, *if there exist positive constants* ζ_i, μ_i, ν_i, c^*, ϑ_i *and matrices* $X_i > 0$ *and* $\bar{K}_i$ $(\forall i \in \mathbb{S})$ *satisfying the following matrix inequalities:*

$$\begin{bmatrix} \Lambda_{11i} & \Lambda_{12i} \\ * & \Lambda_{22i} \end{bmatrix} < 0, \tag{6.42}$$

$$c_1 < c^* < c_2, \tag{6.43}$$

$$\vartheta_i R < X_i < 2R, \tag{6.44}$$

$$\begin{bmatrix} -e^{-\xi_i T}c^* & \sqrt{2c_1 + \delta^2 T\xi_i} \\ * & -\vartheta_i \end{bmatrix} < 0, \tag{6.45}$$

$$2c^* + \delta^2 T\xi_i - e^{-\xi_i T}c_2\vartheta_i < 0, \tag{6.46}$$

where

$$\Lambda_{11i} \triangleq \begin{bmatrix} \mathcal{L}_i & X_i D_i \\ * & D_i^{\mathrm{T}} X_i D_i - \xi_i I \end{bmatrix},$$

$$\mathcal{L}_i \triangleq X_i A_i + A_i^{\mathrm{T}} X_i + \tilde{X}_i + A_i^{\mathrm{T}} X_i A_i - (\xi_i + 2)X_i + \nu_i I$$

$$\Lambda_{12i} \triangleq \begin{bmatrix} \sqrt{\bar{\alpha}}X_i M_i & \sqrt{\bar{\alpha}}\zeta_i N_i^{\mathrm{T}} & A_i^{\mathrm{T}} X_i M_i & \mu_i\bar{\alpha}N_i^{\mathrm{T}} & \sqrt{2}X_i B_i & X_i + (B_i\bar{K}_i)^{\mathrm{T}} \\ 0 & 0 & 0 & 0 & 0 & 0 \end{bmatrix},$$

$$\Lambda_{22i} \triangleq -\mathrm{diag}\{\zeta_i I, \zeta_i I, \mu_i I - M_i^{\mathrm{T}} X_i M_i, \mu_i I, B_i^{\mathrm{T}} X_i B_i, \nu_i I\},$$

the closed-loop system (6.16) *is SFTB with respect to* $\aleph_{[0,T]}$ *over the whole finite-time interval* $[0, T]$. *Moreover, the control gains are* $K_i = \nu_i^{-1}\bar{K}_i$ $(i \in \mathbb{S})$.

Proof Considering the conditions (6.17)–(6.19) in Theorem 6.2 and conditions (6.32)–(6.34) in Theorem 6.3 simultaneously, we have the following inequalities:

$$\begin{bmatrix} \Phi_i & X_i D_i \\ * & D_i^{\mathrm{T}} X_i D_i - \xi_i I \end{bmatrix} < 0, \tag{6.47}$$

$$c_1 < c^* < c_2, \tag{6.48}$$

$$\bar{\sigma}(X_{Ri})c_1 + \delta^2 T \xi_i < e^{-\xi_i T} \underline{\sigma}(X_{Ri})c^*, \tag{6.49}$$

$$\bar{\sigma}(X_{Ri})c^* + \delta^2 T \xi_i < e^{-\xi_i T} \underline{\sigma}(X_{Ri})c_2, \tag{6.50}$$

where

$$\begin{aligned} \Phi_i \triangleq{} & X_i A_i + A_i^{\mathrm{T}} X_i + X_i B_i K_i + (B_i K_i)^{\mathrm{T}} X_i - \xi_i X_i + \tilde{X}_i + A_i^{\mathrm{T}} X_i A_i \\ & + \mu_i \bar{\alpha}^2 N_i^{\mathrm{T}} N_i + \zeta_i^{-1} \bar{\alpha} X_i M_i M_i^{\mathrm{T}} X_i + \zeta_i \bar{\alpha} N_i^{\mathrm{T}} N_i + 2 X_i B_i (B_i^{\mathrm{T}} X_i B_i)^{-1} B_i^{\mathrm{T}} X_i \\ & + A_i^{\mathrm{T}} X_i M_i (\mu_i I - M_i^{\mathrm{T}} X_i M_i)^{-1} M_i^{\mathrm{T}} X_i A_i. \end{aligned}$$

From Lemma 5.1, for a set of sufficiently small scalar ν_i, the matrix inequalities (6.47) hold if and only if the following matrix inequalities hold:

$$\begin{bmatrix} \Xi_i & X_i D_i & X_i + (B_i \bar{K}_i)^{\mathrm{T}} \\ * & D_i^{\mathrm{T}} X_i D_i - \xi_i I & 0 \\ * & * & -\nu_i I \end{bmatrix} < 0, \tag{6.51}$$

where

$$\begin{aligned} \Xi_i \triangleq{} & X_i A_i + A_i^{\mathrm{T}} X_i - 2 X_i + \nu_i I - \xi_i X_i + \tilde{X}_i + A_i^{\mathrm{T}} X_i A_i + \mu_i \bar{\alpha}^2 N_i^{\mathrm{T}} N_i \\ & + \zeta_i^{-1} \bar{\alpha} X_i M_i M_i^{\mathrm{T}} X_i + 2 X_i B_i (B_i^{\mathrm{T}} X_i B_i)^{-1} B_i^{\mathrm{T}} X_i + \zeta_i \bar{\alpha} N_i^{\mathrm{T}} N_i \\ & + A_i^{\mathrm{T}} X_i M_i (\mu_i I - M_i^{\mathrm{T}} X_i M_i)^{-1} M_i^{\mathrm{T}} X_i A_i, \\ \bar{K}_i \triangleq{} & \nu_i K_i, \end{aligned}$$

which can be ensured by (6.42).

By the condition (6.44), one has $\vartheta_i < \underline{\sigma}(X_{Ri}), \bar{\sigma}(X_{Ri}) < 2$. It is easily shown that the inequalities (6.49)–(6.50) can be ensured by (6.45)–(6.46). Thus, it is concluded that the conditions (6.17)–(6.19) in Theorem 6.2 and conditions (6.32)–(6.34) in Theorem 6.3 can be ensured by (6.42)–(6.46). This accomplishes the proof. □

Remark 6.3 In general, the smaller controller gain implies the less energy requirement in control input. Therefore, one can find an ideal ξ_i by the linear searching over $\left(0, \frac{2c_2}{\delta^2 T}\right)$ (the upper bound is obtained from (6.46)) so that the norm of the obtained sliding mode controller gain is the minimum, please refer to [14] for details of the Solving Algorithm.

Table 6.1 Modes of the parameters $\mathcal{M}_i$ and $\mathcal{J}_i$

Mode i	Parameter $\mathcal{M}_i$ (kg)	Parameter $\mathcal{J}_i$ (N·m)
1	0.04	5
2	0.12	6.67
3	0.39	10

6.5 Simulation

Consider a single-link robot arm with the dynamic model as in (5.59), whose system structure diagram is exhibited in Fig. 6.2. The values of parameters m, g and $\mathcal{L}$ are given as $m = 0.02$ kg, $g = 9.8$ m/s^2 and $\mathcal{L} = 1$ m. $\mathcal{D}(t)$ is assumed to be bounded within 2% of its nominal value $\mathcal{D}_0 = 2$, that is, $\mathcal{D}(t) = \mathcal{D}_0 + \Delta\mathcal{D}$ with $|\Delta\mathcal{D}| \leq 0.04$.

The parameters $\mathcal{M}_i$ and $\mathcal{J}_i$ for three different modes are given in Table 6.1 and the TR matrix is given as follows:

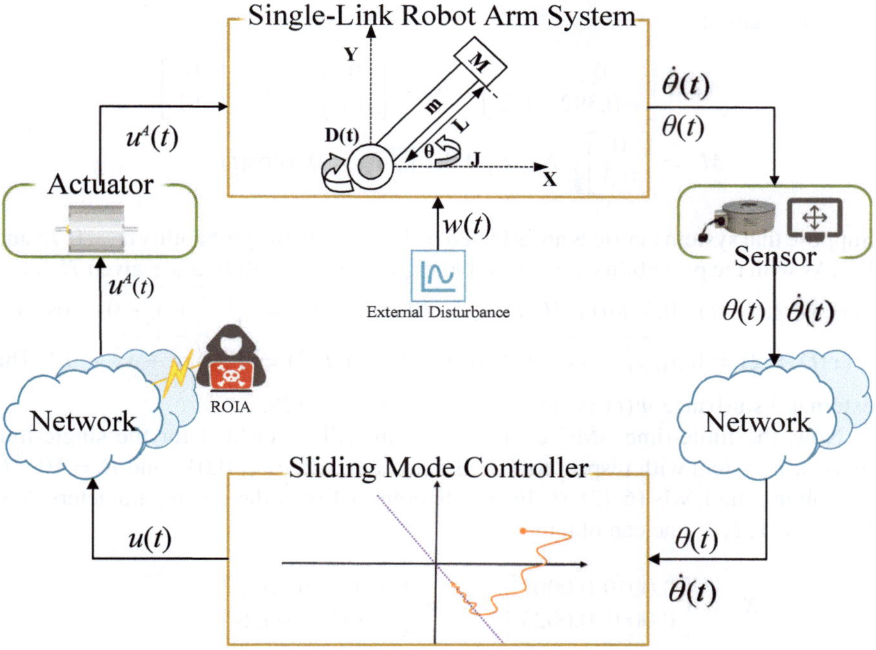

Fig. 6.2 System diagram of a single-link robot arm

$$\Lambda = \begin{bmatrix} -0.1 & 0.02 & 0.08 \\ 0.1 & -0.2 & 0.1 \\ 0.03 & 0.07 & -0.1 \end{bmatrix}.$$

Let $x_1(t) = \theta(t)$ and $x_2(t) = \dot{\theta}(t)$, an then, we can obtain the system matrices as follows:

Subsystem 1:

$$A_1 = \begin{bmatrix} 0 & 1 \\ -0.098 & -0.4 \end{bmatrix}, \ B_1 = \begin{bmatrix} 0 \\ 0.2 \end{bmatrix}, \ D_1 = \begin{bmatrix} 0 \\ 0.2 \end{bmatrix},$$
$$M_1 = \begin{bmatrix} 0 \\ 0.2 \end{bmatrix}, \ N_1 = \begin{bmatrix} 0 & 0.2 \end{bmatrix}, \ F_1 = 0.3\sin(t).$$

Subsystem 2:

$$A_2 = \begin{bmatrix} 0 & 1 \\ -0.196 & -0.3 \end{bmatrix}, \ B_2 = \begin{bmatrix} 0 \\ 0.15 \end{bmatrix}, \ D_2 = \begin{bmatrix} 0 \\ 0.15 \end{bmatrix},$$
$$M_2 = \begin{bmatrix} 0 \\ 0.1 \end{bmatrix}, \ N_2 = \begin{bmatrix} 0 & 0.2 \end{bmatrix}, \ F_2 = 0.4\sin(t).$$

Subsystem 3:

$$A_3 = \begin{bmatrix} 0 & 1 \\ -0.392 & -0.2 \end{bmatrix}, \ B_3 = \begin{bmatrix} 0 \\ 0.1 \end{bmatrix}, \ D_3 = \begin{bmatrix} 0 \\ 0.1 \end{bmatrix},$$
$$M_3 = \begin{bmatrix} 0 \\ 0.3 \end{bmatrix}, \ N_3 = \begin{bmatrix} 0 & 0.2 \end{bmatrix}, \ F_3 = 0.3\cos(t).$$

Suppose that system matrices are affected by ROUs with the probability $\bar{\alpha} = 0.75$ and ROIAs with the probability $\bar{\varphi} = 0.49$. Other parameters of ROIAs are given $H_1(t) = 0.5\cos(t)$, $H_2(t)=0.5\sin(t)$, $H_3(t)=0.5$, $\Psi(x(t), t, 1)=0.5\sqrt{x_1^2 + x_2^2+0.1\cos(t)}$, $\Psi(x(t), t, 2) = 0.5\sqrt{x_1^2 + x_2^2 + 0.1\sin(t)}$, $\Psi(x(t), t, 3) = 0.5\sqrt{x_1^2 + x_2^2 + 0.1}$. The external disturbance $w(t) = \frac{\sqrt{2}}{5}\cos(\pi t)$ with $\delta = 0.2828$.

Now, the finite-time SMC control problem will be tackled for the single-link robot arm system with respect to $T = 3.5$, $c_1 = 0.012$, $c_2 = 0.013$ and $R = 10^{-3}I$. By solving the LMIs (6.42)–(6.46) in Theorem 6.4 with the given parameters $\xi_i = 0.02$, $i = 1, 2, 3$, one can obtain

$$X_1 = \begin{bmatrix} 0.0019 & 0.0001 \\ 0.0001 & 0.0025 \end{bmatrix}, \ X_2 = \begin{bmatrix} 0.0019 & 0.0001 \\ 0.0001 & 0.0026 \end{bmatrix},$$
$$X_3 = \begin{bmatrix} 0.0019 & 0.0001 \\ 0.0001 & 0.0025 \end{bmatrix}, \ K_1 = \begin{bmatrix} -0.4627 & -13.7922 \end{bmatrix},$$
$$K_2 = \begin{bmatrix} -0.6985 & -17.4276 \end{bmatrix}, \ K_3 = \begin{bmatrix} -0.8975 & -25.0305 \end{bmatrix},$$

and $c^* = 0.0122$, based on which the desired SMC law (6.4) may be designed as

$$u(t) = K_i x(t) - \big(\eta_{1i} + \eta_{2i} + (\gamma_{1i} + \gamma_{2i})\|x(t)\|\big)\mathrm{sgn}(s(x(t), i)) \\ - \bar{\varphi}\varsigma_i \psi(x(t), t, i)\mathrm{sgn}(s(x(t), i)),$$

with $\eta_{11} = 4.2186$, $\eta_{12} = 5.6074$, $\eta_{13} = 8.4280$, $\eta_{21} = \eta_{22} = \eta_{23} = 0.2829$, $\gamma_{11} = 19.3413, \gamma_{12} = 24.5179, \gamma_{13} = 35.7352, \gamma_{21} = 0.0010, \gamma_{22} = 0.0045, \gamma_{23} = 0.0021$.

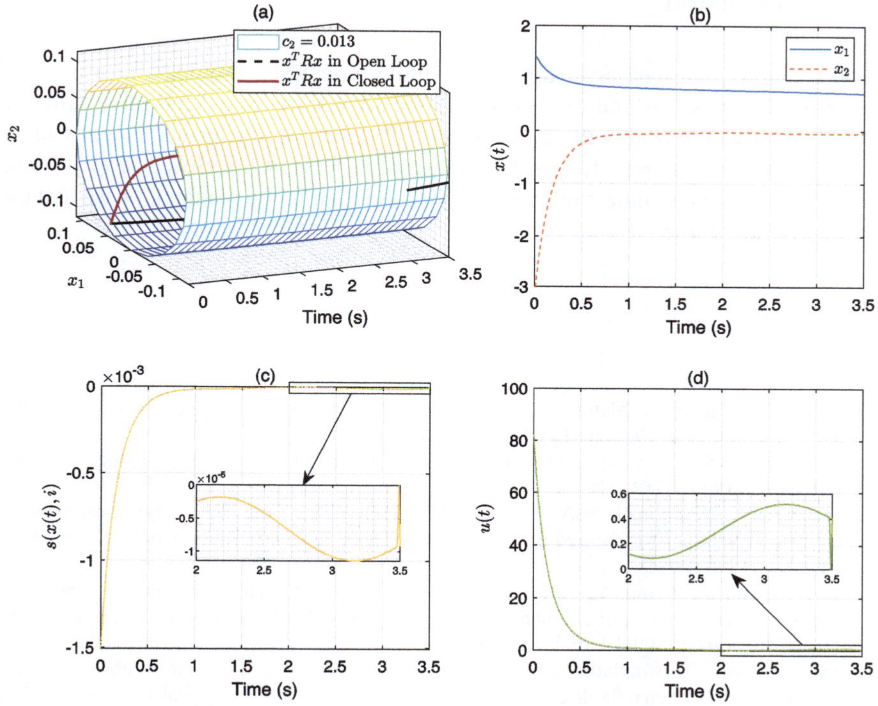

Fig. 6.3 Simulation results of single-link robot arm: **a** evolution of weighted system state variable $x^{\mathrm{T}}(t)Rx(t)$; **b** state responses $x(t)$; **c** sliding variable $s(x(t), i)$; **d** control signal $u(t)$

The simulations are undergone by using the discretization approach with initial state $x(0) = [1.4 \; -3]^{\mathrm{T}}$ and stepsize $\Delta t = 0.01$s. In order to reflect the results in mean square, the simulations are run 50 times, and simulation results as shown in Fig. 6.3. Recalling the simulation results in Chap. 5, there are some notable variations in the control signal $u(t)$, which will influence the transient stationarity of the considered system. To weaken the control signals from chattering, the term $\mathrm{sgn}(s(x(t), i))$ in (6.4) is replaced by $\frac{s(x(t),i)}{(\|s(x(t),i)\|+0.001)}$, then obviously, the variations are weaken in the control signal $u(t)$ as shown in Fig. 6.3d. From the evolution of weighted system state variable $\mathbb{E}\{x^{\mathrm{T}}(t)Rx(t)\}$ in Fig. 6.3a, it is observed that the averaged response of open-loop system exceeds the transient constraint c_2, but dynamic performance of the resultant closed-loop system is satisfactory under the proposed SMC law (6.4)–(6.9) despite existing the ROUs and ROIAs. Figure 6.3b shows the closed-loop response of system states. Furthermore, it can be seen from Fig. 6.3b that the asymptotic stability of controlled states cannot be satisfied during [0, 3.5]s, which implies that the SFTB focuses transient performance.

6.6 Conclusion

In this chapter, we have investigated the finite-time SMC problem for a class of MJSs subject to ROUs and ROIAs, where the adversaries may send false data in a probabilistic way to inject the cyber-attacks. A novel parameter-dependent sliding mode controller has been designed to ensure the SFTB of the closed-loop systems over the whole given finite-time interval $[0, T]$ including the reaching phase and the sliding motion phase.

References

1. Harkat, H., Camarinha-Matos, L.M., Goes, J., Ahmed, H.F.: Cyber-physical systems security: a systematic review. Comput. Ind. Eng., 109891 (2024)
2. Bhadani, U.: Smart grids: a cyber-physical systems perspective. Int. Res. J. Eng. Technol. (IRJET) **11**(06), 801 (2024)
3. Hasan, M.K., Abdulkadir, R.A., Islam, S., Gadekallu, T.R., Safie, N.: A review on machine learning techniques for secured cyber-physical systems in smart grid networks. Energy Rep. **11**, 1268–1290 (2024)
4. Lou, S., Hu, Z., Zhang, Y., Feng, Y., Zhou, M., Lv, C.: Human-cyber-physical system for Industry 5.0: a review from a human-centric perspective. IEEE Trans. Autom. Sci. Eng. (2024). https://doi.org/10.1109/TASE.2024.3360476
5. Panneerselvam, V., Aravinth, N., Sakthivel, R., Hafez, M.G.: Mode-dependent observer-based secured hybrid-driven reliable control for semi-Markov jump cyber-physical systems with DoS attacks. Chaos Solitons Fractals **187**, 115387 (2024)
6. Cheng, G., Liu, H.: Asynchronous finite-time extended dissipative filter on linear neutral semi-Markov jump systems under sensor saturation and dual-channel DoS attacks. Appl. Math. Comput. **475**, 128723 (2024)
7. Zhao, D., Yang, B., Li, Y., Zhang, H.: Replay attack detection for cyber-physical control systems: a dynamical delay estimation method. IEEE Trans. Ind. Electron. **72**(1), 867–875 (2024)
8. Gargoum, S., Yassaie, N., Al-Dabbagh, A.W., Feng, C.: A data-driven framework for verified detection of replay attacks on industrial control systems. IEEE Trans. Autom. Sci. Eng. (2024). https://doi.org/10.1109/TASE.2024.3394315
9. Tajudeen, M.M., Ali, M.S., Thakur, G.K., Priya, B., Perumal, R.: Non-fragile control of discrete-time conic-type nonlinear Markovian jump systems under deception attacks using event-triggered scheme and Its application. Math. Comput. Simul. **227**, 209–225 (2025)
10. Tan, C., Gao, C., Peng, J., Xie, X., Wang, Y.: Neural network-based security control for TS fuzzy system with cooperative event-triggered mechanism. IEEE Trans. Fuzzy Syst. **32**(8), 4633–4645 (2024)

11. Cao, Z., Niu, Y., Zou, Y.: Adaptive neural sliding mode control for singular semi-Markovian jump systems against actuator attacks. IEEE Trans. Syst. Man Cybern. Syst. **51**(3), 1523–1533 (2021)
12. Ni, Z., Li, F., Wang, Y., Shen, H.: Sliding-mode control for 2-D hidden Markov jump Roesser systems with partial information and its application in metal rolling process. IEEE Trans. Autom. Sci. Eng. (2024). https://doi.org/10.1109/TASE.2024.3455570
13. Shen, M., Yan, S., Zhang, G.: Finite-time H_∞ static output control of Markov jump systems with an auxiliary approach. Appl. Math. Comput. **273**, 553–561 (2016)
14. Song, J., Niu, Y., Zou, Y.: Finite-time stabilization via sliding mode control. IEEE Trans. Autom. Control **62**(3), 1478–1483 (2017)
15. Song, J., Niu, Y., Zou, Y.: Finite-time sliding mode control synthesis under explicit output constraint. Automatica **65**, 111–114 (2016)
16. Liu, Y., Ma, Y., Wang, Y.: Reliable sliding mode finite-time control for discrete-time singular Markovian jump systems with sensor fault and randomly occurring nonlinearities. Int. J. Robust Nonlinear Control **28**(2), 381–402 (2018)
17. Zhang, Z., Niu, Y., Song, J.: Input-to-state stabilization of interval type-2 fuzzy systems subject to cyberattacks: An observer-based adaptive sliding mode approach. IEEE Trans. Fuzzy Syst. **28**(1), 190–203 (2020)
18. Chen, B., Niu, Y., Zou, Y.: Adaptive sliding mode control for stochastic Markovian jumping systems with actuator degradation. Automatica **49**(6), 1748–1754 (2013)

Chapter 7
Probability-Dependent Sliding Mode Control Against Sensor Deception Attacks

7.1 Introduction

Due to the reflection, refraction, and diffraction in the process of network transmission, the signal fading phenomena may unavoidably happen, such as, the shadow fading and the multi-path induced fading. For the analog signal, the so-called analog fading can be represented as some typical models, e.g., the memoryless multiplicative fading [1–3], the Lth-order Rice fading [4, 5], and the finite-state Markov fading [6]. Compared with the widely investigated packet dropout phenomena, the above fading phenomena in the form of the multiplicative noise may be seen as a more general case in the mathematical sense [7, 8].

In the fading communication environment, the security issues cannot be ignored. Different from the randomly occurring injection attacks (ROIAs) studied in Chapter 6, the randomly occurring deception attacks (RODAs) are considered in this chapter with the following two distinctions. (1) The former occurs in controller-to-actuator (C/A) channel, which thus is handled by designing an attack-tolerant sliding mode controller with the attack-bound-related robust term. This controller design method is not applicable for dealing with the deception attacks that stochastically occurs in sensor-to-controller (S/C) channel. Thus, a new security control method is needed in this chapter. (2) The ROIAs in Chapter 6 naturally are additive-type attacks, that is, the adversaries do not hinder the transmission of the original signal, but only randomly add packets to the original signal. The RODAs in this chapter are alternate-type attacks, which means that original transmission signal is blocked and only attack signal is transmitted to the receiver. Hence, the defense methods of the two types of attacks are necessarily different.

In view of the above facts, it comes to be a challenging issue to explore a solving scheme for the considered problem in consequence of two fundamental obstacles.

- ► The system model under investigation is quite comprehensive that covers channel fading and RODAs modeled by two random variables. The first difficulty is how

Z. Cao and Y. Niu, *Finite-Time Sliding Mode Control for Markov Jump Systems*,
Studies in Systems, Decision and Control 603,
https://doi.org/10.1007/978-981-96-7651-4_7

to reduce the influences of channel fading and RODAs in reaching and sliding motion phases.

- The channel fading and RODAs cause that the controller can only receive the non-real data that degrades the transient performance of the Markov jump systems (MJSs). The second difficulty is how to construct a feasible sliding mode control (SMC) law to ensure the desired stochastic finite-time boundedness (SFTB) performance of MJSs.

In order to solve above issues, a mode-dependent sliding surface is firstly constructed. Then, a probability-dependent SMC law is developed to adapt the channel fading and RODAs and attain the reachability for the specified sliding surface. Subsequently, sufficient conditions are established for the SFTB of the closed-loop system by means of the partition strategy.

7.2 Communication Problems

Consider the MJS defined on a complete probability space $(\Omega, \mathscr{F}, \Pr)$ as in system (1.6) without parameter uncertainties $\Delta A_i(t) \equiv 0$, in which the matched nonlinear perturbation $f_i(x(t))$ and unmatched external disturbance $w(t)$ satisfy Assumptions 1.3 and 1.4, respectively. The transition probabilities (TPs) are generated by (1.7) with the assumptions that all transition rates (TRs) are completely available.

7.2.1 Channel Fading Model

With the evolution of the communication technique, transmission via network has been a trend shown in Fig. 7.1. In this chapter, the system state is transmitted to controller via a fading channel.

The fading phenomenon might happen according to the following memoryless multiplicative fading model:

$$\bar{x}(t) = \varpi(t)x(t), \tag{7.1}$$

where $\bar{x}(t)$ is the state after fading, and $\varpi(t)$ is a the stochastic variable describing the degree of state fading. Especially, $\varpi(t)$ are mutually independent which take values on the interval $(0, 1]$ with the mathematical expectation $\mathbb{E}\{\varpi(t)\} = \bar{\varpi} \in (0, 1]$.

7.2.2 Deception Attacks in S/C Channel

Due to the fragility of network, it is unavoidable subject to the malicious attack. We consider that the transmitted state is randomly intercepted by adversaries whose model given as

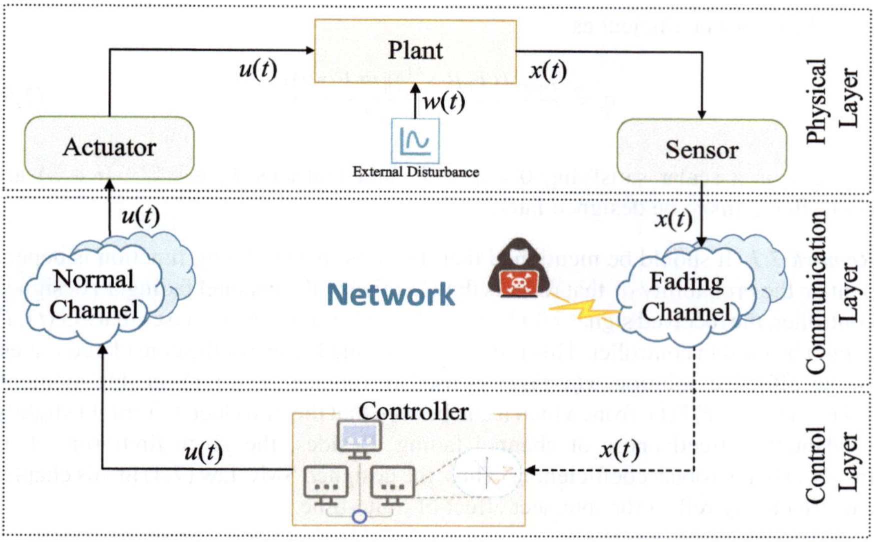

Fig. 7.1 Network control system with channel fading and deception attacks

$$
\begin{aligned}
\tilde{x}(t) &= \alpha(t)\varphi(t) + (1-\alpha(t))\bar{x}(t) \\
&= \alpha(t)\varphi(t) + (1-\alpha(t))\varpi(t)x(t),
\end{aligned} \tag{7.2}
$$

where $\varphi(t)$ represents the injected attack signal sent by attackers satisfying $\|\varphi(t)\| \leq \bar{\varphi}$ with $\bar{\varphi}$ being a known positive scalar; and the stochastic variable $\alpha(t)$ obeys exponential distribution satisfying $\Pr\{\alpha(t)=1\} = \tilde{\alpha}$, $\Pr\{\alpha(t)=0\} = 1-\tilde{\alpha}$ with $\tilde{\alpha} \in [0, 1]$ being a known constant.

7.3 Probability-Dependent Controller Design

Considering the channel fading phenomenon, we construct the following sliding function:

$$
s(x(t), i) = \breve{\varpi} F_i x(t), \tag{7.3}
$$

where $\breve{\varpi} \triangleq 1/\bar{\varpi}$, and the sliding matrices F_i $(i \in \mathbb{S})$ are chosen to ensure that $F_i B_i$ is nonsingular.

Next, the finite-time SMC law is designed as

$$
u(t) = K_i \tilde{x}(t) - \eta_i \operatorname{sgn}(s(\tilde{x}(t), i)) \tag{7.4}
$$

with the robust coefficient as

$$\eta_i \triangleq \frac{\lambda_{\max}((F_iB_i)^{-1})\|\breve{\varpi} F_i x(0)\|}{T-\varphi}, \tag{7.5}$$

where φ is a scalar satisfying $0 < \varphi < T$, and matrices $K_i \in \mathbb{R}^{m\times n}$ $(i \in \mathbb{S})$ are controller gains to be designed later.

Remark 7.1 It should be mentioned that the constructed sliding function is dependent on the probability $\bar{\varpi}$ that reflects the effective of the channel fading. In designed controller, the received signal $\tilde{x}(t)$ instead of $x(t)$ is used, which causes that $s(x(t), i)$ cannot be used in controller. This is the difficulty and feature of the considered issues. As an effective solution, $s(\tilde{x}(t), i)$ is used to replace $s(x(t), i)$, at this moment, $s(\tilde{x}(t), i) = \frac{1}{\bar{\varpi}} F_i\tilde{x}(t)$, from which it can be seen that the introduced $\bar{\varpi}$ might slightly weaken the effectiveness of channel fading. Besides, the given finite-time T is involved in the robust coefficient η_i. Thus, the designed SMC law (7.4) in this chapter can effectively reflect the intersect effect of finite-time.

7.4 Reachability Analysis

In this subsection, we will show that the specified sliding surface $s(x(t), i) = 0$ can be reached in the interval $[0, T^*]$ with $T^* < T$:

Theorem 7.1 *Considering the MJS* (1.6) $(\Delta A_i(t) \equiv 0)$ *under fading channel and RODAs, if there exist scalars* $\zeta_i > 0$, $\iota_i > 0$, *and matrices* K_i $(\forall i \in \mathbb{S})$ *such that the following matrix inequalities hold:*

$$\begin{bmatrix} \Upsilon_i & \breve{\varpi} F_i^{\mathrm{T}} \\ * & -\zeta_i I \end{bmatrix} < 0, \tag{7.6}$$

$$\begin{bmatrix} -\nu_i I & K_i \\ * & -I \end{bmatrix} < 0 \tag{7.7}$$

with

$$\begin{aligned}\Upsilon_i \triangleq\ & \breve{\varpi}^2(F_i^{\mathrm{T}}(F_iB_i)^{-1}F_iA_i + A_i^{\mathrm{T}}F_i^{\mathrm{T}}(F_iB_i)^{-1}F_i) + \breve{\varpi}(1-\bar{\alpha})(F_i^{\mathrm{T}}K_i + K_i^{\mathrm{T}}F_i) \\ & + \zeta_i\breve{\varpi}^2\epsilon_i^2 + \breve{\varpi}^2\sum_{j\in\mathbb{S}}\lambda_{ij}F_j^{\mathrm{T}}(F_jB_j)^{-1}F_j + \iota_i\breve{\varpi}^2F_i^{\mathrm{T}}F_i.\end{aligned}$$

then the SMC law (7.4) *can force the system state into the following sliding domain* Ω *before the instant* T^* *with* $T^* < T$:

$$\Omega \triangleq \left\{ s(x(t), i) \mid \|s(x(t), i)\| \le \frac{2(\bar{\omega}_i + (1+\breve{\varpi}\sqrt{m})\eta_i)}{\iota_i} \right\} \tag{7.8}$$

with

$$\bar{\omega}_i \triangleq \breve{\varpi}\bar{\alpha}\bar{\varphi}\|K_i\| + \breve{\varpi}\delta\|(F_iB_i)^{-1}F_iD_i\|. \tag{7.9}$$

Proof By means of the system model (1.6) with $\Delta A_i(t) \equiv 0$, the sliding function (7.3) and the sliding mode controller (7.4), we have

$$\begin{aligned}\dot{s}(x(t), i) &= \breve{\varpi}F_i(A_ix(t) + B_i(u(t) + f_i(x(t))) + D_iw(t))\\ &= \breve{\varpi}F_i(A_ix(t) + B_i((K_i\tilde{x}(t) - \eta_i\text{sgn}(s(\tilde{x}(t), i))) + f_i(x(t))) + D_iw(t))\\ &= \breve{\varpi}F_i(A_ix(t) + B_i((K_i(\alpha(t)\varphi(t) + (1-\alpha(t))\varpi(t)x(t))\\ &\quad - \eta_i\text{sgn}(s(\tilde{x}(t), i))) + f_i(x(t))) + D_iw(t)).\end{aligned} \tag{7.10}$$

Select the mode-dependent Lyapunov function as:

$$V_1(x(t), i) = s^{\mathrm{T}}(x(t), i)(F_iB_i)^{-1}s(x(t), i). \tag{7.11}$$

Taking the weak infinitesimal operator of $V_1(x(t), i)$, one has

$$\begin{aligned}&\mathscr{L}V_1(x(t), i)\\ &= 2\breve{\varpi}s^{\mathrm{T}}(x(t), i)(F_iB_i)^{-1}F_iA_ix(t) + \sum_{j\in\mathbb{S}}\left(\lambda_{ij}s^{\mathrm{T}}(x(t), j)(F_jB_j)^{-1}s(x(t), j)\right)\\ &\quad + 2\breve{\varpi}s^{\mathrm{T}}(x(t), i)f_i(x(t)) + 2(1-\bar{\alpha})s^{\mathrm{T}}(x(t), i)K_ix(t)\\ &\quad - 2\breve{\varpi}s^{\mathrm{T}}(x(t), i)\eta_i\text{sgn}(s(\tilde{x}(t), i)) + 2\breve{\varpi}\bar{\alpha}s^{\mathrm{T}}(x(t), i)K_i\varphi(t)\\ &\quad + 2\breve{\varpi}s^{\mathrm{T}}(x(t), i)(F_iB_i)^{-1}F_iD_iw(t).\end{aligned} \tag{7.12}$$

For $\zeta_i > 0$, we have

$$2\breve{\varpi}s^{\mathrm{T}}(x(t), i)f_i(x(t)) \le \zeta_i^{-1}\breve{\varpi}^2x^{\mathrm{T}}(t)F_i^{\mathrm{T}}F_ix(t) + \zeta_i\breve{\varpi}^2\epsilon_i^2x^{\mathrm{T}}(t)x(t). \tag{7.13}$$

Combining (7.12) and (7.13), one has

$$\begin{aligned}&\mathscr{L}V_1(x(t), i)\\ &\le \breve{\varpi}^2x^{\mathrm{T}}(t)\text{He}\left(F_i^{\mathrm{T}}(F_iB_i)^{-1}F_iA_i\right)x(t) + \breve{\varpi}^2x^{\mathrm{T}}(t)\sum_{j\in\mathbb{S}}\left(\lambda_{ij}F_j^{\mathrm{T}}(F_jB_j)^{-1}F_j\right)x(t)\\ &\quad + \zeta_i^{-1}\breve{\varpi}^2x^{\mathrm{T}}(t)F_i^{\mathrm{T}}F_ix(t) + \zeta_i\breve{\varpi}^2\epsilon_i^2x^{\mathrm{T}}(t)x(t) + \breve{\varpi}(1-\bar{\alpha})x^{\mathrm{T}}(t)\text{He}\left(F_i^{\mathrm{T}}K_i\right)x(t)\\ &\quad - 2\breve{\varpi}s^{\mathrm{T}}(x(t), i)\eta_i\text{sgn}(s(\tilde{x}(t), i)) + 2\breve{\varpi}\bar{\alpha}s^{\mathrm{T}}(x(t), i)K_i\varphi(t)\\ &\quad + 2\breve{\varpi}s^{\mathrm{T}}(x(t), i)(F_iB_i)^{-1}F_iD_iw(t).\end{aligned} \tag{7.14}$$

Note that

$$
\begin{aligned}
&- \breve{\varpi} s^{\mathrm{T}}(x(t), i)\eta_i \mathrm{sgn}(s(\tilde{x}(t), i)) + \varpi\bar{\alpha} s^{\mathrm{T}}(x(t), i)K_i\varphi(t) \\
&+ \breve{\varpi} s^{\mathrm{T}}(x(t), i)(F_iB_i)^{-1}F_iD_iw(t) \\
&\leq \|s(x(t), i)\|(\breve{\varpi}\bar{\alpha}\bar{\varphi}\|K_i\| + \breve{\varpi}\delta\|(F_iB_i)^{-1}F_iD_i\| + \breve{\varpi}\sqrt{m}\eta_i) \\
&= \bar{\omega}_i\|s(x(t), i)\| + \breve{\varpi}\sqrt{m}\eta_i\|s(x(t), i)\|. \qquad (7.15)
\end{aligned}
$$

From (7.14) and (7.15), one has

$$
\begin{aligned}
\mathscr{L}V_1(x(t), i) &\leq x^{\mathrm{T}}(t)\Psi_i x(t) + 2\bar{\omega}_i\|s(x(t), i)\| \\
&= x^{\mathrm{T}}(t)\Psi_i x(t) + \iota_i\|s(x(t), i)\|^2 - \iota_i\|s(x(t), i)\|^2 + 2\eta_i\|s(x(t), i)\| \\
&\quad - 2\eta_i\|s(x(t), i)\| + 2(\bar{\omega}_i + \breve{\varpi}\sqrt{m}\eta_i)\|s(x(t), i)\| \\
&= x^{\mathrm{T}}(t)\Psi_i x(t) + \iota_i\breve{\varpi}^2 x^{\mathrm{T}}(t)F_i^{\mathrm{T}}F_i x(t) - 2\eta_i\|s(x(t), i)\| \\
&\quad - \iota_i\|s(x(t), i)\|\left(\|s(x(t), i)\| - \frac{2(\bar{\omega}_i + (1 + \breve{\varpi}\sqrt{m})\eta_i)}{\iota_i}\right)
\end{aligned} \tag{7.16}
$$

with $\Psi_i \triangleq \breve{\varpi}^2\mathrm{He}\left(F_i^{\mathrm{T}}(F_iB_i)^{-1}F_iA_i\right) + \breve{\varpi}(1 - \bar{\alpha})\mathrm{He}\left(F_i^{\mathrm{T}}K_i\right) + \zeta_i^{-1}\breve{\varpi}^2F_i^{\mathrm{T}}F_i + \zeta_i\breve{\varpi}^2\epsilon_i^2 I + \breve{\varpi}^2\sum_{j\in\mathbb{S}}\lambda_{ij}F_j^{\mathrm{T}}(F_jB_j)^{-1}F_j$.

When the state trajectories run outside the domain Ω, that is, $\|s(x(t), i)\| > \frac{2(\bar{\omega}_i + (1+\breve{\varpi}\sqrt{m})\eta_i)}{\iota_i}$, we have

$$
\iota_i\|s(x(t), i)\|\left(\|s(x(t), i)\| - \frac{2(\bar{\omega}_i + (1 + \breve{\varpi}\sqrt{m})\eta_i)}{\iota_i}\right) > 0. \tag{7.17}
$$

From (7.16) and (7.17), we obtain

$$
\mathscr{L}V_1(x(t), i) \leq x^{\mathrm{T}}(t)\Psi_i x(t) + \iota_i\breve{\varpi}^2 x^{\mathrm{T}}(t)F_i^{\mathrm{T}}F_i x(t) - 2\eta_i\|s(x(t), i)\|. \tag{7.18}
$$

Obviously, the inequality, $x^{\mathrm{T}}(t)\Psi_i x(t) + \iota_i\breve{\varpi}^2 x^{\mathrm{T}}(t)F_i^{\mathrm{T}}F_i x(t) < 0$, can be guaranteed by the condition (7.6). One has from (7.18)

$$
\mathscr{L}V_1(x(t), i) \leq -2\eta_i\|s(x(t), i)\| < 0. \tag{7.19}
$$

Accordingly, the state trajectories will be driven into the sliding domain Ω around sliding surface $s(x(t), i) = 0$ by the SMC law (7.4).

The following will further give the reaching instant T^*. Similar to the derivation between (4.29)–(4.33), it is easy to see that if there exists an instant T^* satisfying

$$
T^* = \frac{\lambda_{\max}((F_iB_i)^{-1})}{\eta_i}\|\breve{\varpi}F_i x(0)\| \tag{7.20}
$$

then, one has for $t \geq T^*$, $\mathbb{E}\{V_1(x(t), i)\} = 0$, consequently, the sliding domain Ω around sliding surface $s(x(t), i) = 0$ is attained (in the mean square). Thus, together with (7.5) and (7.20), we have $T^* \leq T - \varphi < T$, that is, the state trajectories of the considered MJSs (1.6) with $\Delta A_i(t) \equiv 0$ will enter into the specified sliding region Ω before the instant T^* with $T^* < T$. This fulfills the proof. □

Remark 7.2 Theorem 7.1 provides sufficient conditions for the reachability. It shows from (7.8) and (7.9), $\bar{\omega}_i$, $\iota_i > 0$, and matrices K_i $(i \in \mathbb{S})$ play important roles in sufficient conditions (7.6), (7.7) and the sliding region Ω (7.8), which implies that channel fading and deception attacks affect the feasibility of the solution and the SMC performance. For example, a larger attack occurrence probability would result in a larger sliding bound.

7.5 SFTB Analysis

According to the theory of SMC method, the moving process of system states is composed of two phases, that is, reaching phase and sliding motion phase. Since it is already proven in Theorem (7.1) that the state trajectories of MJS (1.6) ($\Delta A_i(t) \equiv 0$) will enter into the sliding domain before the instant $T^* < T$, the above two phases will occur within the interval $[0, T^*]$ and the interval $[T^*, T]$, respectively. Hence, we will give the analysis process of the SFTB during the interval $[0, T^*]$ (i.e., the reaching phase) and the interval $[T^*, T]$ (i.e., the sliding motion phase), simultaneously, by which it is shown that the SFTB of considered system will be attained for the overall interval $[0, T]$.

7.5.1 SFTB over Reaching Phase

It is known that the system state over the reaching phase will be controlled by the designed sliding mode controller. Based on this, we substitute the sliding mode controller (7.4) into the system (1.6) ($\Delta A_i(t) \equiv 0$) to obtain the closed-loop system as

$$\dot{x}(t) = \breve{A}x(t) + \alpha(t)B_i K_i \varphi(t) + B_i(f_i(x(t)) - \eta_i \mathrm{sgn}(s(\tilde{x}(t), i))) + D_i w(t) \tag{7.21}$$

with $\breve{A}_i \triangleq A_i + (1 - \alpha(t))\varpi(t)B_i K_i$.

It notices that since the SMC law (7.4) is designed based on the corrupted faded state $\tilde{x}(t)$ involving in the random fading and attacks, two random variables $\alpha(t)$ and $\varpi(t)$ influence the closed-loop MJS (7.21).

In what follows, for the MJS (7.21), its SFTB will be analyzed to deduce the sufficient conditions during the interval $[0, T^*]$.

Theorem 7.2 *The closed-loop MJS* (7.21) *is SFTB with respect to* $\aleph_{[0,T^*]}$, *for the given* $\psi_i > 0$, *if there exist scalars* $\varepsilon_i > 0$, $\zeta_i > 0$, $c^* > 0$, *and matrices* $P_i > 0$ ($\forall i \in \mathbb{S}$), *satisfying the following conditions:*

$$\begin{bmatrix} \Xi_{1i} & P_i D_i \\ * & -\psi_i I \end{bmatrix} < 0, \tag{7.22}$$

$$c_1 < c^* < c_2, \tag{7.23}$$

$$\frac{\bar{\varsigma}(P_i)c_1 + \delta^2 T\psi_i + \frac{C_i}{\psi_i}T}{\underline{\varsigma}(P_i)} < e^{-\psi_i T}c^* \tag{7.24}$$

with

$$\Xi_{1i} \triangleq \mathrm{He}(P_i A_i) - \psi_i P_i + \sum_{j=1}^{N}\lambda_{ij}P_j + \bar{\varpi}(1-\bar{\alpha})P_i B_i K_i + \bar{\varpi}(1-\bar{\alpha})(B_i K_i)^{\mathrm{T}}P_i$$
$$+ \zeta_i^{-1}P_i B_i B_i^{\mathrm{T}}P_i + \zeta_i\epsilon_i^2 I + \varepsilon_i^{-1}P_i B_i B_i^{\mathrm{T}}P_i,$$
$$\bar{\varsigma}(P_i) \triangleq \max_{i\in\mathbb{S}}\{\lambda_{\max}(R^{-\frac{1}{2}}P_i R^{-\frac{1}{2}})\},\ \underline{\varsigma}(P_i) \triangleq \min_{i\in\mathbb{S}}\{\lambda_{\min}(R^{-\frac{1}{2}}P_i R^{-\frac{1}{2}})\},$$
$$C_i \triangleq 2\varepsilon_i(\bar{\alpha}^2\bar{\varphi}^2\|K_i\|^2 + \eta_i^2 m).$$

Proof Choose the mode-dependent Lyapunov function as follows:

$$V_2(x(t), i) = x^{\mathrm{T}}(t)P_i x(t). \tag{7.25}$$

It follows from the closed-loop MJS (7.21) that

$$\mathscr{L}V_2(x(t), i) \tag{7.26}$$
$$= x^{\mathrm{T}}(t)\left(\mathrm{He}\,(P_i A_i + \bar{\varpi}(1-\bar{\alpha})P_i B_i K_i) + \sum_{j=1}^{N}\lambda_{ij}P_j\right)x(t) + 2x^{\mathrm{T}}(t)P_i B_i f_i(x(t))$$
$$+ w^{\mathrm{T}}(t)D_i^{\mathrm{T}}P_i^{\mathrm{T}}x(t) + x^{\mathrm{T}}(t)P_i D_i w(t) + 2\bar{\alpha}x^{\mathrm{T}}(t)P_i B_i K_i\varphi(t)$$
$$- 2x^{\mathrm{T}}(t)P_i B_i\eta_i\mathrm{sgn}(s(\tilde{x}(t), i)). \tag{7.27}$$

It is known that we have for $\zeta_i > 0$ ($i \in \mathbb{S}$),

$$2x^{\mathrm{T}}(t)P_i B_i f_i(x(t)) \le \zeta_i^{-1}x^{\mathrm{T}}(t)P_i B_i B_i^{\mathrm{T}}P_i x(t) + \zeta_i\epsilon_i^2 x^{\mathrm{T}}(t)x(t) \tag{7.28}$$

and for $\varepsilon_i > 0$ ($i \in \mathbb{S}$),

$$2x^{\mathrm{T}}(t)P_i B_i(\bar{\alpha}K_i\varphi(t) - \eta_i\mathrm{sgn}(s(\tilde{x}(t), i)))$$
$$\le \varepsilon_i^{-1}x^{\mathrm{T}}(t)P_i B_i B_i^{\mathrm{T}}P_i x(t) + 2\varepsilon_i(\bar{\alpha}^2\bar{\varphi}^2\|K_i\|^2 + \eta_i^2 m)$$
$$= \varepsilon_i^{-1}x^{\mathrm{T}}(t)P_i B_i^{\mathrm{T}}B_i^{\mathrm{T}}P_i x(t) + C_i. \tag{7.29}$$

Then, by means of (7.28) and (7.29), it can be obtained from (7.27) that

$$\begin{aligned}\mathscr{L}V_2(x(t),i) \leq x^{\mathrm{T}}(t)\bigg(&\mathrm{He}\,(P_iA_i+\bar{\varpi}(1-\bar{\alpha})P_iB_iK_i)+\sum_{j=1}^{N}\lambda_{ij}P_j+\zeta_i\epsilon_i^2 I\\&+\zeta_i^{-1}P_iB_iB_i^{\mathrm{T}}P_i+\varepsilon_i^{-1}P_iB_iB_i^{\mathrm{T}}P_i\bigg)x(t)\\&+w^{\mathrm{T}}(t)D_i^{\mathrm{T}}P_i^{\mathrm{T}}x(t)+x^{\mathrm{T}}(t)P_iD_iw(t)+C_i.\end{aligned} \tag{7.30}$$

For scalars $\psi_i > 0$ $(i \in \mathbb{S})$, define the following auxiliary functions:

$$\Pi(x(t),i) \triangleq \mathscr{L}V_2(x(t),i)-\psi_i V_2(x(t),i)-\psi_i w^{\mathrm{T}}(t)w(t)-C_i. \tag{7.31}$$

It is obtained that $\Pi(x(t),i) < 0$ can be guaranteed by the conditions (7.22), from which we have

$$\mathscr{L}V_2(x(t),i) < \psi_i V_2(x(t),i)+\psi_i w^{\mathrm{T}}(t)w(t)+C_i. \tag{7.32}$$

Multiplying both sides of expression (7.32) by $e^{-\psi_i t}$ and integrating (7.32) from 0 to t with $t \in [0, T^*]$, we obtain

$$\begin{aligned}&e^{-\psi_i t}\mathbb{E}\{V_2(x(t),i)\}\\&< V_2(x(0),r_0)+\psi_i\int_0^t e^{-\psi_i\tau}w^{\mathrm{T}}(\tau)w(\tau)\mathrm{d}\tau+\psi_i\int_0^t e^{-\psi_i\tau}\frac{C_i}{\psi_i}\mathrm{d}\tau\\&\leq \bar{\varsigma}(P_i)c_1+\delta^2T\psi_i+\frac{C_i}{\psi_i}T.\end{aligned} \tag{7.33}$$

Besides, it is obtained from (7.25) that

$$\mathbb{E}\{V_2(x(t),i)\} \geq \underline{\varsigma}(P_i)\mathbb{E}\{x^{\mathrm{T}}(t)Rx(t)\}. \tag{7.34}$$

By means of the expressions (7.23), (7.24), (7.34), and (7.35), we obtain

$$\mathbb{E}\{x^{\mathrm{T}}(t)Rx(t)\} < \frac{(\bar{\varsigma}(P_i)c_1+\delta^2T\psi_i+\frac{C_i}{\psi_i}T)e^{\psi_i T}}{\underline{\varsigma}(P_i)}. \tag{7.35}$$

If the condition (7.24) holds, the right term of the above inequality is less than c^*. So we have $\mathbb{E}\{x^{\mathrm{T}}(t)Rx(t)\} < c^*$ with $t \in [0, T^*]$, which, according to Definition 1.5, means the SFTB of the closed-loop system (7.21) with respect to $\aleph_{[0,T^*]}$. This fulfills the proof. □

7.5.2 SFTB over Sliding Motion Phase

We further analyze SFTB of sliding motion during the interval $[T^*, T]$. According to the SMC theory, on the sliding surface $s(x(t), i) = 0$, we can obtain the following equivalent control law from (1.6) ($\Delta A_i(t) \equiv 0$) and (7.3)

$$u_{eq}(t) = -(F_i B_i)^{-1}(F_i A_i(t)x(t) + F_i D_i w(t)) - f_i(x(t)). \tag{7.36}$$

Substituting (7.36) into (1.6) ($\Delta A_i(t) \equiv 0$) yields the sliding mode dynamics as:

$$\dot{x}(t) = \grave{A}_i x(t) + \grave{D}_i w(t) \tag{7.37}$$

with $\grave{A}_i \triangleq Q_i A_i$, $\grave{D}_i \triangleq Q_i D_i$, $Q_i \triangleq I - B_i(F_i B_i)^{-1} F_i$.

Theorem 7.3 *The sliding mode dynamics* (7.37) *are SFTB with respect to* $\aleph_{[T^*,T]}$, *for given scalar* $\psi_i > 0$, *if there exist scalars* $\xi_i > 0$, $c^* > 0$ *and matrices* $P_i > 0$ ($\forall i \in \mathbb{S}$) *satisfying* (7.23) *and the following conditions*

$$\begin{bmatrix} \Xi_{2i} & P_i D_i - P_i B_i (F_i B_i)^{-1} F_i D_i \\ * & -\psi_i I \end{bmatrix} < 0, \tag{7.38}$$

$$\frac{\bar{\varsigma}(P_i)c^* + \delta^2 T \psi_i}{\underline{\varsigma}(P_i)} < e^{-\psi_i T} c_2 \tag{7.39}$$

with

$$\Xi_{2i} \triangleq \text{He}\left(P_i A_i - P_i B_i (F_i B_i)^{-1} F_i A_i\right) - \psi_i P_i + \sum_{j=1}^{N} \lambda_{ij} P_j.$$

Proof For the Lyapunov function (7.25), its weak infinitesimal operator along the sliding mode dynamics (7.37) is given as

$$\begin{aligned}
&\mathscr{L} V_2(x(t), i) \\
&= x^{\mathrm{T}}(t)\left(P_i \grave{A}_i + \grave{A}_i^{\mathrm{T}} P_i + \sum_{j=1}^{N} \lambda_{ij} P_j\right)x(t) + w^{\mathrm{T}}(t)\grave{D}_i^{\mathrm{T}} P_i x(t) + x^{\mathrm{T}}(t) P_i \grave{D}_i w(t) \\
&= x^{\mathrm{T}}(t)\left(\text{He}\left(P_i A_i - P_i B_i (F_i B_i)^{-1} F_i A_i\right) + \sum_{j=1}^{N} \lambda_{ij} P_j\right)x(t) \\
&\quad - x^{\mathrm{T}}(t) P_i B_i (F_i B_i)^{-1} F_i D_i w(t) - w^{\mathrm{T}}(t) D_i^{\mathrm{T}} F_i^{\mathrm{T}} (F_i B_i)^{-1} B_i^{\mathrm{T}} P_i x(t) \\
&\quad + w^{\mathrm{T}}(t) D_i^{\mathrm{T}} P_i^{\mathrm{T}} x(t) + x^{\mathrm{T}}(t) P_i D_i w(t).
\end{aligned} \tag{7.40}$$

For some scalars $\psi_i > 0$, define the following auxiliary functions:

$$\check{\Pi}(x(t), i) = \mathscr{L}V_2(x(t), i) - \psi_i V_2(x(t), i) - \psi_i w^{\mathrm{T}}(t)w(t). \tag{7.41}$$

Then, if the condition (7.22) holds, we have $\check{\Pi}(x(t), i) < 0$. Thus, one has

$$\mathscr{L}V_2(x(t), i) < \psi_i V_2(x(t), i) + \psi_i w^{\mathrm{T}}(t)w(t). \tag{7.42}$$

From Theorem 7.2, one has $\mathbb{E}\{x^{\mathrm{T}}(T^*)Rx(T^*)\} < c^*$. Integrating 7.42 from T^* to t with $t \in [T^*, T]$, we have

$$\begin{aligned} e^{-\psi_i \tau}\mathbb{E}\{V_2(x(t), i)\} &< e^{-\psi_i \tau}\mathbb{E}\{V_2(x(T^*), r_{T^*})\} + \psi_i \int_{T^*}^{t} e^{-\psi_i \tau} w^{\mathrm{T}}(\tau)w(\tau)\mathrm{d}\tau \\ &< \bar{\varsigma}(P_i)c^* + \delta^2 T\psi_i. \end{aligned} \tag{7.43}$$

Besides, we can get from the Lyapunov function (7.25):

$$\mathbb{E}\{V_2(x(t), i)\} \geq \underline{\varsigma}(P_i)\mathbb{E}\{x^{\mathrm{T}}(t)Rx(t)\}, \tag{7.44}$$

which, together with (7.43), implies

$$\mathbb{E}\{x^{\mathrm{T}}(t)Rx(t)\} < \frac{(\bar{\varsigma}(P_i)c^* + \delta^2 T\psi_i)e^{\psi_i t}}{\underline{\varsigma}(P_i)}. \tag{7.45}$$

Obviously, if there exists a scalar c^* satisfying (7.23) and (7.24), it can be obtained from the condition (7.45) that $\mathbb{E}\{x^{\mathrm{T}}(t)Rx(t)\} < c_2$ for $\forall t \in [T^*, T]$. Then, the SFTB for the sliding mode dynamics (7.37) are ensured with respect to $\aleph_{[T^*,T]}$. The proof is finished. □

Remark 7.3 By applying the partition strategy, we have proven in Theorems 7.2 and 7.3 the SFTB of the closed-loop MJS (7.21) with respect to $\aleph_{[0,T^*]}$, and the SFTB of the sliding mode dynamics (7.37) with respect to $\aleph_{[T^*,T]}$, respectively. Besides, due to $c_1 < c^* < c_2$, the results in Theorems 7.2 and 7.3 can attain the desirable boundedness of the MJS (5) during the overall interval $[0, T]$ with respect to $\aleph_{[0,T]}$. However, different from the results in previous chapters where the system state can be successfully sent to the controller side, the existing channel fading and attacks in this chapter result in the unavailability of the raw system state $x(t)$, which also leads to more difficulties in analyzing the finite time SMC design. As a result, in Theorem 7.1, the system state cannot be strictly ensured to attain the sliding surface, only can into a bounded region around the desirable sliding surface. This just reflects the effect of the unreliable network involving fading and attacks on the SMC performance. Similarly, this effect still exists in analyzing the SFTB of the closed-loop MJS (7.21). To deal with this difficulty, some mode-dependent scalars ζ_i and ε_i are employed to deduce the results in Theorem 3 for ensuring the SFTB of closed-loop system (7.21).

7.5.3 Solving Algorithm

It should be noted that the SFTB of MJS (1.6) ($\Delta A_i(t) \equiv 0$) over the overall given interval $[0, T]$ can be guaranteed only when the conditions in Theorems 7.2 and 7.3 can be satisfied simultaneously. Meanwhile, the conditions in Theorem 7.1 should also be ensured for achieving the reachability of the specified sliding surface. To this end, we further give the sufficient conditions in the following Theorem 7.4 to guarantee the results in Theorems 7.1–7.3 such that the SFTB of the MJS (7.21) during $[0, T]$ can be achieved.

Theorem 7.4 *Given positive scalars ψ_i, ϑ_i, ρ_i, δ_i, and parameters (c_1, c_2, T, R, δ), if for any $i \in \mathbb{S}$, there exist scalars $\xi_i > 0$, $\chi_i > 0$, $\varepsilon_i > 0$, and matrices $P_i > 0$ and K_i ($\forall i \in \mathbb{S}$) satisfying conditions* (7.6), (7.7), (7.23), (7.38), *and the following conditions:*

$$\begin{bmatrix} \Gamma_{11i} & \Gamma_{12i} \\ * & \Gamma_{22i} \end{bmatrix} < 0, \tag{7.46}$$

$$\vartheta_i R < P_i < \rho_i R, \tag{7.47}$$

$$\begin{bmatrix} -e^{-\psi_i T} c^* \vartheta_i I & \sqrt{\rho_i c_1 + \delta^2 T \psi_i} I & \varepsilon_i \eta_i \sqrt{\frac{2mT}{\psi_i}} I & \bar{\alpha}\bar{\varphi}\varepsilon_i \sqrt{\frac{2\upsilon_i T}{\psi_i}} I \\ * & -I & 0 & 0 \\ * & * & -\varepsilon_i I & 0 \\ * & * & 0 & -\varepsilon_i I \end{bmatrix} < 0, \tag{7.48}$$

$$\rho_i c^* + \delta^2 T \psi_i - e^{-\psi_i T} c_2 \vartheta_i < 0 \tag{7.49}$$

with

$$\Gamma_{11i} \triangleq P_i A_i + A_i^{\mathrm{T}} P_i - \psi_i P_i + \sum_{j=1}^{N} \lambda_{ij} P_j + \zeta_i \epsilon_i^2 I,$$

$$\Gamma_{12i} \triangleq \begin{bmatrix} P_i D_i & P_i B_i & P_i B_i & P_i + B_i K_i \end{bmatrix},$$

$$\Gamma_{22i} \triangleq -\operatorname{diag}\{\psi_i I, \zeta_i I, \varepsilon_i I, \frac{1}{\varpi(1-\bar{\alpha})} I\},$$

then the closed-loop MJS (7.21) *is SFTB with respect to $\aleph_{[0,T]}$ during the interval* $[0, T]$.

Proof By Schur Complement (Lemma 1.3), (7.22) can be ensured by (7.46). In addition, it shows from the condition (7.47) that $\vartheta_i < \underline{\varsigma}(P_i)$, $\bar{\varsigma}(P_i) < \rho_i$. Then, (7.24) can be ensured by the following inequality

$$\rho_i c_1 + \delta^2 T \psi_i + \frac{C_i}{\psi_i} T - e^{-\psi_i T} \vartheta_i c^* < 0. \tag{7.50}$$

Recalling $C_i = 2\varepsilon_i(\bar{\alpha}^2\bar{\varphi}^2\|K_i\|^2 + \eta_i^2 m)$, we have from (7.6)

$$C_i \le 2\varepsilon_i\bar{\alpha}^2\bar{\varphi}^2\nu_i + 2\varepsilon_i\eta_i^2 m. \tag{7.51}$$

Based on (7.51), the following inequality can ensure that (7.50) holds:

$$\rho_i c_1 + \delta^2 T\psi_i + \frac{2\varepsilon_i\bar{\alpha}^2\bar{\varphi}^2\nu_i + 2\varepsilon_i\eta_i^2 m}{\psi_i}T - e^{-\psi_i T}\vartheta_i c^* < 0, \tag{7.52}$$

which can be ensured by the condition (7.48). Similarly, the condition (7.39) in Theorem 7.3 can be ensured by the inequality (7.49). Thus, if the conditions in Theorem 7.4 hold, it can ensure the SFTB of the MJS (7.21) with respect to $\aleph_{[0,T]}$ over the whole interval $[0, T]$. This finishes the proof. □

Remark 7.4 Due to the effect of channel fading and network attacks, the analysis on the dynamic characteristic including the sliding motion and the closed-loop system will become more difficult, which is mainly due to the unavailability of the system state $x(t)$. That is, only the faded and corrupted signal $\tilde{x}(t)$ can be utilized to design an effective controller for ensuring practical implementation. The above-considered network issues and the resulting difficulties just generate the main contributions of this present chapter. First, two kinds of random network phenomena, that is, fading and attacks, are simultaneously investigated for MJSs, and a sliding surface is constructed based on the system mode and fading coefficient. Then, by only using the faded and corrupted system state, a feasible SMC law is designed to drive the system state into a convergence domain around the sliding surface, which is different from some existing works on finite-time SMC under the unconstrained communication environment. Besides, during two stages, that is, reaching phase and sliding motion phase, the SFTB of the closed-loop MJS and sliding motion is ensured respectively.

7.6 Simulation

Consider the following 2-mode MJSs with parameters:

$$A_1 = \begin{bmatrix} 0.4 & -0.1 \\ 0.2 & -0.2 \end{bmatrix}, \; B_1 = \begin{bmatrix} 0.6 \\ 0.2 \end{bmatrix}, \; D_1 = \begin{bmatrix} 0.2 \\ 0.3 \end{bmatrix},$$
$$A_2 = \begin{bmatrix} -0.5 & 0.2 \\ 0.4 & -0.3 \end{bmatrix}, \; B_2 = \begin{bmatrix} 0.5 \\ 0.8 \end{bmatrix}, \; D_2 = \begin{bmatrix} 0.1 \\ 0.5 \end{bmatrix}.$$

and the TR matrix is assumed to be

$$\Lambda = \begin{bmatrix} -0.7 & 0.7 \\ 1.2 & -1.2 \end{bmatrix}.$$

Suppose the nonlinear functions as $f_i(x(t)) = \epsilon_i \sin\sqrt{x_1^2 + x_2^2}$ $(i = 1, 2)$ with $\epsilon_1 = 0.05, \epsilon_2 = 0.1$, and the external disturbances as $w(t) = \delta e^{-0.1t}\cos(0.2\pi t)$ with $\delta = 0.05$. Besides, the transmitted state signal might be attacked by the deception signal $\varphi(t) = 0.05\sqrt{2}\left[\tanh(x_1^2); \tanh(x_2^2)\right]$ with the upper bound $\bar{\varphi} = 0.1$ and the probability $\tilde{\alpha} = 0.15$, meanwhile, the mathematical expectation of the fading coefficient $\varpi(t)$ is $\bar{\varpi} = 0.9$.

For given parameters of finite time SMC issue as $T = 6, c_1 = 0.45, c_2 = 1.2$, and $R = 0.5I$, choosing sliding matrices $F_1 = [0.5, 1.0]$ and $F_2 = [0.6, 0.8]$ satisfying $\det(F_i B_i) \neq 0$, and parameters $\delta_1 = 8$, $\delta_2 = 5$, $\psi_1 = 0.01$, $\psi_2 = 0.05$, $\vartheta_1 = 0.1$, $\vartheta_2 = 0.2$, $\rho_1 = 2$, $\rho_2 = 2.5$, the following matrices can be obtained by solving the conditions (7.6), (7.7), (7.23), (7.38), and (7.46)–(7.49) in Theorem (7.4):

$$P_1 = \begin{bmatrix} 0.6843 & -0.2383 \\ -0.2383 & 0.5877 \end{bmatrix}, \quad P_2 = \begin{bmatrix} 0.6075 & -0.1567 \\ -0.1567 & 0.6778 \end{bmatrix},$$
$$K_1 = \begin{bmatrix} -1.3586 & -0.1614 \end{bmatrix}, \quad K_2 = \begin{bmatrix} -0.5161 & -0.7550 \end{bmatrix}.$$

The simulation results are plotted in Figs. 7.2 and 7.3 with the initial condition $x(0) = [0.8, 0.5]^{\mathrm{T}}$. Under one trial, the system mode jumping follows the sequence in Fig. 7.3b. The deception attack occurrence and state fading coefficient variation are exhibited in Fig. 7.2. It can be seen from Fig. 7.3a that the open-loop system cannot attain SFTB, but the closed-loop system can be ensured to be SFTB under the proposed SMC scheme with the control signal shown in Fig. 7.3d. Therefore, the proposed SMC scheme is effective to realize the SFTB of an MJS during the given time interval despite the negative effects of the deception attack and channel fading.

To show the robustness of the proposed control scheme, we further consider that the above system suffers from a *worse* fading situation with $\bar{\varpi} = 0.6$ and the occurrence probability of deception attack $\tilde{\alpha} = 0.5$. The Lyapunov matrices and controller gains can be obtained by solving the conditions in Theorem 7.4:

$$P_1 = \begin{bmatrix} 0.9606 & -0.3689 \\ -0.3689 & 0.6601 \end{bmatrix}, \quad P_2 = \begin{bmatrix} 0.6659 & -0.2871 \\ -0.2871 & 0.7335 \end{bmatrix},$$
$$K_1 = \begin{bmatrix} -2.4527 & 0.1729 \end{bmatrix}, \quad K_2 = \begin{bmatrix} -0.6755 & -0.7399 \end{bmatrix}.$$

Meanwhile, the simulation results are plotted in Figs. 7.4 and 7.5. Compared with Figs. 7.2 and 7.4, it can be seen that when the attack occurrence probability increases, the number of attacks increases significantly, and when the fading expectation decrease, the degree of state decay increases. As shown in Fig. 7.4a, the state of the closed-loop systems still can attain the SFTB, which verifies the effectiveness of the proposed finite-time SMC scheme. However, when the system state is transmitted over a worse network environment with smaller $\bar{\varpi}$ and larger $\tilde{\alpha}$, it can be observed from Fig. 7.4c and Fig. 7.4d that the SMC performance will degrade, which is mainly due to the fact that less information is obtained by the controller.

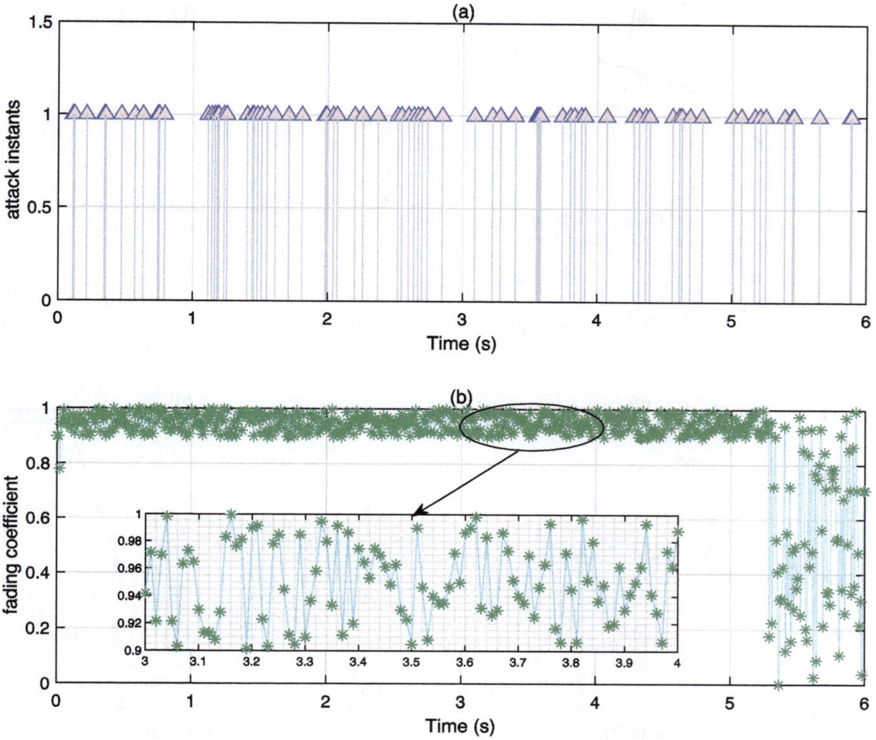

Fig. 7.2 Attack and fading phenomena: **a** attack instants under probability $\tilde{\alpha} = 0.15$; **b** fading coefficient variation with expectation $\bar{\varpi} = 0.9$

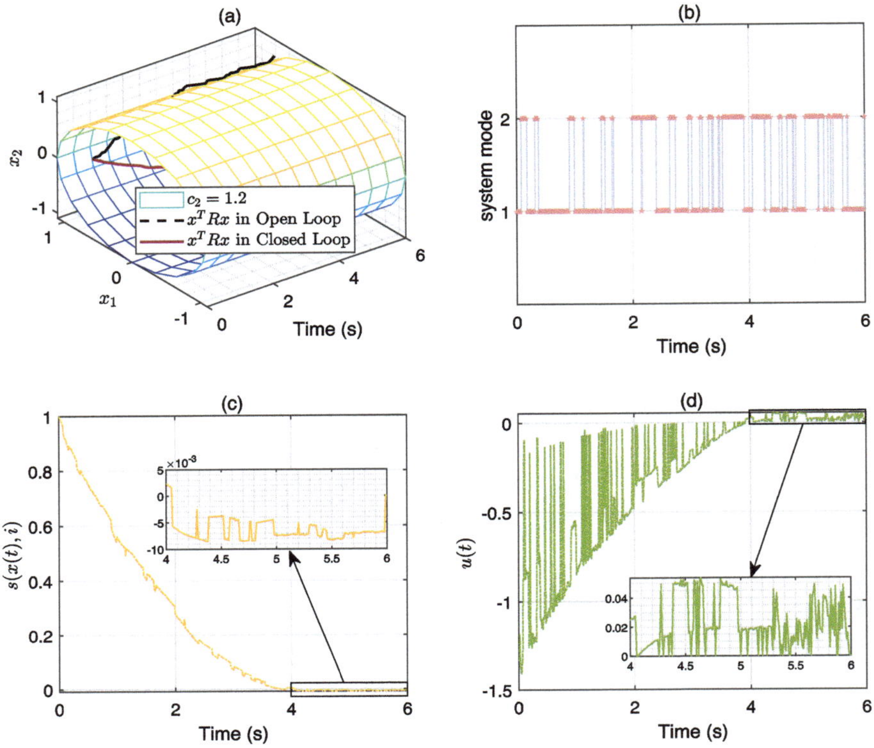

Fig. 7.3 Simulation results of the controlled system under $\bar{\varpi} = 0.9$, $\tilde{\alpha} = 0.15$: **a** evolution of weighted system state variable $x^{\mathrm{T}}(t)Rx(t)$; **b** possible mode transition sequence r_t; **c** sliding variable $s(x(t), i)$; **d** control signal $u(t)$

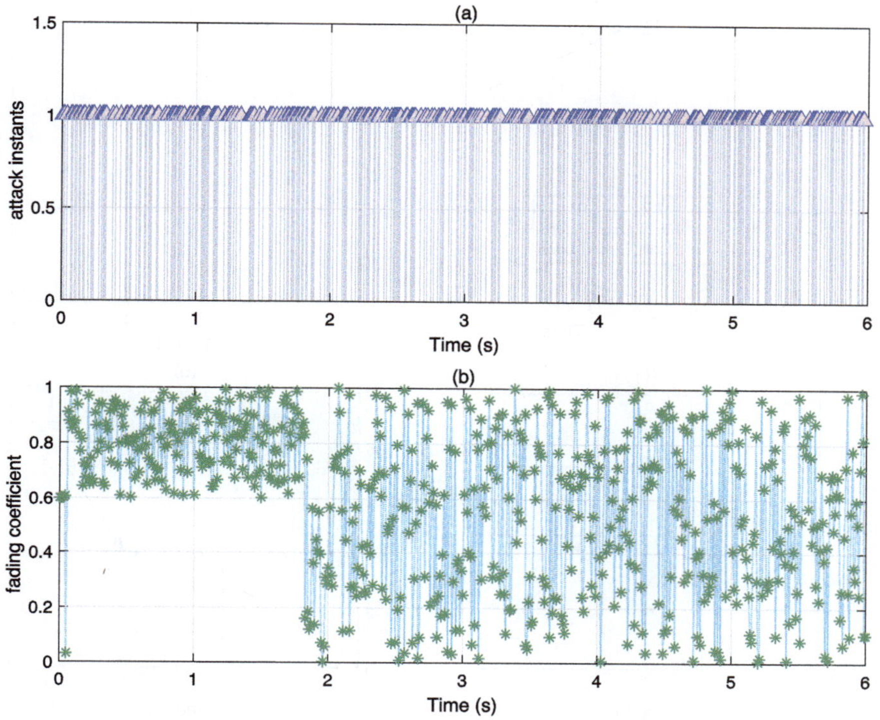

Fig. 7.4 Attack and fading phenomena: **a** attack instants under probability $\tilde{\alpha} = 0.5$; **b** fading coefficient variation with expectation $\bar{\varpi} = 0.6$

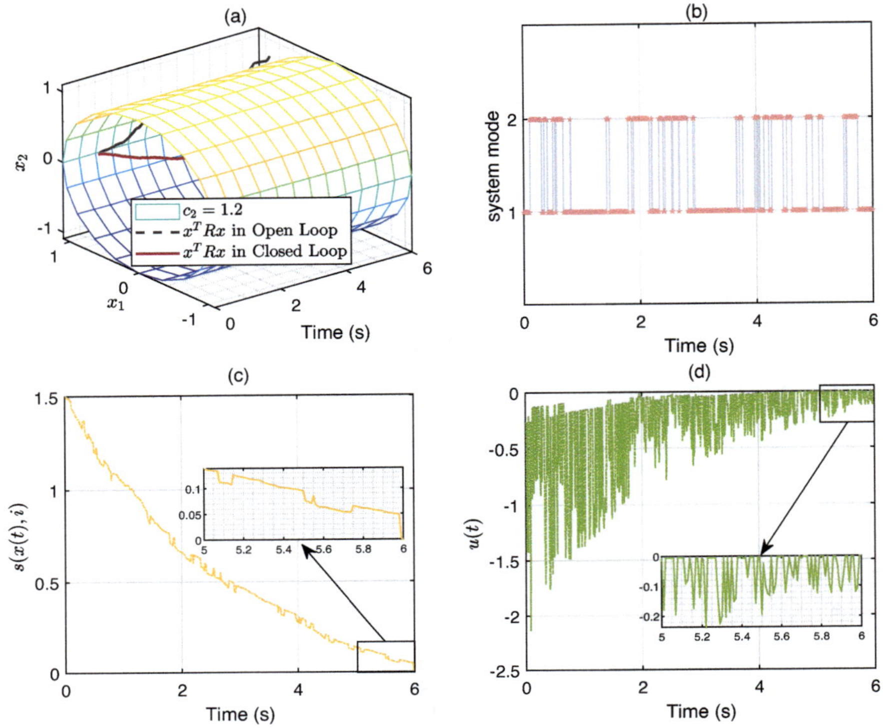

Fig. 7.5 Simulation results of the controlled system under $\bar{\varpi} = 0.6$, $\tilde{\alpha} = 0.5$: **a** evolution of weighted system state variable $x^{\mathrm{T}}(t)Rx(t)$; **b** possible mode transition sequence r_t; **c** sliding variable $s(x(t), i)$; **d** control signal $u(t)$

7.7 Conclusion

This chapter has addressed the security SMC issue for MJSs subject to channel fading and deception attacks over the S/C channel. Two independent stochastic variables have been introduced to describe the fading coefficient and the occurrence of deception attacks. By designing the mode-dependent sliding function involving the fading coefficient and using only the corrupted states with channel fading, a security SMC law has been developed for guaranteeing the reachability of the sliding surface and the stochastic finite time boundedness of the closed-loop system. Meanwhile, in the presence of the fading phenomenon and deception attacks, sufficient conditions have been derived for the boundedness over the reaching phase and sliding motion phase. The simulation and some comparisons on different network circumstances have been made to verify the proposed security finite time SMC scheme.

References

1. Su, L., Chesi, G.: On the design of output feedback controllers for LTI systems over fading channels. IEEE Trans. Autom. Control **63**(5), 1503–1508 (2017)
2. Zhang, Z., Su, S.F., Niu, Y.: Dynamic event-triggered control for interval type-2 fuzzy systems under fading channel. IEEE Trans. Cybern. **51**(11), 5342–5351 (2021)
3. Song, J., Niu, Y., Wang, S.: Robust finite-time dissipative control subject to randomly occurring uncertainties and stochastic fading measurements. J. Frankl. Inst. **354**(9), 3706–3723 (2017)
4. Li, J., Niu, Y.: Sliding mode control subject to rice channel fading. IET Control Theory Appl. **13**(16), 2529–2537 (2019)
5. Yang, Y., Niu, Y., Zhang, Z.: Dynamic event-triggered sliding mode control for interval type-2 fuzzy systems with fading channels. ISA Trans. **110**, 53–62 (2021)
6. Song, J., Wang, Z., Niu, Y., Dong, H.: Genetic-algorithm-assisted sliding-mode control for networked state-saturated systems over hidden Markov fading channels. IEEE Trans. Cybern. **51**(7), 3664–3675 (2021)
7. Zhang, S., Wang, Z., Ding, D., Wei, G., Alsaadi, F.E., Hayat, T.: A gain-scheduling approach to non-fragile H_∞ fuzzy control subject to fading channels. IEEE Trans. Fuzzy Syst. **26**(1), 142–154 (2016)
8. Han, K.Y., Lee, S.W., Lim, J.S., Sung, K.M.: Channel estimation for OFDM with fast fading channels by modified Kalman filter. IEEE Trans. Consum. Electron. **50**(2), 443–449 (2004)

Chapter 8
Saturated Sliding Mode Control Under Injection Attacks

8.1 Introduction

When the designed control signals are transmitted via the open network to the actuator, both physical and communication problems will be met. From the physical constraint perspective, due to the requirement of the actuator structure and its safe operation, the actuator saturation phenomenon occurs inevitably, which brings strong interference and destructive effects to the system performance. Generally, there are two main methods for dealing with actuator saturations. One approach is to consistently consider the saturation characteristics during the controller design process, ensuring that the system performance can either meet or withstand the constraints of actuator saturation, thereby guaranteeing system stability [1–3]. Another method is to construct the controller, ensuring that the closed-loop system maintains good performance during the unsaturated phase of the actuator. When saturation occurs, an anti-windup compensator is designed to input the difference between the control signal and the actuator signal back into the controller, indirectly handling the constraints of saturation.

From the communication constraint perspective, the control signals might be distorted by cyber attacks due to the openness and vulnerability of the communication network. Among various types of attacks, injection attacks have stirred considerable interest on account of breaking the authenticity of control signals. A kind of randomly occurring injection attack has been studied for Markov jump systems (MJSs) in [4] with an assumption that both constant and nonlinear bounds of injection attacks could be achieved and used in control parts. An adaptive method has been further proposed in [5] to estimate the constant bound of injection attacks, while their nonlinear bounds have been still assumed to be known. Although this assumption is effective for control design, it counters two problems, that is, it might be unreliable in practical scenarios due to the hard acquirement of attack signals. How to relax this assumption is one of the motivations of this chapter.

Z. Cao and Y. Niu, *Finite-Time Sliding Mode Control for Markov Jump Systems*,
Studies in Systems, Decision and Control 603,
https://doi.org/10.1007/978-981-96-7651-4_8

In this chapter, an actuator saturation problem is considered for MJSs, and a neural network method is used to approximate the attack nonlinear bound of the injection attacks in the controller-to-actuator (C/A) channel. Then, a mode-dependent sliding function is established, based on which, the adaptive rules of neural network parameters are designed according to the assigned sliding variable, and the sliding mode control (SMC) law is constructed to realize the finite-time transient performance. Furthermore, by analyzing the stochastic finite-time boundedness (SFTB) of the closed-loop system during the reaching phase and the SFTB of sliding mode dynamics during the sliding motion phase, respectively, sufficient conditions are derived for solving controller gains.

8.2 Physical and Network Constraints

Consider a class of MJSs on a complete probability space $(\Omega, \mathscr{F}, \Pr)$ as in system (1.6) ($f_i(x(t)) \equiv 0$) with randomly occurring uncertainties (ROUs) and cyber attacks. The transition probabilities (TPs) are generated by (1.7) with the assumptions that all transition rates (TRs) are completely available. Accordingly, for mode $r_t = i \in \mathbb{S}$, the considered MJS with ROUs and cyber attacks is expressed as

$$\dot{x}(t) = (A_i + \Delta A_i(t))x(t) + B_i\Phi(u^A(t)) + D_i w(t) \tag{8.1}$$

where $x(t) \in \mathbb{R}^n$ is the state, $\Phi(u^A(t)) \in \mathbb{R}^m$ stands for the nonlinear saturation characteristic with $u^A(t)$ being the input signal from the controller under injection attacks, $\Delta A_i(t)$ represents the parameter uncertainty satisfying Assumption 1.2, and $w(t)$ is the external disturbance satisfying Assumption 1.4.

8.2.1 *Actuator Saturation Model*

In addition, we define $u^A(t) \triangleq [u_1^A(t)\ u_2^A(t)\ \cdots\ u_m^A(t)]^{\mathrm{T}}$ and constrain the maximum control input as $u_{l\max}$ ($u_{l\max} > 0$) for each control input $u_l^A(t)$ ($l = 1, 2, \ldots, m$). The nonlinear saturation function $\Phi(\cdot)$ owns the following characteristic [2, 3]:

$$\Phi(u^A(t)) \triangleq \phi(u^A(t))u^A(t) \tag{8.2}$$

where

$$\phi(u^A(t)) \triangleq \mathrm{diag}\{\phi_1(u_1^A(t)), \phi_2(u_2^A(t)), \ldots, \phi_m(u_m^A(t))\} \tag{8.3}$$

with

$$\phi_l(u_l^A(t)) \triangleq \begin{cases} u_{l\max}/u_l^A(t), & \text{if } u_l^A(t) > u_{l\max} \\ 1, & \text{if } |u_l^A(t)| \le u_{l\max} \\ -u_{l\max}/u_l^A(t), & \text{if } u_l^A(t) < -u_{l\max} \end{cases} \tag{8.4}$$

Remark 8.1 The coefficient $\phi_l(\cdot) \in [0, 1]$ is an indicator of the saturation level of the lth control input. If $u_{l\,\max} = 0$, then $\phi_l(\cdot) = 0$, which indicates no feedback from the lth control signal $u_l(t)$. Conversely, when $\phi_l(\cdot) = 1$, it signifies that there is no saturation in the lth control signal $u_l(t)$.

A real number has the density property [11], according to which since $0 < \phi_l(u_l(t)) \leq 1$, there exists a known constant ξ_l meeting $0 < \xi_l \leq \phi_l(u_l(t)) \leq 1$ for $l = 1, 2, \ldots, m$.Thus, we have

$$0 < \Xi \leq \phi(u^A(t)) \leq I \ \ (l = 1, 2, \ldots, m) \tag{8.5}$$

with $\Xi \triangleq \text{diag}\{\xi_1, \xi_2, \ldots, \xi_m\}$.

8.2.2 Injection Attacks in C/A Channel

Suppose that the injection attacks may occur in C/A network, in which the false information $\Psi(x(t), t)$ injected by the adversaries affects the authenticity of the control signal $u(t)$. As the above description, the attack model can be expressed as

$$u^A(t) = u(t) + \Psi(x(t), t) \tag{8.6}$$

where $\|\Psi(x(t), t)\| \leq \psi(x(t), t)$ with $\psi(x(t), t)$ an unknown nonlinear function related to the available system states.

Remark 8.2 From [6], we can know that the considered injection attacks (8.6) may inject the desired signals into the control input to realize the stealthy and deception purpose. In order to consume the low cost of attacks, the adversaries launch the same attacks against different subsystems. Additionally, the bound of injected information usually obeys a nonlinear function on the system state $x(t)$ captured by the adversaries. Actually, although we don't know the information, we can employ the neural network technique to estimate the attack nonlinear bound $\psi(x(t), t)$, which can allow for more accurate information about the attack, rather than the constant upper bound of the attack as in [7, 8]

Consequently, as shown in Fig. 8.1, considering the actuator saturation (8.2) and injection attacks (8.6) simultaneously, system (8.1) can be further expressed as

$$\dot{x}(t) = (A_i + \Delta A_i(t))x(t) + B_i\phi(u^A(t))(u(t) + \Psi(x(t), t)) + D_i w(t) \tag{8.7}$$

The objective of this chapter is: (1) to utilize the neural network technique and design the corresponding adaptive parameters to estimate the attack nonlinear bound; and (2) to construct a sliding mode controller dependent on the estimated attack information such that the SFTB of the MJS (8.7) can be attained despite actuator saturation and injection attacks.

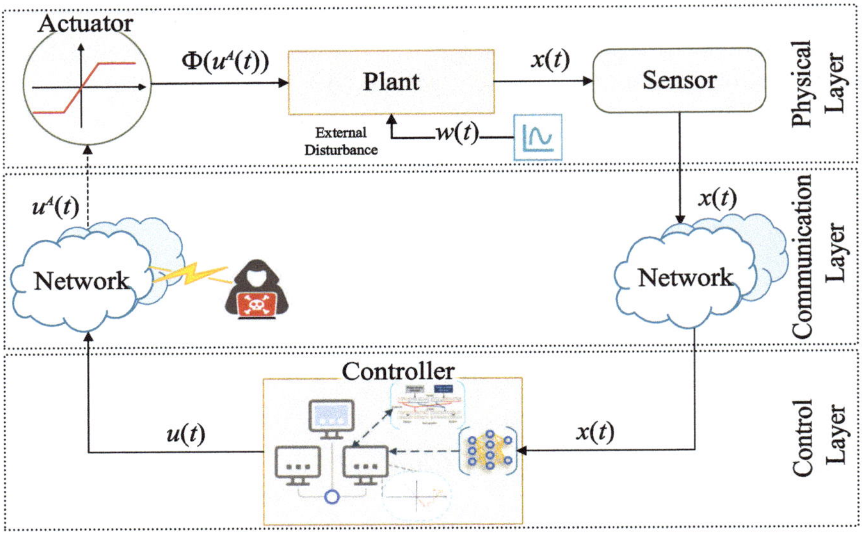

Fig. 8.1 Block diagram of MJSs under injection attacks

8.3 Neural Network Approximation

In virtue of neural network technique, the attack nonlinear bound $\psi(x(t), t)$ can be approximated by

$$\psi(x(t), t) = \mathcal{W}^{\mathrm{T}} \varsigma(\mathcal{V}^{\mathrm{T}} \vartheta(t)) + \varepsilon(x) \ (\forall x(t) \in \mho \subset \mathbb{R}^n) \tag{8.8}$$

with $\mho$ being a compact set and the symbols as in Table 8.1.

Due to unknown $\mathcal{W}$ and $\mathcal{V}$, an adaptive approach will be proposed to estimate them as $\hat{\mathcal{W}}$ and $\hat{\mathcal{V}}$, respectively. Denote the estimation errors as $\tilde{\mathcal{W}}$ and $\tilde{\mathcal{V}}$, where $\tilde{\mathcal{W}}$ indicates the difference between the estimated value $\hat{\mathcal{W}}$ and the actual value $\mathcal{W}$,

Table 8.1 Symbols in neural networks

Notations	Explanations
$\mathcal{W} \in \mathbb{R}^{l \times 1}$	Optimization weight of the hidden-to-output layer
$\mathcal{V} \in \mathbb{R}^{(n+1) \times l}$	Optimization weight of input-to-hidden layer
$\varepsilon(x) \in \mathbb{R}$	Error vector satisfying $\|\varepsilon(x)\| \leq \mu$ $(\mu > 0)$
$l \in \mathbb{R}$	Number of hidden neurons
$\vartheta(t) \triangleq \mathrm{col}\{x(t), -1\}$	Augmented input vector for a neural network, including -1 bias
$\varsigma(\cdot) \triangleq \mathrm{col}\{\varsigma_1(x_1), \ldots, \varsigma_l(x_l)\}$	Activation function: $\varsigma_s(x_s) = \frac{1}{1+e^{-a_s x_s}}$ $(a_s > 0,\ s = 1, 2, \ldots, l)$

and $\tilde{\mathcal{V}}$ indicates the difference between the estimated value $\hat{\mathcal{V}}$ and the actual value $\mathcal{V}$. Specifically, $\tilde{\mathcal{W}} \triangleq \hat{\mathcal{W}} - \mathcal{W}$ and $\tilde{\mathcal{V}} \triangleq \hat{\mathcal{V}} - \mathcal{V}$. For the purpose of subsequent derivation, the following lemma is introduced.

Lemma 8.1 *Let $\hat{\psi}(x(t), t)$ as the estimated approximation of the nonlinear function $\psi(x(t), t)$. The error $\tilde{\psi}(\cdot) = \hat{\psi}(\cdot) - \psi(\cdot)$ is expressed as*

$$\tilde{\psi}(x(t), t) = \tilde{\mathcal{W}}^{\mathrm{T}}(\hat{\varsigma} - \hat{\varsigma}'\hat{\mathcal{V}}^{\mathrm{T}}\vartheta(t)) + \hat{\mathcal{W}}^{\mathrm{T}}\hat{\varsigma}'\tilde{\mathcal{V}}^{\mathrm{T}}\vartheta(t) + \iota(x)$$

where $\iota(x)$ is the residual error and its boundary is $\|\iota(x)\| < \alpha^T z$, with $\alpha \in \mathbb{R}^4$ being the vector of unascertained parameter and $z \triangleq \mathrm{col}\{1, \|\vartheta(t)\|, \|\vartheta(t)\| \|\hat{\mathcal{W}}\|_F, \|\vartheta(t)\| \|\hat{\mathcal{V}}\|_F\}$.

Denote the estimation value as $\hat{\alpha}$ and the estimation error of parameter α as $\tilde{\alpha} \triangleq \hat{\alpha} - \alpha$. Then, adaptive laws for the parameters $\hat{\mathcal{W}}$, $\hat{\mathcal{V}}$, and $\hat{\alpha}$ will be processed later.

8.4 Saturated Controller Design and Reachability Analysis

Choose the sliding functions as follows:

$$s(x(t), i) = G_i x(t) \tag{8.9}$$

where matrix $G_i \triangleq B_i^T P_i$ with $P_i > 0$ to be designed later for each $i \in \mathbb{S}$. Obviously, $G_i B_i > 0$ is nonsingular.

For a given finite-time interval $[0, T]$, an appropriate sliding mode controller will be constructed so that the state trajectories of the system can be pushed into the specified sliding surfaces $s(x(t), i) = 0$ before a finite instant T^* with $T^* < T$ and sustain there within the interval $[T^*, T]$. Therefore, we provide the following theorem:

Theorem 8.1 *For a given finite-time interval $[0, T]$, if the following SMC law for MJS (8.7) is designed with the sliding function (8.9):*

$$\begin{aligned} u(t) = K_i x(t) &- \left(\hat{\mathcal{W}}^{\mathrm{T}}\varsigma(\hat{\mathcal{V}}^{\mathrm{T}}\vartheta(t)) + \hat{\alpha}^{\mathrm{T}} z\right)\Xi^{-1}\mathrm{sgn}(s(x(t), i)) \\ &- \rho_i(t)\Xi^{-1}\mathrm{sgn}(s(x(t), i)) \end{aligned} \tag{8.10}$$

where

$$\dot{\hat{\mathcal{W}}} \triangleq \Gamma_1(\hat{\varsigma} - \hat{\varsigma}'\hat{\mathcal{V}}^{\mathrm{T}}\vartheta(t))\|s(x(t), i)\|, \ \dot{\hat{\mathcal{V}}} \triangleq \Gamma_2\vartheta(t)\hat{\mathcal{W}}^{\mathrm{T}}\hat{\varsigma}'\|s(x(t), i)\| \tag{8.11}$$

$$\dot{\hat{\alpha}} \triangleq \Gamma_3 z\|s(x(t), i)\| \tag{8.12}$$

$$\rho_i(t) \triangleq \varrho_i + \gamma_i\|x(t)\| + \|(G_i B_i)^{-1} G_i D_i\|\delta + \frac{\eta_i\|x(t)\|^2}{2\|s(x(t), i)\|} \tag{8.13}$$

$$\gamma_i \triangleq \|(G_i B_i)^{-1} G_i M_i\| \|N_i\| + \|(G_i B_i)^{-1} G_i A_i\| + \|K_i\| \tag{8.14}$$

$$\eta_i \triangleq \| \sum_{j=1}^{N} \lambda_{ij} G_j^{\mathrm{T}} (G_j B_j)^{-1} G_j \|, \ \varrho_i \geq \frac{\lambda_{\max}((G_i B_i)^{-1})}{T - \chi} \|G_i x(0)\| \tag{8.15}$$

with $0 < \chi < T$ *a given small constant, the state trajectories of the system can be pushed into the sliding surfaces* $s(x(t), i) = 0$ *within the interval* $[0, T^*]$ *with* $T^* < T$.

Proof We select the following Lyapunov function as

$$\begin{aligned} V_1(x(t), i) &= s^{\mathrm{T}}(x(t), i)(G_i B_i)^{-1} s(x(t), i) \\ &\quad + \mathrm{tr}(\tilde{\mathcal{W}}^{\mathrm{T}} \Gamma_1^{-1} \tilde{\mathcal{W}}) + \mathrm{tr}(\tilde{\mathcal{V}}^{\mathrm{T}} \Gamma_2^{-1} \tilde{\mathcal{V}}) + \tilde{\alpha}^{\mathrm{T}} \Gamma_3^{-1} \tilde{\alpha} \end{aligned} \tag{8.16}$$

Subsequently, the infinitesimal operator [10] can be obtained as

$$\begin{aligned} &\mathscr{L} V_1(x(t), i) \\ &= 2s^{\mathrm{T}}(x(t), i)((G_i B_i)^{-1} G_i M_i F_i(t) N_i + (G_i B_i)^{-1} G_i A_i) x(t) \\ &\quad + 2s^{\mathrm{T}}(x(t), i)\phi(u^A(t))(u(t) + \Psi(x(t), t)) \\ &\quad + 2s^{\mathrm{T}}(x(t), i)(G_i B_i)^{-1} G_i D_i w(t) + x^{\mathrm{T}}(t) \sum_{j=1}^{N} \lambda_{ij} G_j^{\mathrm{T}} (G_j B_j)^{-1} G_j x(t) \\ &\quad + 2\mathrm{tr}(\tilde{\mathcal{W}}^{\mathrm{T}} \Gamma_1^{-1} \dot{\hat{\mathcal{W}}}) + 2\mathrm{tr}(\tilde{\mathcal{V}}^{\mathrm{T}} \Gamma_2^{-1} \dot{\hat{\mathcal{V}}}) + 2\tilde{\alpha}^{\mathrm{T}} \Gamma_3^{-1} \dot{\hat{\alpha}} \end{aligned} \tag{8.17}$$

Substituting (8.10)–(8.15) into (8.17) and perceiving that $\|s(x(t), i)\| < |s(x(t), i)|$ gives

$$\begin{aligned} &\mathbb{E}\{\mathscr{L} V_1(x(t), i)\} \\ &\leq 2\|s(x(t), i)\| \Big(\|(G_i B_i)^{-1} G_i M_i\| \|N_i\| \|x(t)\| + \|(G_i B_i)^{-1} G_i A_i\| \|x(t)\| \\ &\quad + \|K_i\| \|x(t)\| + \|(G_i B_i)^{-1} G_i D_i\| \|w(t)\| \\ &\quad + \frac{\|x(t)\|^2}{2\|s(x(t), i)\|} \| \sum_{j=1}^{N} \lambda_{ij} G_j^{\mathrm{T}} (G_j B_j)^{-1} G_j \| + \psi(x(t), t) \Big) \\ &\quad - 2s^{\mathrm{T}}(x(t), i)\phi(u^A(t)) \Big(\hat{\mathcal{W}}^{\mathrm{T}} \varsigma(\hat{\mathcal{V}}^{\mathrm{T}} \vartheta(t)) + \hat{\alpha}^T z \Big) \Xi^{-1} \mathrm{sgn}(s(x(t), i)) \\ &\quad + 2\mathrm{tr}(\tilde{\mathcal{W}}^{\mathrm{T}} \Gamma_1^{-1} \dot{\hat{\mathcal{W}}}) + 2\mathrm{tr}(\tilde{\mathcal{V}}^{\mathrm{T}} \Gamma_2^{-1} \dot{\hat{\mathcal{V}}}) + 2\tilde{\alpha}^{\mathrm{T}} \Gamma_3^{-1} \dot{\hat{\alpha}} \\ &\quad - 2s^{\mathrm{T}}(x(t), i)\phi(u^A(t)) \Xi^{-1} \rho_i(t) \mathrm{sgn}(s(x(t), i)) \end{aligned} \tag{8.18}$$

In fact, from (8.5), one has

$$
\begin{aligned}
& s^{\mathrm{T}}(x(t), i)\phi(u^A(t))\Xi^{-1}\mathrm{sgn}(s(x(t), i)) \\
= & \left[s_1(x(t), i) \cdots s_m(x(t), i)\right] \mathrm{diag}\{\phi_1(u_1^A(t))/\xi_1, \ldots, \phi_m(u_1^A(t))/\xi_m\} \\
& \times \left[\mathrm{sgn}(s_1(x(t), i)) \cdots \mathrm{sgn}(s_m(x(t), i))\right]^{\mathrm{T}} \\
= & \sum_{l=1}^{m} \phi_l(u_l^A(t))/\xi_l s_l(x(t), i)\mathrm{sgn}(s_l(x(t), i)) \\
\geq & \sum_{l=1}^{m} s_l(x(t), i)\mathrm{sgn}(s_l(x(t), i)) \geq |s(x(t), i)| \geq \|s(x(t), i)\|
\end{aligned} \tag{8.19}
$$

According to (8.8), (8.10) and (8.19), we have

$$
\begin{aligned}
& 2\|s(x(t), i)\|\psi(x(t), t) + 2\mathrm{tr}(\tilde{\mathcal{W}}^{\mathrm{T}}\Gamma_1^{-1}\dot{\hat{\mathcal{W}}}) + 2\mathrm{tr}(\tilde{\mathcal{V}}^{\mathrm{T}}\Gamma_2^{-1}\dot{\hat{\mathcal{V}}}) + 2\tilde{\alpha}^{\mathrm{T}}\Gamma_3^{-1}\dot{\hat{\alpha}} \\
& - 2s^{\mathrm{T}}(x(t), i)\phi(u^A(t))\Big(\hat{\mathcal{W}}^{\mathrm{T}}\varsigma(\hat{\mathcal{V}}^{\mathrm{T}}\vartheta(t)) + \hat{\alpha}^T z\Big)\Xi^{-1}\mathrm{sgn}(s(x(t), i)) \\
\leq & 2\|s(x(t), i)\|\Big(- \tilde{\mathcal{W}}^{\mathrm{T}}(\hat{\varsigma} - \hat{\varsigma}'\hat{\mathcal{V}}^{\mathrm{T}}\vartheta(t)) - \hat{\mathcal{W}}^{\mathrm{T}}\hat{\varsigma}'\tilde{\mathcal{V}}^{\mathrm{T}}\vartheta(t) - \iota(x) - \hat{\alpha}^T z\Big) \\
& + 2\mathrm{tr}(\tilde{\mathcal{W}}^{\mathrm{T}}\Gamma_1^{-1}\dot{\hat{\mathcal{W}}}) + 2\mathrm{tr}(\tilde{\mathcal{V}}^{\mathrm{T}}\Gamma_2^{-1}\dot{\hat{\mathcal{V}}}) + 2\tilde{\alpha}^{\mathrm{T}}\Gamma_3^{-1}\dot{\hat{\alpha}}
\end{aligned} \tag{8.20}
$$

By utilizing adaptive adjustment rules (8.11)–(8.12), the following can be obtained from (8.20):

$$
\begin{aligned}
& - 2\|s(x(t), i)\|\tilde{\mathcal{W}}^{\mathrm{T}}(\hat{\varsigma} - \hat{\varsigma}'\hat{\mathcal{V}}^{\mathrm{T}}\vartheta(t)) + 2\mathrm{tr}(\tilde{\mathcal{W}}^{\mathrm{T}}\Gamma_1^{-1}\dot{\hat{\mathcal{W}}}) \\
\leq & - 2\mathrm{tr}\Big(\|s(x(t), i)\|\tilde{\mathcal{W}}^{\mathrm{T}}(\hat{\varsigma} - \hat{\varsigma}'\hat{\mathcal{V}}^{\mathrm{T}}\vartheta(t))\Big) \\
& + 2\mathrm{tr}\Big(\tilde{\mathcal{W}}^{\mathrm{T}}(\hat{\varsigma} - \hat{\varsigma}'\hat{\mathcal{V}}^{\mathrm{T}}\vartheta(t))\|s(x(t), i)\|\Big) = 0
\end{aligned} \tag{8.21}
$$

$$
\begin{aligned}
& - 2\|s(x(t), i)\|\hat{\mathcal{W}}^{\mathrm{T}}\hat{\varsigma}'\tilde{\mathcal{V}}^{\mathrm{T}}\vartheta(t) + 2\mathrm{tr}(\tilde{\mathcal{V}}^{\mathrm{T}}\Gamma_2^{-1}\dot{\hat{\mathcal{V}}}) \\
\leq & - 2\mathrm{tr}\Big(\|s(x(t), i)\|\hat{\mathcal{W}}^{\mathrm{T}}\hat{\varsigma}'\tilde{\mathcal{V}}^{\mathrm{T}}\vartheta(t)\Big) \\
& + 2\mathrm{tr}\Big(\tilde{\mathcal{V}}^{\mathrm{T}}\vartheta(t)\hat{\mathcal{W}}^{\mathrm{T}}\hat{\varsigma}'\|s(x(t), i)\|\Big) = 0
\end{aligned} \tag{8.22}
$$

$$
\begin{aligned}
& - 2\|s(x(t), i)\|l(x) + 2\tilde{\alpha}^{\mathrm{T}}\Gamma_3^{-1}\dot{\hat{\alpha}} - 2\|s(x(t), i)\|\hat{\alpha}^T z \\
\leq & 2\|s(x(t), i)\|\alpha^T z - 2\|s(x(t), i)\|\hat{\alpha}^T z + 2\tilde{\alpha}^T z\|s(x(t), i)\| = 0
\end{aligned} \tag{8.23}
$$

Therefore, the following formula is established:

$$
\mathbb{E}\{\mathscr{L}V_1(x(t), i)\} \leq -2\varrho_i\|s(x(t), i)\| < 0 \tag{8.24}
$$

In summary, by the SMC law (8.10)–(8.15), the state trajectories will be pushed into the specified sliding surfaces $s(x(t), i) = 0$.

Below, we will further provide the reaching instant T^*. One has from (8.16) that $V_1(x(t), i) \leq \lambda_{\max}((G_i B_i)^{-1})\|s(x(t), i)\|^2$. Then, referring to Chap. 6, the state trajectories will be pushed into the specified sliding surfaces $s(x(t), i) = 0$ within finite time T^* with $T^* < T$. This completes the proof. □

Remark 8.3 From (8.4), we can see that the function $\phi_l(\cdot)$ tends to zero as $u_l(t)$ converges to get close to infinity [2], that is, the constant ξ_l is impossible to exist and the diagonal matrix Ξ in (8.5) may also not exist. However, according to (8.20) and (8.24), the signals have boundary in finite time, and then, the $u(t)$ has boundary in finite time, which means the existence of Ξ, as $\phi_l(\cdot)$ converges to zero. Therefore, the assumption (8.5) is feasible.

8.5 SFTB Analysis

Within this segment, by analyzing SFTB problem of the MJS (8.7) against actuator saturation (8.2) and the injection attacks (8.6) during the reaching and sliding motion phases respectively, the SFTB problem of the MJS (8.7) will be investigated and sufficient conditions will be proposed to guarantee SFTB over the specified finite-time interval $[0, T]$.

►**SFTB of closed-loop MJS over the reaching phase** $[0, T^*]$

Combining the SMC law (8.10) and the system (8.7), the following closed-loop system can be generated:

$$\begin{aligned}\dot{x}(t) &= (A_i + \Delta A_i(t) + B_i\phi(u^A(t))K_i)x(t) - B_i\phi(u^A(t))\rho_{si}(t)\\ &\quad + B_i\phi(u^A(t))\Psi(x(t), t) + D_i w(t)\end{aligned} \tag{8.25}$$

where $\rho_{si}(t) \triangleq \left(\hat{W}^{\mathrm{T}}\varsigma(\hat{V}^{\mathrm{T}}\vartheta(t)) + \hat{\alpha}^T z + \rho_i(t)\right)\Xi^{-1}\mathrm{sgn}(s(x(t), i))$.

In the following theorem, the sufficient conditions are proposed to ensure SFTB over the reaching phase $[0, T^*]$.

Theorem 8.2 *For the given parameters* (c_1, c_2, T, R, δ) *and a feasible scalar* $\varphi_i > 0$, *consider the closed-loop system* (8.25), *if there are constants* $\sigma_i > 0$, $c^* > 0$, *and matrices* $P_i > 0$ *and* K_i $(\forall i \in \mathbb{S})$ *satisfying*

$$\begin{bmatrix} \Lambda_{11i} & \Lambda_{12i} \\ * & \Lambda_{22i} \end{bmatrix} < 0 \tag{8.26}$$

$$c_1 < c^* < c_2 \tag{8.27}$$

$$\frac{\bar{\varpi}_P c_1 + \varphi_i \delta^2 T}{\underline{\varpi}_P} < e^{-\varphi_i T} c^* \tag{8.28}$$

where

$$\Lambda_{11i} \triangleq \begin{bmatrix} \text{He}(P_i\bar{A}_i(t)) - \varphi_i P_i + \sum_{j=1}^{N} \lambda_{ij} P_j & P_i D_i \\ * & -\varphi_i I \end{bmatrix}$$

$$\Lambda_{12i} \triangleq \begin{bmatrix} P_i M_i & \sigma_i N_i^{\text{T}} \\ 0 & 0 \end{bmatrix}, \ \Lambda_{22i} \triangleq -\text{diag}\{\sigma_i I, \sigma_i I\}, \ \bar{A}_i(t) \triangleq A_i + B_i \phi(u^A(t)) K_i$$

$$\bar{\varpi}_P \triangleq \max_{i \in \mathbb{S}} \{\lambda_{\max}[R^{-\frac{1}{2}} P_i R^{-\frac{1}{2}}]\}, \ \underline{\varpi}_P \triangleq \min_{i \in \mathbb{S}} \{\lambda_{\min}[R^{-\frac{1}{2}} P_i R^{-\frac{1}{2}}]\}$$

the closed-loop system (8.25) *is SFTB with respect to* $\aleph_{[0,T^*]}$.

Proof Select the following Lyapunov function as

$$V_2(x(t), i) = x^{\text{T}}(t) P_i x(t) + \text{tr}(\tilde{\mathcal{W}}^{\text{T}} \Gamma_1^{-1} \tilde{\mathcal{W}}) + \text{tr}(\tilde{\mathcal{V}}^{\text{T}} \Gamma_2^{-1} \tilde{\mathcal{V}}) + \tilde{\alpha}^{\text{T}} \Gamma_3^{-1} \tilde{\alpha} \quad (8.29)$$

whose infinitesimal operator is given as follows:

$$\begin{aligned} &\mathscr{L} V_2(x(t), i) \\ &= x^{\text{T}}(t) \left(\text{He}(P_i A_i + P_i B_i \phi(u^A(t)) K_i) + \sum_{j=1}^{N} \lambda_{ij} P_j \right) x(t) \\ &+ w^{\text{T}}(t) D_i^T P_i x(t) + x^{\text{T}}(t) P_i D_i w(t) - 2x^{\text{T}}(t) P_i B_i \phi(u^A(t)) \rho_{si}(t) \\ &+ 2x^{\text{T}}(t) P_i \Delta A_i x(t) + 2x^{\text{T}}(t) P_i B_i \phi(u^A(t)) \Psi(x(t), t) \\ &+ 2\text{tr}(\tilde{\mathcal{W}}^{\text{T}} \Gamma_1^{-1} \dot{\tilde{\mathcal{W}}}) + 2\text{tr}(\tilde{\mathcal{V}}^{\text{T}} \Gamma_2^{-1} \dot{\tilde{\mathcal{V}}}) + 2\tilde{\alpha}^{\text{T}} \Gamma_3^{-1} \dot{\tilde{\alpha}} \end{aligned} \quad (8.30)$$

In virtue of (8.19)–(8.23), we have

$$\begin{aligned} &2x^{\text{T}}(t) P_i B_i \phi(u^A(t)) \Psi(x(t), t) - 2x^{\text{T}}(t) P_i B_i \phi(u^A(t)) \rho_{si}(t) \\ &\leq -2\rho_i(t) \|s(x(t), i)\| < 0 \end{aligned}$$

Then, we have for $\sigma_i > 0$

$$\begin{aligned} \mathscr{L} V_2(x(t), i) \leq x^{\text{T}}(t) \Big(&\text{He}(P_i \bar{A}_i(t)) + \sum_{j=1}^{N} \lambda_{ij} P_j + \sigma_i^{-1} P_i M_i M_i^T P_i \\ &+ \sigma_i N_i^{\text{T}} N_i \Big) x(t) + w^{\text{T}}(t) D_i^T P_i x(t) + x^{\text{T}}(t) P_i D_i w(t) \end{aligned} \quad (8.31)$$

Define the auxiliary function:

$$J_1(x(t), i) \triangleq \mathscr{L} V_2(x(t), i) - \varphi_i V_2(x(t), i) - \varphi_i w^{\text{T}}(t) w(t) \quad (8.32)$$

By means of Schur Complement (Lemma 1.3), the condition (8.26) implies $J_1(x(t), i) < 0$, meanwhile, one can get

$$\mathscr{L} V_2(x(t), i) < \varphi_i V_2(x(t), i) + \varphi_i w^{\text{T}}(t) w(t) \quad (8.33)$$

Taking the expression (8.33), multiplying both sides by $e^{-\varphi_i t}$, and integrating the result from 0 to t with $t \in [0, T^*]$, we have

$$\begin{aligned} e^{-\varphi_i t}\mathbb{E}\{V_2(x(t), r_0)\} &< \mathbb{E}\{V_2(x(0), i)\} + \varphi_i \int_0^t e^{-\varphi_i \tau} w^{\mathrm{T}}(\tau)w(\tau)\mathrm{d}\tau \\ &\leq \bar{\varpi}_P c_1 + \varphi_i \delta^2 T \end{aligned} \tag{8.34}$$

Furthermore, it follows from (8.29) that

$$e^{-\varphi_i t}\mathbb{E}\{V_2(x(t), i)\} \geq e^{-\varphi_i t}x^{\mathrm{T}}(t)P_i x(t) \geq e^{-\varphi_i T}\underline{\varpi}_P \mathbb{E}\{x^{\mathrm{T}}(t)Rx(t)\} \tag{8.35}$$

It yields from (8.34) and (8.35) that

$$\mathbb{E}\{x^{\mathrm{T}}(t)Rx(t)\} < \frac{\bar{\varpi}_P c_1 + \varphi_i \delta^2 T}{e^{-\varphi_i T}\underline{\varpi}_P} \tag{8.36}$$

By employing (8.28), it follows from (8.36) that $\mathbb{E}\{x^{\mathrm{T}}(t)Rx(t)\} < c^*$ for all $t \in [0, T^*]$. Therefore, as stated in Definition 1.5, the resulting closed-loop system (8.25) is SFTB with respect to $\aleph_{[0,T^*]}$, which completes the proof. □

►SFTB of sliding mode dynamics over sliding motion phase $[T^*, T]$

Presently, it will be analyzed that the sliding mode dynamics are SFTB within the interval $[T^*, T]$. As dictated by the SMC theory, when the state trajectory of the system remains within the sliding surface $s(x(t), i) = 0$, we have an equivalent controller as follows:

$$\begin{aligned} u_{eq}(t) = &- \phi^{-1}(u^A(t))(G_i B_i)^{-1}G_i(A_i + \Delta A_i(t))x(t) \\ &- \Psi(x(t), t) - \phi^{-1}(u^A(t))(G_i B_i)^{-1}G_i D_i w(t) \end{aligned} \tag{8.37}$$

Upon substituting (8.37) into the model (8.1), we get the following sliding mode dynamics:

$$\dot{x}(t) = \tilde{G}_i(A_i + \Delta A_i(t))x(t) + \tilde{G}_i D_i w(t) \tag{8.38}$$

where $\tilde{G}_i \triangleq I - B_i(G_i B_i)^{-1}G_i$.

Theorem 8.3 *For the given the parameters* (c_1, c_2, T, R, δ) *and feasible scalar* $\varphi_i > 0$, *consider the sliding mode dynamics* (8.38). *If there are constants* $\zeta_i > 0$, $\sigma_i > 0$, $c^* > 0$ *and matrices* $P_i > 0$ $(\forall i \in \mathbb{S})$ *satisfying*

$$\begin{bmatrix} \Omega_{11i} & \Omega_{12i} \\ * & \Omega_{22i} \end{bmatrix} < 0 \tag{8.39}$$

$$c_1 < c^* < c_2 \tag{8.40}$$

$$\frac{\bar{\varpi}_P c^* + \varphi_i \delta^2 T}{\underline{\varpi}_P} < e^{-\varphi_i T} c_2 \tag{8.41}$$

where

$$\Omega_{11i} \triangleq \begin{bmatrix} \mathrm{He}(P_iA_i) + A_i^T P_i A_i - \varphi_i P_i + \sum_{j=1}^N \lambda_{ij} P_j & P_i D_i \\ * & D_i^T P_i D_i - \varphi_i I \end{bmatrix}$$

$$\Omega_{12i} \triangleq \begin{bmatrix} P_iM_i & \sigma_i N_i^{\mathrm{T}} & A_i^T P_i M_i & \zeta_i N_i^{\mathrm{T}} & P_iB_i \\ 0 & 0 & 0 & 0 & 0 \end{bmatrix}$$

$$\Omega_{22i} \triangleq -\mathrm{diag}\{\sigma_i I, \sigma_i I, \zeta_i I - M_i^T P_i M_i, \zeta_i I, B_i^T P_i B_i\}$$

the sliding mode dynamics (8.38) *are SFTB with respect to* $\aleph_{[T^*,T]}$.

Proof Choose the following Lyapunov functional candidate:

$$V_3(x(t), i) = x^{\mathrm{T}}(t) P_i x(t) \tag{8.42}$$

We obtain the infinitesimal operator as

$$\begin{aligned}
&\mathscr{L}V_3(x(t), i) \\
&= x^{\mathrm{T}}(t)(\mathrm{He}(P_iA_i))x(t) + 2x^{\mathrm{T}}(t)P_i\Delta A_i(t)x(t) \\
&\quad - 2x^{\mathrm{T}}(t)P_iB_i(G_iB_i)^{-1}G_i\Delta A_i(t)x(t) - 2x^{\mathrm{T}}(t)P_iB_i(G_iB_i)^{-1}G_iA_i(t)x(t) \\
&\quad + w^{\mathrm{T}}(t)D_i^T P_i x(t) + x^{\mathrm{T}}(t)P_iD_iw(t) \\
&\quad - 2x^{\mathrm{T}}(t)P_iB_i(G_iB_i)^{-1}G_iD_iw(t) + x^{\mathrm{T}}(t)\sum_{j=1}^N \lambda_{ij}P_jx(t)
\end{aligned} \tag{8.43}$$

For $\zeta_i > 0$, one has

$$\begin{aligned}
&- 2x^{\mathrm{T}}(t)P_iB_i(B_i^TP_iB_i)^{-1}B_i^TP_i(A_i + \Delta A_i(t))x(t) \\
\leq &x^{\mathrm{T}}(t)\Big(P_iB_i(B_i^TP_iB_i)^{-1}B_i^TP_i + A_i^TP_iA_i + \zeta_iN_i^TN_i \\
&+ A_i^TP_iM_i(\zeta_iI - M_i^TP_iM_i)^{-1}M_i^{\mathrm{T}}P_iA_i\Big)x(t) \\
&- 2x^{\mathrm{T}}(t)P_iB_i(B_i^TP_iB_i)^{-1}B_i^TP_iD_iw(t) \\
\leq &x^{\mathrm{T}}(t)P_iB_i(B_i^TP_iB_i)^{-1}B_i^TP_ix(t) + w^{\mathrm{T}}(t)D_i^TP_iD_iw(t)
\end{aligned}$$

Define an auxiliary function as follows:

$$J_2(x(t), i) \triangleq \mathscr{L}V_3(x(t), i) - \varphi_iV_3(x(t), i) - \varphi_iw^{\mathrm{T}}(t)w(t) \tag{8.44}$$

The condition (8.39) implies $J_2(x(t), i) < 0$ by Schur Complement (Lemma 1.3). Then, we obtain

$$\mathscr{L}V_3(x(t), i) < \varphi_iV_3(x(t), i) + \varphi_iw^{\mathrm{T}}(t)w(t) \tag{8.45}$$

According to the result in the Theorem 8.2, one has $\mathbb{E}\{x^{\mathrm{T}}(T^*)Rx(T^*)\} < c^*$. Thus, by multiplying both sides of expression (8.45) by $e^{-\varphi_i t}$ and integrating the obtained inequality from T^* to t with $t \in [T^*, T]$, we have

$$
\begin{aligned}
e^{-\varphi_i t}\mathbb{E}\{V_3(x(t), i)\} <& e^{-\varphi_i T^*}\mathbb{E}\{V_3(x(T^*), r_{T^*})\} + \varphi_i \int_{T^*}^{t} e^{-\varphi_i \tau} w^{\mathrm{T}}(\tau)w(\tau)\mathrm{d}\tau \\
<& \bar{\varpi}_P \mathbb{E}\{x^{\mathrm{T}}(T^*)Rx(T^*)\} + \varphi_i \delta^2 T \\
<& \bar{\varpi}_P c^* + \varphi_i \delta^2 T
\end{aligned} \tag{8.46}
$$

In addition, it follows from (8.42) that

$$
e^{-\varphi_i t}\mathbb{E}\{V_3(x(t), i)\} \geq e^{-\varphi_i T} \underline{\varpi}_P \mathbb{E}\{x^{\mathrm{T}}(t)Rx(t)\} \tag{8.47}
$$

Putting together (8.46) and (8.47), we have

$$
\mathbb{E}\{x^{\mathrm{T}}(t)Rx(t)\} < \frac{\bar{\varpi}_P c^* + \varphi_i \delta^2 T}{e^{-\varphi_i T} \underline{\varpi}_P} \tag{8.48}
$$

Further, if there exists a scalar c^* satisfying (8.41), (8.27), we have from (8.48) that $\mathbb{E}\{x^{\mathrm{T}}(t)Rx(t)\} < c_2$ for all $t \in [T^*, T]$, which completes the proof. □

►SFTB over the whole phase $[0, T]$

It is worthy of pointing out that the control gain K_i should be designed to ensure the conditions in Theorems 8.2 and 8.3 to be feasible simultaneously. However, it is noted that the matrix inequalities in Theorems 8.2 and 8.3 are nonlinear, which are difficult to solve. Following this, we will put a sufficient condition formulated using the linear matrix inequalities (LMIs), which can certify that the requirements in Theorems 8.2 and 8.3 are met concurrently.

Theorem 8.4 *For the given the parameters* (c_1, c_2, T, R, δ) *and a feasible scalar* $\varphi_i > 0$, *consider the closed-loop system* (8.25). *If there are constants* $\sigma_i > 0$, $\zeta_i > 0$, $\kappa_i > 0$, $\theta_i > 0$, $c^* > 0$ *and matrices* $P_i > 0$ *and* $\mathcal{K}_i$ $(\forall i \in \mathbb{S})$ *satisfying*

$$
\begin{bmatrix} \bar{\Lambda}_{11i} & \Lambda_{12i} & \Lambda_{13i} & \Lambda_{14i} \\ * & \Lambda_{22i} & 0 & 0 \\ * & * & \Lambda_{33i} & 0 \\ * & * & * & \Lambda_{44i} \end{bmatrix} < 0 \tag{8.49}
$$

$$
c_1 < c^* < c_2 \tag{8.50}
$$

$$
\theta_i R < P_i < 2R \tag{8.51}
$$

$$
\begin{bmatrix} -e^{-\varphi_i T} c^* & \sqrt{2c_1 + \varphi_i \delta^2 T} \\ * & -\theta_i \end{bmatrix} < 0 \tag{8.52}
$$

$$
-e^{-\varphi_i T} c_2 \theta_i + 2c^* + \varphi_i \delta^2 T < 0 \tag{8.53}
$$

where

$$\bar{\Lambda}_{11i} \triangleq \begin{bmatrix} \mathrm{He}(P_iA_i) + A_i^T P_i A_i - \varphi_i P_i + \sum_{j=1}^{N} \lambda_{ij} P_j & P_i D_i \\ * & D_i^T P_i D_i - \varphi_i I \end{bmatrix}$$

$$\Lambda_{13i} \triangleq \begin{bmatrix} A_i^T P_i M_i & \zeta_i N_i^{\mathrm{T}} & P_i B_i \\ 0 & 0 & 0 \end{bmatrix}, \quad \Lambda_{33i} \triangleq -\mathrm{diag}\{\zeta_i I - M_i^T P_i M_i, \zeta_i I, B_i^T P_i B_i\}$$

$$\Lambda_{14i} \triangleq \begin{bmatrix} P_i B_i & \mathcal{K}_i^{\mathrm{T}} \\ 0 & 0 \end{bmatrix}, \quad \Lambda_{44i} \triangleq -\mathrm{diag}\{\kappa_i I, \kappa_i I\}$$

the closed-loop system (8.25) *is SFTB with respect to* $\aleph_{[0,T^*]}$ *over the whole finite-time interval* $[0, T]$ *and the sliding mode controller gain in* (8.9) *is given as* $K_i = \kappa_i^{-1}\mathcal{K}_i$.

Proof Simultaneously considering the conditions (8.26)–(8.28) along with the conditions (8.39)–(8.41), it is easily shown according to Schur Complement (Lemma 1.3) that conditions (8.26) and (8.39) can be ensured by (8.49). Then, according to condition (8.51), one has $\theta_i < \underline{\varpi}_P$, $\bar{\varpi}_P < 2$. Thus, it is easy to indicate that the inequalities (8.28) and (8.41) can be guaranteed by (8.52) and (8.53). This completes the proof.

□

8.6 Simulation

Consider an MJS with 4 modes as

$$A_1 = \begin{bmatrix} -1.2 & 0.8 \\ 0.5 & -2.4 \end{bmatrix}, B_1 = \begin{bmatrix} 1.4 \\ 2.5 \end{bmatrix}, D_1 = \begin{bmatrix} 0.03 \\ 0.01 \end{bmatrix}, M_1 = \begin{bmatrix} 0.01 \\ 0.02 \end{bmatrix}, N_1 = \begin{bmatrix} 0.3 \\ 0 \end{bmatrix}^{\mathrm{T}}$$

$$A_2 = \begin{bmatrix} -1.0 & 0.7 \\ 0.6 & -1.5 \end{bmatrix}, B_2 = \begin{bmatrix} 1.5 \\ 1.4 \end{bmatrix}, D_2 = \begin{bmatrix} 0.02 \\ 0.03 \end{bmatrix}, M_2 = \begin{bmatrix} 0.02 \\ 0.01 \end{bmatrix}, N_2 = \begin{bmatrix} 0 \\ 0.2 \end{bmatrix}^{\mathrm{T}}$$

$$A_3 = \begin{bmatrix} -1.2 & 0.5 \\ 0.8 & -0.6 \end{bmatrix}, B_3 = \begin{bmatrix} 0.5 \\ 1.4 \end{bmatrix}, D_3 = \begin{bmatrix} 0.01 \\ 0.02 \end{bmatrix}, M_3 = \begin{bmatrix} 0.01 \\ 0.01 \end{bmatrix}, N_3 = \begin{bmatrix} 0.2 \\ 0 \end{bmatrix}^{\mathrm{T}}$$

$$A_4 = \begin{bmatrix} -1.4 & 0 \\ 0.6 & -2.4 \end{bmatrix}, B_4 = \begin{bmatrix} 1.8 \\ 2.1 \end{bmatrix}, D_4 = \begin{bmatrix} 0.02 \\ 0.01 \end{bmatrix}, M_4 = \begin{bmatrix} 0.01 \\ 0.02 \end{bmatrix}, N_4 = \begin{bmatrix} 0 \\ 0.3 \end{bmatrix}^{\mathrm{T}}$$

$$F_1(t) = 0.3\sin(t), F_2(t) = 0.4\sin(t), F_3(t) = 0.3\cos(t), F_4(t) = 0.4\cos(t)$$

whoes TR matrix is given as

$$\Lambda = \begin{bmatrix} -1.5 & 0.2 & 0.5 & 0.8 \\ 0.4 & -0.8 & 0.1 & 0.3 \\ 0.4 & 1.2 & -2.5 & 0.9 \\ 0.1 & 0.3 & 0.5 & -0.9 \end{bmatrix}$$

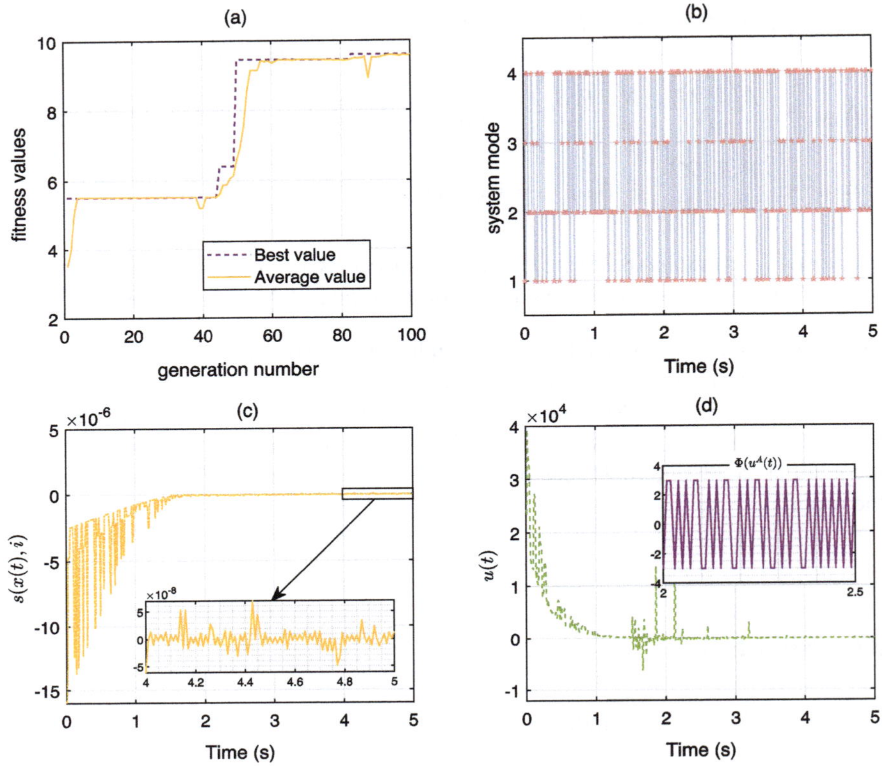

Fig. 8.2 Simulation results of the controlled system: **a** fitness values of each generation in GA; **b** possible mode transition sequence r_t; **c** sliding variable $s(x(t), i)$; **d** control signal $u(t)$

Let $T = 5$, $c_1 = 1.01$, $c_2 = 15$, $R = 10^{-7} I_2$. In this example, we set an optimization problem (**OP**) as follows:

$$\begin{aligned} \textbf{OP} &: \min_{\mathbb{G}} C \\ &\text{subject to } (8.49) - (8.53) \end{aligned}$$

with $\mathbb{G} \triangleq \{\sigma_i,\ \zeta_i,\ \kappa_i,\ \theta_i,\ c^*,\ P_i,\ \mathcal{K}_i\}$ and then adopt genetic algorithm (GA) as in [12, 13] to solve this **OP**. After 100-generation evolution, the best fitness values are shown in Fig. 8.2a, and the controller gains are

$$\begin{aligned} K_1 &= \begin{bmatrix} 0.0260 & -0.0002 \end{bmatrix}, \quad K_2 = \begin{bmatrix} -0.0460 & -0.0413 \end{bmatrix} \\ K_3 &= \begin{bmatrix} -0.0299 & -0.0225 \end{bmatrix}, \quad K_4 = \begin{bmatrix} 0.0241 & 0.0593 \end{bmatrix} \end{aligned}$$

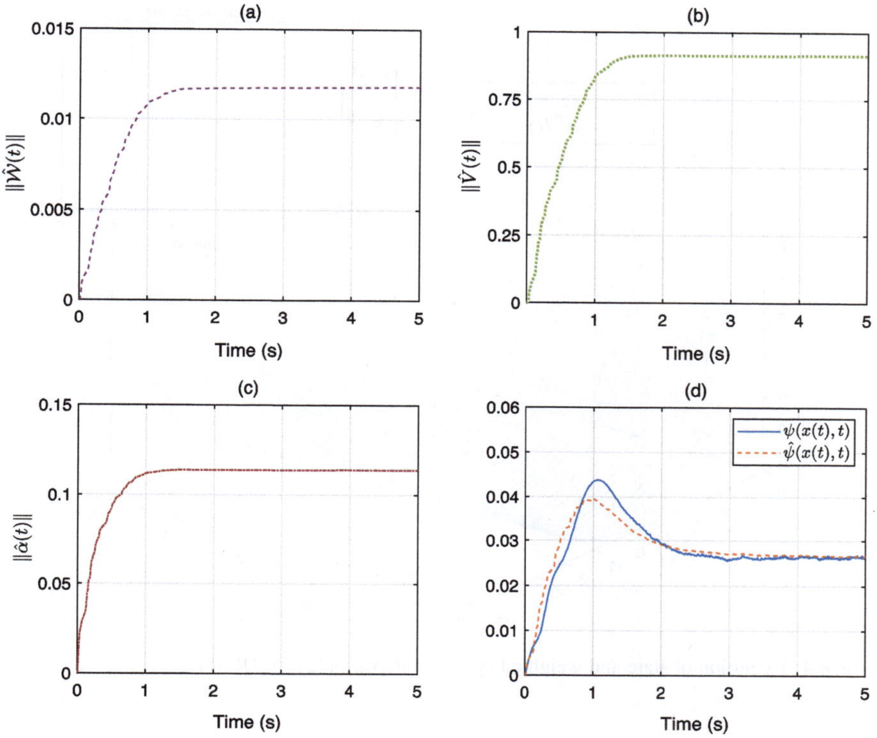

Fig. 8.3 Estimated attack nonlinear bound: **a** adaptive parameter $\|\hat{\mathcal{W}}(t)\|$; **b** adaptive parameter $\|\hat{\mathcal{V}}(t)\|$; **c** adaptive parameter $\|\hat{\alpha}(t)\|$; **d** attack bound and its estimation

For given initial state $x(0) = \begin{bmatrix}-40 & 30\end{bmatrix}^{\mathrm{T}}$, the simulation results are exhibited in Figs. 8.3 and 8.4. Figure 8.2b depicts the mode sequence in one trial, from which it can be seen that system mode jumps among 4 subsystems. Figure 8.2c and Fig. 8.2d, respectively, show the sliding variable and control signal, and especially, we can find that the attacked control input is always constrained in the given saturation bounds $[-3, 3]$. Figure 8.3 provides the evolution of adaptive parameters. From Fig. 8.3d, it is shown that the proposed neural network estimation approach is effective in approximating the attack nonlinear bound. In Fig. 8.4, by operating the neural network algorithm and designed controller, the closed-loop system is attained to be SFTB, which further illustrates the effectiveness of the presented neural network security control method for countering actuator saturation and injection attack.

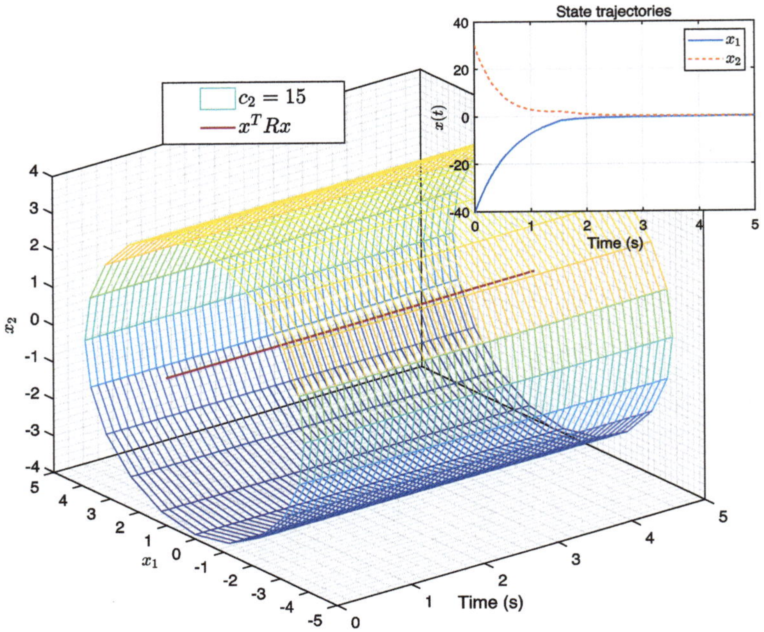

Fig. 8.4 Evolution of state and weighted system state variable $x^{\mathrm{T}}(t)Rx(t)$

8.7 Conclusion

A security control problem has been investigated in this chapter for MJSs subject to actuator saturation and injection attack. A neural network approximation technique is used to estimate the attack nonlinear bound, and the adaptive rules of neural network parameters are designed. Based on this, a sliding mode controller is constructed to ensure the reachability of the specified sliding surface within given finite time T. By analyzing the SFTB of the arrival and sliding phases, respectively, the solving conditions are obtained to ensure the SFTB of the closed-loop system over the whole finite phase despite actuator saturation.

References

1. Deng, Y., Lu, H., Zhou, W.: Security event-triggered control for Markovian jump neural networks against actuator saturation and hybrid cyber attacks. J. Frankl. Inst. **358**(14), 7096–7118 (2021)
2. Zhu, Z., Xia, Y., Fu, M.: Adaptive sliding mode control for attitude stabilization with actuator saturation. IEEE Trans. Ind. Electron. **58**(10), 4898–4907 (2011)
3. Li, H., Wang, J., Shi, P.: Output-feedback based sliding mode control for fuzzy systems with actuator saturation. IEEE Trans. Fuzzy Syst. **24**(6), 1282–1293 (2016)
4. Cao, Z., Niu, Y., Song, J.: Finite-time sliding-mode control of Markovian jump cyber-physical systems against randomly occurring injection attacks. IEEE Trans. Autom. Control **65**(3), 1264–1271 (2019)
5. Zhang, Z., Niu, Y., Song, J.: Input-to-state stabilization of interval type-2 fuzzy systems subject to cyberattacks: an observer-based adaptive sliding mode approach. IEEE Trans. Fuzzy Syst. **28**(1), 190–203 (2020)
6. Ao, W., Song, Y., Wen, C.: Adaptive cyber-physical system attack detection and reconstruction with application to power systems. IET Control Theory Appl. **10**(12), 1458–1468 (2016)
7. An, L., Yang, G.-H.: Improved adaptive resilient control against sensor and actuator attacks. Inf. Sci. **423**, 145–156 (2018)
8. Jin, X., Haddad, W.M., Yucelen, T.: An adaptive control architecture for mitigating sensor and actuator attacks in cyber-physical systems. IEEE Trans. Autom. Control **62**(11), 6058–6064 (2017)
9. Song, J., Niu, Y., Zou, Y.: Asynchronous output feedback control of time-varying Markovian jump systems within a finite-time interval. J. Frankl. Inst. **354**(15), 6747–6765 (2017)
10. Skorokhod, A.V.: Asymptotic methods in the theory of stochastic differential equation. Am. Math. Soc. (1989)
11. Hu, Q.: Robust adaptive sliding mode attitude maneuvering and vibration damping of three-axis-stabilized flexible spacecraft with actuator saturation limits. Nonlinear Dyn. **55**(4), 301–321 (2009)
12. Song, J., Niu, Y.: Co-design of 2-D event generator and sliding mode controller for 2-D Roesser model via genetic algorithm. IEEE Trans. Cybern. **51**(9), 4581–4590 (2020)
13. Chen, B., Niu, Y., Liu, H.: Input-to-state stabilization of stochastic Markovian jump systems under communication constraints: genetic algorithm-based performance optimization. IEEE Trans. Cybern. **52**(10), 10379–10392 (2021)

Chapter 9
Hybrid Sliding Mode Control Scheme with Transition Rate Synthesis

9.1 Introduction

As a special class of stochastic systems, continuous-time Markov jump systems (MJSs) feature the transition rates (TRs) that govern the rule of the subsystem switching. Obviously, the TRs play key roles in the steady-state and dynamic performances of MJSs. For example, an MJS can be stabilized by an appropriate TR matrix under the case that no subsystems are stable. However, in the existing research, there exists a potential assumption that the TRs are priori whether fully or partially known. Based on the discussion above, the TRs can regulate switching/jumping among subsystems and affect the stability of the whole MJS. Thus, an interesting and meaningful issue is that the engineers might choose/design an appropriate TR matrix to ensure the desirable performance of MJSs. Pioneering research on TR synthesis has been investigated in [1], in which the means-quare stability of the forced system was ensured by the hybrid-designed controller and transition rates. Motivated by [1], some TR synthesis methods for MJSs have been developed to realize the stochastic stability [2] and exponential mean-square stability [3]. However, to the authors' best knowledge, the problem of synthesizing sliding mode control (SMC) and the TR matrix simultaneously has not received attention yet, especially, under the requirement of stochastic finite-time boundedness (SFTB) performance. This is the primary motivation of the present chapter.

On the other hand, in many practical systems, including power systems, image processing systems, and mechanical systems, the control design process is often constrained by the physical structure and the requirement for safe operation, inevitably leading to saturation phenomena [4–7]. From the perspective of components that occurs saturations, saturation phenomena can be divided into actuator saturation [4], state saturation [5, 6], and sensor saturation [7], and of course, multiple variables may saturate simultaneously. Among them, actuator saturation is the most common type of non-linear constraint, and it has strong interference and destructive effects. The

Z. Cao and Y. Niu, *Finite-Time Sliding Mode Control for Markov Jump Systems*, Studies in Systems, Decision and Control 603,
https://doi.org/10.1007/978-981-96-7651-4_9

deviation between the control signal and the actuator signal results in the deterioration of certain performance of the closed-loop system, and may even lead to system instability, causing irreversible effects and losses in practical engineering. Therefore, providing appropriate control strategies for handling the actuator saturation constraints is of great theoretical and practical significance.

Currently, there are two main methods for dealing with actuator saturations. One approach is to consistently consider the saturation characteristics during the controller design process, ensuring that the system performance can either meet or withstand the constraints of actuator saturation, thereby guaranteeing system stability [8]. Another method is to design the controller, ensuring that the closed-loop system maintains good performance during the unsaturated stage of the actuator. When saturation occurs, an anti-windup compensator is designed to input the difference between the control signal and the actuator signal back into the controller, indirectly handling the constraints of saturation.

It is worth noting that, although many significant achievements have been made in the SMC design of MJSs under actuator saturations, the control objectives of existing research works have focused on ensuring the asymptotic stability of closed-loop systems. Deficiently, there has been a lack of research on how to design desired finite-time sliding mode controllers for facilitating the closed-loop systems to achieve specified finite-time transient performance in the presence of actuator saturation, especially considering TR design. This further motivates this chapter.

In this chapter, we shall investigate the SFTB control problem of uncertain MJSs by designing a hybrid control scheme involving in TR matrix and SMC law. Apparently, this will be a challenging task due to the following three essential difficulties:

Q1. For the unforced MJSs, how to design the TR matrix to ensure the SFTB of the unforced systems?

Q2. For a given short time interval $[0, T]$, how to ensure the reachability of the sliding surface during the interval $[0, T^*]$ with $T^* < T$?

Q3. How to synthesize a hybrid control scheme especially subject to actuator saturation, in which both the TR matrix and SMC law are to be designed, to guarantee the SFTB of the resultant closed-loop MJSs over the finite-time interval composed of reaching phase and sliding motion phase?

The present method can not only provide the freedom for engineers to design appropriate TR matrices, but also take the *the transient performance* of the controlled MJSs into consideration. This chapter highlights the necessity for robust control strategies capable of accommodating TRs, thereby enhancing the reliability and performance of MJSs in practical applications. The exploration of co-design methodologies not only contributes to theoretical advancements but also offers practical solutions to real-world challenges faced by systems characterized by stochastic transitions.

9.2 TR Matrix Design

Consider a *unforced* MJS defined on a complete probability space $(\Omega, \mathscr{F}, \Pr)$ as in system (1.6) with $u(t) \equiv 0$, in which the parameter uncertainty $\Delta A_i(t)$, matched nonlinear perturbation $f_i(x(t))$ and unmatched external disturbance $w(t)$ satisfy Assumptions 1.2–1.4, respectively. The transition probabilities (TPs) are generated by (1.7). In this section, the SFTB problem for the *unforced* MJS will be considered by designing the TR matrix.

9.2.1 SFTB Conditions with Specified TR Matrix

First, we give a sufficient condition on the SFTB of the unforced MJSs (1.6) ($u(t) \equiv 0$) with a specified TR matrix Λ.

Lemma 9.1 *Given the scalars $\varphi_i > 0$ and TR matrix $\Lambda = [\lambda_{ij}]_{N\times N}$ ($i, j \in \mathbb{S}$), if there exist matrices $P_i > 0$, scalars $\gamma_i > 0$, $\zeta_i > 0$ ($i \in \mathbb{S}$) satisfying*

$$\begin{bmatrix} \mho_i & P_iD_i & P_iM_i & P_iB_i \\ * & -\varphi_i I & 0 & 0 \\ * & * & -\gamma_i I & 0 \\ * & * & * & -\zeta_i I \end{bmatrix} < 0, \tag{9.1}$$

$$\bar{\sigma}_P c_1 + \delta^2 T \varphi_i < e^{-\varphi_i T} \underline{\sigma}_P c_2, \tag{9.2}$$

where

$$\mho_i \triangleq \mathrm{He}(P_iA_i) + \gamma_i N_i^{\mathrm{T}} N_i + \zeta_i \epsilon_i^2 I + \sum_{j=1}^{N} \lambda_{ij} P_j - \varphi_i P_i, \tag{9.3}$$

$$\bar{\sigma}_P \triangleq \max_{i\in\mathbb{S}}\{\lambda_{\max}(R^{-\frac{1}{2}} P_i R^{-\frac{1}{2}})\},\ \underline{\sigma}_P \triangleq \min_{i\in\mathbb{S}}\{\lambda_{\min}(R^{-\frac{1}{2}} P_i R^{-\frac{1}{2}})\}, \tag{9.4}$$

the unforced MJS (1.6) ($u(t) \equiv 0$) is SFTB with respect to $\aleph_{[0,T]}$.

Proof Choose the following Lyapunov function:

$$V(x(t), i) = x^{\mathrm{T}}(t) P_i x(t). \tag{9.5}$$

For the unforced MJS (1.6) ($u(t) \equiv 0$), the infinitesimal operator of function $V(x(t), i)$ from Definition 1.4 is given as

$$\mathscr{L}V(x(t), i) = x^{\mathrm{T}}(t)\Big(P_iA_i + A_i^{\mathrm{T}}P_i + \sum_{j=1}^{N} \lambda_{ij} P_j\Big)x(t) + 2x^{\mathrm{T}}(t) P_i B_i f_i(x(t))$$

$$+ 2x^{\mathrm{T}}(t)P_i\Delta A_i(t)x(t) + x^{\mathrm{T}}(t)P_iD_iw(t) + w^{\mathrm{T}}(t)D_i^{\mathrm{T}}P_ix(t). \tag{9.6}$$

For $\gamma_i > 0$ and $\zeta_i > 0$, one can obtain

$$\begin{aligned}\mathscr{L}V(x(t), i) \leq x^{\mathrm{T}}(t)\Big(&P_iA_i + A_i^{\mathrm{T}}P_i + \sum_{j=1}^{N}\lambda_{ij}P_j + \gamma_i^{-1}P_iM_iM_i^{\mathrm{T}}P_i \\ &+ \gamma_iN_i^{\mathrm{T}}N_i + \zeta_i\epsilon_i^2 I + \zeta_i^{-1}P_iB_iB_i^{\mathrm{T}}P_i\Big)x(t) \\ &+ x^{\mathrm{T}}(t)P_iD_iw(t) + w^{\mathrm{T}}(t)D_i^{\mathrm{T}}P_ix(t).\end{aligned} \tag{9.7}$$

Define an auxiliary function as follows:

$$J(x(t), i) = \mathscr{L}V(x(t), i) - \varphi_iV(x(t), i) - \varphi_iw^{\mathrm{T}}(t)w(t). \tag{9.8}$$

Apparently, $J(x(t), i) < 0$ can be ensured by (9.1). Thus, it follows from $J(x(t), i) < 0$ that

$$\mathscr{L}V(x(t), i) < \varphi_iV(x(t), i) + \varphi_iw^{\mathrm{T}}(t)w(t). \tag{9.9}$$

Multiplying both sides of expression (9.9) by $e^{-\varphi_it}$ and integrating the obtained inequality from 0 to t with $t \in [0, T]$, we have

$$\begin{aligned}e^{-\varphi_it}\mathbb{E}\{V(x(t), i)\} &< V(x(0), r_0) + \varphi_i\int_0^t e^{-\varphi_i\tau}w^{\mathrm{T}}(\tau)w(\tau)\mathrm{d}\tau \\ &\leq \bar{\sigma}_Pc_1 + \delta^2T\varphi_i.\end{aligned} \tag{9.10}$$

On the other hand, it follows from (9.5) that

$$\begin{aligned}e^{-\varphi_it}\mathbb{E}\{V(x(t), i)\} &= e^{-\varphi_it}\mathbb{E}\{x^{\mathrm{T}}(t)P_ix(t)\} \\ &\geq e^{-\varphi_iT}\underline{\sigma}_P\mathbb{E}\{x^{\mathrm{T}}(t)Rx(t)\}.\end{aligned} \tag{9.11}$$

Combining (9.10) and (9.11) derives that

$$\mathbb{E}\{x^{\mathrm{T}}(t)Rx(t)\} < \frac{\bar{\sigma}_Pc_1 + \delta^2T\varphi_i}{e^{-\varphi_iT}\underline{\sigma}_P}. \tag{9.12}$$

It yields from (9.2) and (9.12) that $\mathbb{E}\{x^{\mathrm{T}}(t)Rx(t)\} < c_2$ for all $t \in [0, T]$. Thus, according to Definition 1.5, the unforced MJS (1.6) ($u(t) \equiv 0$) is SFTB with respect to $\aleph_{[0,T]}$. This finishes the proof. □

It is noted from (9.4) that the inequality (9.2) in Lemma 9.1 is infeasible due to $\bar{\sigma}_P$ and $\underline{\sigma}_P$. To this end, a sufficient condition will be further given in the following lemma to ensure the unforced MJS (1.6) ($u(t) \equiv 0$) to be SFTB.

Lemma 9.2 *Given scalars $\varphi_i > 0$, $\varpi_i > 0$ and the TR matrix $\Lambda = [\lambda_{ij}]_{N\times N}$ $(i, j \in \mathbb{S})$, if there exist matrices $P_i > 0$, scalars $\vartheta_i > 0$, $\gamma_i > 0$, $\zeta_i > 0$ $(i \in \mathbb{S})$ satisfying the following linear matrix inequalities (LMIs):*

$$
\begin{bmatrix} \mho_i & P_iD_i & P_iM_i & P_iB_i \\ * & -\varphi_i I & 0 & 0 \\ * & * & -\gamma_i I & 0 \\ * & * & * & -\zeta_i I \end{bmatrix} < 0, \tag{9.13}
$$

$$
\vartheta_i R < P_i < \varpi_i R, \tag{9.14}
$$

$$
-e^{-\varphi_i T} c_2 \vartheta_i + \varpi_i c_1 + \delta^2 T \varphi_i < 0, \tag{9.15}
$$

where $\mho_i$ is defined as (9.3) in Lemma 9.1, the unforced MJS (1.6) ($u(t) \equiv 0$) is SFTB with respect to $\aleph_{[0,T]}$.

Proof It is easily seen that the inequality (9.13) is equivalent to (9.1) in Lemma 9.1. On the other hand, if the condition (9.14) holds, one has $\vartheta_i < \underline{\sigma}_P$, $\bar{\sigma}_P < \varpi_i$. Then, the inequality (9.2) can be ensured by (9.15). Hence, the conditions (9.1)–(9.2) can be ensured by the conditions (9.13)–(9.15). □

Lemma 9.2 proposes a feasible SFTB condition based on LMIs for the unforced MJS (1.6) ($u(t) \equiv 0$), which will be used in Simulation part later to check whether the open-loop MJS with a given TR matrix is SFTB.

9.2.2 Existing Conditions of TR Matrix

Lemma 9.1 provides an SFTB condition based on the given TR matrix Λ. In the sequel, a sufficient condition on the existence of the TR matrix Λ will be further given to ensure that SFTB of the unforced MJS (1.6) ($u(t) \equiv 0$).

Theorem 9.1 *Given scalars $\varphi_i > 0$, $\varpi_i > 0$ and $\nu > 0$, if there exist matrices $P_i > 0$, scalars $\vartheta_i > 0$, $\gamma_i > 0$, $\zeta_i > 0$, and a set of scalars $\bar{\lambda}_{ij} > 0$ $(i, j \in \mathbb{S}, i \neq j)$ satisfying*

$$
\begin{bmatrix} \bar{\mho}_i & P_iD_i & P_iM_i & P_iB_i & \Gamma_i \\ * & -\varphi_i I & 0 & 0 & 0 \\ * & * & -\gamma_i I & 0 & 0 \\ * & * & * & -\zeta_i I & 0 \\ * & * & * & * & -\Theta_i \end{bmatrix} < 0, \tag{9.16}
$$

$$
\vartheta_i R < P_i < \varpi_i R, \tag{9.17}
$$

$$- e^{-\varphi_i T} c_2 \vartheta_i + \varpi_i c_1 + \delta^2 T \varphi_i < 0, \tag{9.18}$$

where

$$\begin{aligned}
\bar{\mho}_i &\triangleq \mathrm{He}(P_i A_i) + \gamma_i N_i^{\mathrm{T}} N_i + \zeta_i \epsilon_i^2 I - \varphi_i P_i \\
&\quad + \nu^2 \sum_{j=1, j \neq i}^{N} \varepsilon_j I + 2\nu (N-1) P_i - 2\nu \sum_{j=1, j \neq i}^{N} P_j, \\
\Gamma_i &\triangleq [P_1 - P_i + \bar{\lambda}_{i1} I \ \cdots \ P_{i-1} - P_i + \bar{\lambda}_{i(i-1)} I \\
&\qquad P_{i+1} - P_i + \bar{\lambda}_{i(i+1)} I \ \cdots \ P_N - P_i + \bar{\lambda}_{iN} I], \\
\Theta_i &\triangleq \mathrm{diag}\{\varepsilon_1 I, \ldots, \varepsilon_{j-1} I, \varepsilon_{j+1} I, \ldots, \varepsilon_N I\},
\end{aligned}$$

the unforced MJS (1.6) ($u(t) \equiv 0$) is SFTB with respect to $\aleph_{[0,T]}$. In this case, $\bar{\Lambda} \triangleq [\bar{\lambda}_{ij}]_{N \times N}$ with $\bar{\lambda}_{ii} \triangleq -\varepsilon_i \sum_{j=1, j \neq i}^{N} \varepsilon_j^{-1} \bar{\lambda}_{ij}$. Thus, the designed TR matrix is given as $\Lambda = 2\bar{\Lambda}\Xi^{-1}$ with $\Xi \triangleq \mathrm{diag}\{\varepsilon_1, \varepsilon_2, \ldots, \varepsilon_N\}$.

Proof According to Lemma 9.1, if conditions (9.1)–(9.2) are solvable, the SFTB of the unforced MJS (1.6) ($u(t) \equiv 0$) can be attained. Next, it will be shown that conditions (9.1)–(9.2) can be guaranteed by conditions (9.16)–(9.18). Thus, when conditions (9.16)–(9.18) hold, the unforced MJS (1.6) ($u(t) \equiv 0$) is SFTB.

From (9.1), we have

$$\mho_i = P_i A_i + A_i^{\mathrm{T}} P_i + \gamma_i N_i^{\mathrm{T}} N_i + \zeta_i \epsilon_i^2 I + \sum_{j=1}^{N} \lambda_{ij} P_j - \varphi_i P_i. \tag{9.19}$$

Noticing the fact, $\lambda_{ii} = -\sum_{j=1, j \neq i}^{N} \lambda_{ij}$, it is easily seen that $\mho_i < \hat{\mho}_i$ for $\varepsilon_i > 0$, where

$$\begin{aligned}
\hat{\mho}_i &\triangleq P_i A_i + A_i^{\mathrm{T}} P_i + \sum_{j=1, j \neq i}^{N} \lambda_{ij} (P_j - P_i) + \gamma_i N_i^{\mathrm{T}} N_i \\
&\quad + \zeta_i \epsilon_i^2 I - \varphi_i P_i + \sum_{j=1, j \neq i}^{N} \frac{\varepsilon_j \lambda_{ij}^2}{4} \\
&= P_i A_i + A_i^{\mathrm{T}} P_i + \gamma_i N_i^{\mathrm{T}} N_i + \zeta_i \epsilon_i^2 I - \varphi_i P_i \\
&\quad + \sum_{j=1, j \neq i}^{N} (P_j - P_i + \bar{\lambda}_{ij})^{\mathrm{T}} \varepsilon_i^{-1} I (P_j - P_i + \bar{\lambda}_{ij}) \\
&\quad - \sum_{j=1, j \neq i}^{N} \varepsilon_j^{-1} (P_j - P_i)(P_j - P_i)
\end{aligned} \tag{9.20}$$

with $\bar{\lambda}_{ij} \triangleq \frac{\varepsilon_j}{2} \lambda_{ij}$.

Noting $\varepsilon_j > 0$, we have for an adjustable scalar $\nu > 0$ and $\forall i, j \in \mathbb{S}$,

$$\begin{aligned}&[(P_j - P_i) - \nu\varepsilon_j I](\varepsilon_j^{-1} I)[(P_j - P_i) - \nu\varepsilon_j I]\\ &= \varepsilon_j^{-1}(P_j - P_i)(P_j - P_i) - 2\nu(P_j - P_i) + \nu^2\varepsilon_j I \geq 0,\end{aligned} \tag{9.21}$$

which gives

$$-\varepsilon_j^{-1}(P_j - P_i)(P_j - P_i) \leq -2\nu(P_j - P_i) + \nu^2\varepsilon_j I.$$

Then, it yields from (9.20) and (9.21) that $\hat{\mho}_i \leq \tilde{\mho}_i$, where

$$\begin{aligned}\tilde{\mho}_i \triangleq\ & P_i A_i + A_i^{\mathrm{T}} P_i + \gamma_i N_i^{\mathrm{T}} N_i + \zeta_i \epsilon_i^2 I - \varphi_i P_i\\ &+ \sum_{j=1, j\neq i}^{N} (P_j - P_i + \bar{\lambda}_{ij})^{\mathrm{T}} \varepsilon_i^{-1} I (P_j - P_i + \bar{\lambda}_{ij})\\ &+ \nu^2 \sum_{j=1, j\neq i}^{N} \varepsilon_j I + 2\nu(N-1)P_i - 2\nu \sum_{j=1, j\neq i}^{N} P_j.\end{aligned} \tag{9.22}$$

Thus, if the following inequality holds, the condition (9.1) holds:

$$\begin{bmatrix} \tilde{\mho}_i & P_i D_i & P_i M_i & P_i B_i \\ * & -\varphi_i I & 0 & 0 \\ * & * & -\gamma_i I & 0 \\ * & * & * & -\zeta_i I \end{bmatrix} < 0, \tag{9.23}$$

which can be ensured by (9.16), that is, the condition (9.1) be ensured by the condition (9.16).

On the other hand, by the condition (9.17), one has $\vartheta_i < \underline{\sigma}_P$, $\bar{\sigma}_P < \varpi_i$. Thus, it is easily shown that the inequality (9.2) can be ensured by (9.18). This completes the proof. □

Theorem 9.1 provides a design scheme on the TR matrix to achieve the SFTB of the *unforced* MJS (1.6) ($u(t) \equiv 0$), which answers **Q1**.

Remark 9.1 The key idea of the common parameter approach in [1] is to introduce a sufficiently small mode-independent scalar $\varepsilon > 0$ into some coupled Lyapunov inequalities. It is noted that the conditions obtained in [1] are solvable only when the parameter ε satisfies all modes. That is, there must exist a parameter ε satisfying the obtained conditions in [1] for all modes $i \in \mathbb{S}$, then, the concerned problem only become feasible, which brings great difficulty in solving the parameter ε. In order to increase the freedom of solution, a set of mode-dependent sufficiently small scalars $\varepsilon_i > 0$ is introduced in the inequality (9.19), which can find the feasible solutions more effectively than the common parameter approach in [1, 2].

Remark 9.2 In this work, the problem of finding N^2 scalars λ_{ij} for $i, j \in \mathbb{S}$ with a constraint condition $\sum_{j=1}^{N} \lambda_{ij} = 0$ is changed into the problem of determining $N^2 - N$ scalars λ_{ij} for $i \neq j, i, j \in \mathbb{S}$. Thus, by calculating $\lambda_{ii} = -\sum_{j=1, j\neq i}^{N} \lambda_{ij}$, all λ_{ij} for $i, j \in \mathbb{S}$ can be obtained. This is quite different from the work [2], in which N^2 scalars λ_{ij} need to be solved. Moreover, the conditions in Theorem 9.1 contain the performance information of TRs in MJSs directly and explicitly, which was not reflected in corresponding conditions in [2].

9.2.3 *Two-Dimensional Search Algorithm*

Notice that in the matrix inequality (9.16), the decision variable ν is exploited to reduce conservatism by introducing two extra degrees of freedom. By applying the following two-dimensional search over $\nu > 0$ and $\varpi_i > 0$, the feasibility of LMIs (9.16)–(9.18) can be checked.

Algorithm 9.1 Searching parameters ν and ϖ_i simultaneously

Step 1. Give the scalar $\nu > 0$ (for example $\nu = 0.001$). By a linear searching over $(0.1, 100)$, if at least a set of values $\{\varpi_i, i \in \mathbb{S}\}$ is found such that LMIs (9.16)–(9.18) are feasible, go to Step 2; else, modify the scalar ν (for example $\nu = \nu + 0.001$), and then, restart the above search for ϖ_i.

Step 2. Calculate $\bar{\lambda}_{ii}$ by $\bar{\lambda}_{ii} = -\varepsilon_i \sum_{j=1, j\neq i}^{N} \varepsilon_j^{-1} \bar{\lambda}_{ij}$. The designed TR matrix is given as $\Lambda = 2\bar{\Lambda}\Xi^{-1}$, where Ξ is defined in Theorem 9.1.

Step 3. The TR matrix Λ will be obtained.

9.2.4 *Simulation Example 1*

This example is given to verify a possible case that a *unforced* MJS with a given TR matrix may not be SFTB, but a designed TR matrix via Theorem 9.1 can ensure the SFTB of the *unforced* MJS. Consider MJS (1.6) ($u(t) \equiv 0$) containing two subsystems with

$$A_1 = \begin{bmatrix} -1 & 1 \\ 3 & -5 \end{bmatrix}, \; A_2 = \begin{bmatrix} -1 & 0 \\ 0 & -2 \end{bmatrix}, \; B_1 = \begin{bmatrix} 1 \\ -1 \end{bmatrix},$$
$$B_2 = \begin{bmatrix} -2 \\ -3 \end{bmatrix}, \; D_1 = \begin{bmatrix} 0.1 \\ 0.2 \end{bmatrix}, \; D_2 = \begin{bmatrix} 0.1 \\ 0.3 \end{bmatrix},$$

$f_i(x(t)) = \epsilon_i \sin(\|x(t)\|)$ with $\epsilon_1 = 0.2$ and $\epsilon_2 = 0.3$, and the external disturbance satisfy

$$w(t) = \begin{cases} 0.5\sin^2(0.84t)\cos(t), & 0 \le t < \frac{11}{12}s, \\ 0, & \frac{11}{12}s \le t < 5\,s. \end{cases}$$

Obviously, $\delta = 0.5$.

► **Case 1** For the given TR matrix as $\Lambda = \begin{bmatrix} -2 & 2 \\ 23 & -23 \end{bmatrix}$, there are no feasible solution to ensure the SFTB of the considered system.

► **Case 2** Here, we shall design the TR matrix Λ according to Theorem 9.1. By means of Algorithm 9.1, when setting the adjustable scalar $\nu_1 = 0.001$, we can obtain the TR matrix as $\Lambda = \begin{bmatrix} -0.0091 & 0.0091 \\ 0.0095 & -0.0095 \end{bmatrix}$ with $\varpi_1 = \varpi_2 = 0.07$. For the initial state $x(0) = [-0.5\ \ 0.2]^{\mathrm{T}}$, given the finite time $T = 5$, parameters $c_1 = 1.05$, $c_2 = 5.05$, weight matrix $R = \mathrm{diag}\{4, 1\}$.

Both the simulation results in Cases 1.1 and 1.2 are shown in Fig. 9.1, from which it is exhibited that the given TR matrix cannot ensure the SFTB of the unforced MJS, while the designed TR matrix via Lemma 9.1 can stabilize the unforced MJS to be SFTB.

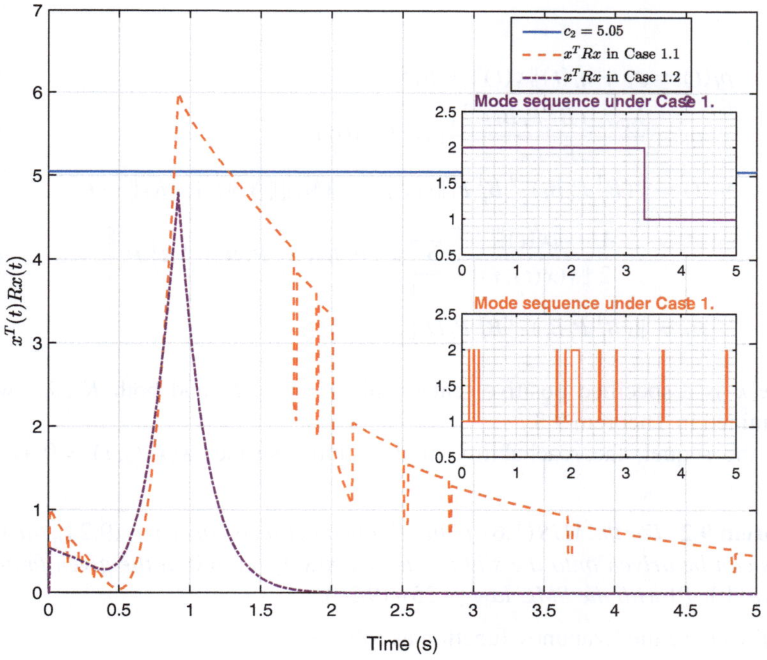

Fig. 9.1 Simulation results in Example 1

9.3 Hybrid Control Scheme

In this section, we further consider the hybrid design problem of sliding mode controller $u(t)$ and a corresponding TR matrix $\Lambda = [\lambda_{ij}]_{N\times N}$ for *forced* MJS (1.6).

9.3.1 Sliding Mode Controller Establishment

First, we construct the following sliding function:

$$s(x(t), i) = B_i^{\mathrm{T}} P_i x(t), \tag{9.24}$$

where $P_i > 0$ will be designed later. Due to the full column of B_i, the nonsingular of $B_i^{\mathrm{T}} P_i B_i$ can be ensured.

Now, we establish the following SMC law by using the TRs:

$$u(t) = K_i x(t) - \rho_i(t)\mathrm{sgn}(s(x(t), i)) \tag{9.25}$$

with

$$\rho_i(t) = \varrho_i + \eta_i(t)\|x(t)\| + \mu_i\delta, \tag{9.26}$$

$$\varrho_i \geq \frac{\lambda_{\max}[(B_i^{\mathrm{T}} P_i B_i)^{-1}]}{T-\kappa}\|B_i^{\mathrm{T}} P_i x(0)\|, \tag{9.27}$$

$$\begin{aligned}\eta_i(t) = {} & \|(B_i^{\mathrm{T}} P_i B_i)^{-1} B_i^{\mathrm{T}} P_i\|(\|A_i\| + \|M_i\|\|N_i\|) + \|K_i\| + \epsilon_i \\ & + \frac{1}{2}\frac{\|x(t)\|}{\|s(x(t), i)\|}\left\|\sum_{j=1}^{N}\lambda_{ij} P_j B_j (B_j^{\mathrm{T}} P_j B_j)^{-1} B_j^{\mathrm{T}} P_j\right\|,\end{aligned} \tag{9.28}$$

$$\mu_i = \|(B_i^{\mathrm{T}} P_i B_i)^{-1} B_i^{\mathrm{T}} P_i D_i\|, \tag{9.29}$$

where κ is a specified positive scalar with $0 < \kappa < T$, and both K_i, λ_{ij} will be determined in Theorem 9.3.

In the sequel, the reachability of the sliding surface $s(x(t), i) = 0$ is firstly analyzed.

Theorem 9.2 *For the MJS* (1.6), *if the sliding function is chosen as* (9.24), *the system states can be driven onto the sliding surface* $s(x(t), i) = 0$ *in the given finite-time interval* $[0, T]$ *with the SMC law* (9.25)–(9.29).

Proof Choose the Lyapunov function as follows:

$$U(x(t), i) = \frac{1}{2} s^{\mathrm{T}}(x(t), i)(B_i^{\mathrm{T}} P_i B_i)^{-1} s(x(t), i). \tag{9.30}$$

By Definition 1.4, we obtain the infinitesimal operator of $U(x(t), i)$ as

$$
\begin{aligned}
&\mathscr{L}U(x(t), i) \\
&= s^{\mathrm{T}}(x(t), i)(B_i^{\mathrm{T}} P_i B_i)^{-1} B_i^{\mathrm{T}} P_i \Big((A_i + \Delta A_i(t))x(t) + B_i u(t) + B_i f_i(x(t)) \\
&\quad + D_i w(t) \Big) + \frac{1}{2} \sum_{j=1}^{N} \Big[\lambda_{ij} s^{\mathrm{T}}(x(t), j)(B_j^{\mathrm{T}} P_j B_j)^{-1} s(x(t), j) \Big] \\
&= s^{\mathrm{T}}(x(t), i)(B_i^{\mathrm{T}} P_i B_i)^{-1} B_i^{\mathrm{T}} P_i \Big((A_i + \Delta A_i(t))x(t) \\
&\quad + D_i w(t) \Big) + s^{\mathrm{T}}(x(t), i) K_i x(t) + s^{\mathrm{T}}(x(t), i) f_i(x(t)) \\
&\quad + \frac{1}{2} \sum_{j=1}^{N} \Big[\lambda_{ij} s^{\mathrm{T}}(x(t), j)(B_j^{\mathrm{T}} P_j B_j)^{-1} s(x(t), j) \Big] \\
&\quad - s^{\mathrm{T}}(x(t), i) \rho_i(t) \mathrm{sgn}(s(x(t), i)).
\end{aligned} \tag{9.31}
$$

Noting the fact $\|s(x(t), i)\| \leq |s(x(t), i)|$ and $\|w(t)\| \leq \delta$, one has from (9.19)

$$
\begin{aligned}
\mathscr{L}U(x(t), i) \leq\ & \|s(x(t), i)\| \Big(\|(B_i^{\mathrm{T}} P_i B_i)^{-1} B_i^{\mathrm{T}} P_i A_i\| + \|K_i\| \\
& + \|(B_i^{\mathrm{T}} P_i B_i)^{-1} B_i^{\mathrm{T}} P_i M_i\| \|N_i\| + \epsilon_i \Big) \|x(t)\| \\
& + \|s(x(t), i)\| \|(B_i^{\mathrm{T}} P_i B_i)^{-1} B_i^{\mathrm{T}} P_i D_i\| \|w(t)\| \\
& + \frac{1}{2} \|x(t)\| \Big\| \sum_{j=1}^{N} \lambda_{ij} P_j B_j (B_j^{\mathrm{T}} P_j B_j)^{-1} B_j^{\mathrm{T}} P_j \Big\| \|x(t)\| \\
& - \|s(x(t), i)\|_1 (\varrho_i + \eta_i \|x(t)\| + \mu_i \delta) \\
\leq & - \varrho_i \|s(x(t), i)\| < 0,
\end{aligned} \tag{9.32}
$$

where the expressions (9.27)–(9.29) are utilized. Thus, we obtain that the state trajectories will be driven onto the specified sliding surface $s(x(t), i) = 0$ by the SMC law (9.25)–(9.29).

The following will further give the reaching instant T^*. One has from (9.30) that $U(x(t), i) \leq \lambda_{\max}((B_i^{\mathrm{T}} P_i B_i)^{-1}) \|s(x(t), i)\|^2$. Then, referring to Chapter 6, the state trajectories will be pushed into the specified sliding surfaces $s(x(t), i) = 0$ within finite time T^* with $T^* < T$. This fulfills the proof. □

Theorem 9.2 proves that the proposed sliding mode controller can guarantee the reachability of the specified sliding surface $s(x(t), i) = 0$ in a finite time T^* with $T^* < T$, which answers **Q2**.

9.3.2 SFTB Analysis and Hybrid Design

► **SFTB analysis during reaching phase**. By substituting SMC law (9.25) into the MJS (1.6), we obtain the following closed-loop system:

$$\dot{x}(t) = (\bar{A}_i + \Delta A_i(t))x(t) - B_i\rho_i(t)\text{sgn}(s(x(t), i)) + B_i f_i(x(t)) + D_i w(t), \tag{9.33}$$

where $\bar{A}_i \triangleq A_i + B_i K_i$.

Lemma 9.3 *Given the parameters* (c_1, c_2, T, R, δ), *scalars* $\varphi_i > 0$ *and TR matrix* Λ, *if there exist matrices* $P_i > 0$, K_i, *scalars* $\gamma_i > 0$, $\zeta_i > 0$, $c^* > 0$ $(i \in \mathbb{S})$ *satisfying*

$$\begin{bmatrix} \Upsilon_i & P_i D_i & P_i M_i & P_i B_i \\ * & -\varphi_i I & 0 & 0 \\ * & * & -\gamma_i I & 0 \\ * & * & * & -\zeta_i I \end{bmatrix} < 0, \tag{9.34}$$

$$c_1 < c^* < c_2, \tag{9.35}$$

$$\bar{\sigma}_P c_1 + \delta^2 T \varphi_i < e^{-\varphi_i T} \underline{\sigma}_P c^*, \tag{9.36}$$

where $\bar{\sigma}_P$ *and* $\underline{\sigma}_P$ *are defined as* (9.4) *in Lemma 9.1, and*

$$\Upsilon_i \triangleq \text{He}(P_i A_i + P_i B_i K_i) + \gamma_i N_i^{\mathrm{T}} N_i + \zeta_i \epsilon_i^2 I + \sum_{j=1}^{N} \lambda_{ij} P_j - \varphi_i P_i,$$

the closed-loop system (9.33) *is SFTB with respect to* $\aleph_{[0,T^*]}$.

Proof Choose the Lyapunov function $V(x(t), i)$ as the one in (9.5). For the closed-loop system (9.33), we have the infinitesimal operator on $V(x(t), i)$ as

$$\begin{aligned} &\mathscr{L}V(x(t), i) \\ &\le x^{\mathrm{T}}(t)\Big(\text{He}(P_i A_i + P_i B_i K_i) + \sum_{j=1}^{N} \lambda_{ij} P_j + \gamma_i^{-1} P_i M_i M_i^{\mathrm{T}} P_i \\ &\quad + \gamma_i N_i^{\mathrm{T}} N_i + \zeta_i \epsilon_i^2 I + \zeta_i^{-1} P_i B_i B_i^{\mathrm{T}} P_i\Big)x(t) \\ &\quad + x^{\mathrm{T}}(t) P_i D_i w(t) + w^{\mathrm{T}}(t) D_i^{\mathrm{T}} P_i x(t). \end{aligned} \tag{9.37}$$

By Schur Complement (Lemma 1.3), the auxiliary function $J(x(t), i)$ as (9.8) can be ensured to be $J(x(t), i) < 0$ by (9.34).

Furthermore, $J(x(t), i) < 0$ implies

$$\mathscr{L}V(x(t), i) < \varphi_i V(x(t), i) + \varphi_i w^{\mathrm{T}}(t) w(t). \tag{9.38}$$

Multiplying both sides of expression (9.38) by $e^{-\varphi_i t}$ and integrating the obtained inequality from 0 to t with $t \in [0, T^*]$, we have

$$
\begin{aligned}
e^{-\varphi_i t}\mathbb{E}\{V(x(t), r_0)\} &< V(x(0), i) + \varphi_i \int_0^t e^{-\varphi_i \tau} w^{\mathrm{T}}(\tau) w(\tau) \mathrm{d}\tau \\
&\leq \bar{\sigma}_P c_1 + \delta^2 T \varphi_i .
\end{aligned} \tag{9.39}
$$

On the other hand, it follows from (9.5) and (9.36) that

$$
\mathbb{E}\{x^{\mathrm{T}}(t) R x(t)\} < \frac{\bar{\sigma}_P c_1 + \delta^2 T \varphi_i}{e^{-\varphi_i T} \underline{\sigma}_P}. \tag{9.40}
$$

According to (9.36) and (9.40), one has $\mathbb{E}\{x^{\mathrm{T}}(t) R x(t)\} < c^*$ for all $t \in [0, T^*]$. Thus, according to Definition 1.5, the closed-loop MJS (9.33) is SFTB with respect to $\aleph_{[0,T^*]}$. This accomplishes the proof. □

► **SFTB analysis during sliding motion phase**. According to SMC theory, when the state trajectories maintain in the sliding surface $s(x(t), i) = 0$ for each $i \in \mathbb{S}$, we have the following equivalent controller:

$$
\begin{aligned}
u_{eq}(t) = &- (B_i^{\mathrm{T}} P_i B_i)^{-1} B_i^{\mathrm{T}} P_i (A_i + \Delta A_i(t)) x(t) - f(x(t), t, i) \\
&- (B_i^{\mathrm{T}} P_i B_i)^{-1} B_i^{\mathrm{T}} P_i D_i w(t).
\end{aligned} \tag{9.41}
$$

Then, when the equivalent controller (9.41) is applied to the MJS (1.6), the sliding mode dynamics can be obtained as

$$
\dot{x}(t) = [I - B_i (B_i^{\mathrm{T}} P_i B_i)^{-1} B_i^{\mathrm{T}} P_i]\Big((A_i + \Delta A_i(t)) x(t) + D_i w(t)\Big). \tag{9.42}
$$

Now, a sufficient condition for the existence of sliding mode controller and TR matrix is given in the following theorem to ensure the SFTB of the sliding mode dynamics (9.42).

Lemma 9.4 *Given the parameters* (c_1, c_2, T, R, δ), *scalars* $\varphi_i > 0$ *and TR matrix* Λ, *if there exist matrices* $P_i > 0$, K_i, *scalars* $\gamma_i > 0$, $\xi_i > 0$, $c^* > 0$ $(i \in \mathbb{S})$ *satisfying*

$$
\begin{bmatrix}
\Omega_i & P_i D_i & P_i M_i & \sqrt{3} P_i B_i & 0 & 0 \\
* & D_i^{\mathrm{T}} P_i D_i - \varphi_i I & 0 & 0 & 0 & 0 \\
* & * & -\gamma_i I & 0 & 0 & 0 \\
* & * & * & -B_i^{\mathrm{T}} P_i B_i & 0 & 0 \\
* & * & * & * & 0 & 0 \\
* & * & * & * & -P_i & P_i M_i \\
* & * & * & * & * & -\xi_i I
\end{bmatrix} < 0, \tag{9.43}
$$

$$
c_1 < c^* < c_2, \tag{9.44}
$$

$$
\bar{\sigma}_P c^* + \delta^2 T \varphi_i < e^{-\varphi_i T} \underline{\sigma}_P c_2, \tag{9.45}
$$

where $\bar{\sigma}_P$ *and* $\underline{\sigma}_P$ *are defined as* (9.4) *in Lemma 9.1, and*

$$\Omega_i \triangleq \mathrm{He}(P_iA_i) + A_i^{\mathrm{T}}P_iA_i + \gamma_iN_i^{\mathrm{T}}N_i + \xi_iN_i^{\mathrm{T}}N_i + \sum_{j=1}^{N}\lambda_{ij}P_j - \varphi_iP_i,$$

the sliding mode dynamics (9.42) *are SFTB with respect to* $\aleph_{[T^*,T]}$.

Proof Choose the Lyapunov function defined as (9.5). For the sliding mode dynamics (9.42), we have the infinitesimal operator from Definition 1.4

$$\begin{aligned}
&\mathscr{L}V(x(t), i)\\
&\le x^{\mathrm{T}}(t)\Big(\mathrm{He}(P_iA_i) + \sum_{j=1}^{N}\lambda_{ij}P_j + \gamma_i^{-1}P_iM_iM_i^{\mathrm{T}}P_i + \gamma_iN_i^{\mathrm{T}}N_i\Big)x(t)\\
&\quad + x^{\mathrm{T}}(t)P_iB_i(B_i^{\mathrm{T}}P_iB_i)^{-1}B_i^{\mathrm{T}}P_ix(t) + x^{\mathrm{T}}(t)\Delta A_i^{\mathrm{T}}(t)P_i\Delta A_i(t)x(t)\\
&\quad + x^{\mathrm{T}}(t)P_iB_i(B_i^{\mathrm{T}}P_iB_i)^{-1}B_i^{\mathrm{T}}P_ix(t) + x^{\mathrm{T}}(t)A_i^{\mathrm{T}}P_iA_ix(t)\\
&\quad + x^{\mathrm{T}}(t)P_iD_iw(t) + w^{\mathrm{T}}(t)D_i^{\mathrm{T}}P_ix(t).
\end{aligned} \tag{9.46}$$

By Schur Complement (Lemma 1.3), the auxiliary function $J(x(t), i) < 0$ defined as (9.8) can be ensured by the following matrix inequalities:

$$\begin{bmatrix}
\bar{\Omega}_i & P_iD_i & P_iM_i & \sqrt{3}P_iB_i & \Delta A_i^{\mathrm{T}}(t)P_i\\
* & D_i^{\mathrm{T}}P_iD_i - \varphi_iI & 0 & 0 & 0\\
* & * & -\gamma_iI & 0 & 0\\
* & * & * & -B_i^{\mathrm{T}}P_iB_i & 0\\
* & * & * & * & 0\\
* & * & * & * & -P_i
\end{bmatrix} < 0, \tag{9.47}$$

where

$$\bar{\Omega}_i \triangleq \mathrm{He}(P_iA_i) + A_i^{\mathrm{T}}P_iA_i + \gamma_iN_i^{\mathrm{T}}N_i + \sum_{j=1}^{N}\lambda_{ij}P_j - \varphi_iP_i,$$

which can be guaranteed by (9.43) for $\xi_i > 0$.

Furthermore, $J(x(t), i) < 0$ implies

$$\mathscr{L}V(x(t), i) < \varphi_iV(x(t), i) + \varphi_iw^{\mathrm{T}}(t)w(t). \tag{9.48}$$

According to the result in Lemma 9.3, one has $\mathbb{E}\{x^{\mathrm{T}}(T^*)Rx(T^*)\} < c^*$. Thus, by multiplying both sides of expression (9.48) by $e^{-\varphi_it}$ and integrating the obtained inequality from T^* to t with $t \in [T^*, T]$, we have

$$\begin{aligned}
e^{-\varphi_it}\mathbb{E}\{V(x(t), i)\} &< e^{-\varphi_iT^*}\mathbb{E}\{V(x(T^*), r_{T^*})\} + \varphi_i\int_{T^*}^{t}e^{-\varphi_i\tau}w^{\mathrm{T}}(\tau)w(\tau)\mathrm{d}\tau\\
&< \bar{\sigma}_P\mathbb{E}\{x^{\mathrm{T}}(T^*)Rx(T^*)\} + \delta^2T\varphi_i
\end{aligned}$$

$$< \bar{\sigma}_P c^* + \delta^2 T \varphi_i. \tag{9.49}$$

In addition, it follows from (9.5) and (9.49) that

$$\mathbb{E}\{x^{\mathrm{T}}(t)Rx(t)\} < \frac{\bar{\sigma}_P c^* + \delta^2 T \varphi_i}{e^{-\varphi_i T}\underline{\sigma}_P}. \tag{9.50}$$

Further, if there exists a scalar c^* satisfying (9.44), (9.45), we have from (9.50) that $\mathbb{E}\{x^{\mathrm{T}}(t)Rx(t)\} < c_2$ for all $t \in [T^*, T]$. Thus, the sliding mode dynamics (9.42) are SFTB with respect to $\aleph_{[T^*,T]}$. The proof is completed. □

► **Hybrid design on controller gains $\mathbf{K_i}$ ($\mathbf{i} \in \mathbb{S}$) and TR matrix $\mathbf{\Lambda}$.** It is noted that the SFTB of the closed-loop MJS over the whole interval $[0, T]$ is attained only when the conditions in Lemmas 9.3 and 9.4 hold simultaneously. This means that the matrices P_i, K_i and Λ should be designed to ensure the conditions in Lemmas 9.3 and 9.4 to be feasible simultaneously. In the sequel, we shall give a sufficient condition in Theorem 9.3, which can guarantee that the conditions in Lemmas 9.3 and 9.4 hold simultaneously.

Theorem 9.3 *Given the parameters (c_1, c_2, T, R, δ) and scalars $\varphi_i > 0$, $\varpi_i > 0$, $\nu_1 > 0$, $\nu_2 > 0$, if there exist matrices $P_i > 0$, $\bar{K}_i$, scalars $\varepsilon_i > 0$, $\gamma_i > 0$, $\zeta_i > 0$, $\xi_i > 0$, $c^* > 0$, $\vartheta_i > 0$, and a set of scalars $\bar{\lambda}_{ij} > 0$ $(i, j \in \mathbb{S}, i \neq j)$ satisfying*

$$\begin{bmatrix} \Psi_{11i} & \Psi_{12i} & \Psi_{13i} \\ * & \Psi_{22i} & 0 \\ * & * & \Psi_{33i} \end{bmatrix} < 0, \tag{9.51}$$

$$c_1 < c^* < c_2, \tag{9.52}$$

$$\vartheta_i R < P_i < \varpi_i R, \tag{9.53}$$

$$\begin{bmatrix} -e^{-\varphi_i T} c^* & \sqrt{\varpi_i c_1 + \delta^2 T \varphi_i} \\ * & -\vartheta_i \end{bmatrix} < 0, \tag{9.54}$$

$$\varpi_i c^* + \delta^2 T \varphi_i - e^{-\varphi_i T} c_2 \vartheta_i < 0, \tag{9.55}$$

where

$$\Psi_{11i} \triangleq \mathrm{diag}\{\Pi_{1i}, \Pi_{2i}\},$$

$$\Pi_{1i} \triangleq \mathrm{He}(P_i A_i) + \gamma_i N_i^{\mathrm{T}} N_i + \zeta_i \epsilon_i^2 I - \varphi_i P_i + \nu_2^2 \varepsilon_i I - 2\nu_2 P_i + \nu_1^2 \sum_{j=1, j\neq i}^{N} \varepsilon_j I + 2\nu_1 (N-1) P_i - 2\nu_1 \sum_{j=1, j\neq i}^{N} P_j,$$

$$\Pi_{2i} \triangleq \mathrm{He}(P_i A_i) + A_i^{\mathrm{T}} P_i A_i + \gamma_i N_i^{\mathrm{T}} N_i + \xi_i N_i^{\mathrm{T}} N_i - \varphi_i P_i + \nu_1^2 \sum_{j=1, j\neq i}^{N} \varepsilon_j I$$

$$+ 2\nu_1(N-1)P_i - 2\nu_1 \sum_{j=1, j\neq i}^{N} P_j,$$

$$\Psi_{12i} \triangleq \begin{bmatrix} P_i D_i & 0 & P_i M_i & 0 & P_i B_i & 0 \\ 0 & P_i D_i & 0 & P_i M_i & 0 & \sqrt{3} P_i B_i \end{bmatrix},$$

$$\Psi_{13i} \triangleq \begin{bmatrix} 0 & 0 & P_i + \bar{K}_i^{\mathrm{T}} B_i^{\mathrm{T}} & \Gamma_i & 0 \\ 0 & 0 & 0 & 0 & \Gamma_i \end{bmatrix},$$

$$\Psi_{22i} \triangleq -\operatorname{diag}\{\varphi_i I, -D_i^{\mathrm{T}} P_i D_i + \varphi_i I, \gamma_i I, \gamma_i I, \zeta_i I, B_i^{\mathrm{T}} P_i B_i\},$$

$$\Psi_{33i} \triangleq \begin{bmatrix} -P_i & P_i M_i & 0 & 0 & 0 \\ * & -\xi_i I & 0 & 0 & 0 \\ * & * & -\varepsilon_i I & 0 & 0 \\ * & * & * & -\Theta_i & 0 \\ * & * & * & * & -\Theta_i \end{bmatrix}$$

with Γ_i and Θ_i are defined in Theorem 9.1, the closed-loop system is SFTB with respect to $\aleph_{[0,T]}$. In this case, the controller gain is given as $K_i = \varepsilon_i^{-1} \bar{K}_i$ and the designed TR matrix is given as $\Lambda = 2\bar{\Lambda}\Xi^{-1}$, where Ξ is defined in Theorem 9.1.

Proof Taking the conditions (9.34)–(9.36) in Lemma 9.3 and conditions (9.43)–(9.45) in Lemma 9.4 simultaneously, we have the following inequalities:

$$\begin{bmatrix} \bar{\Psi}_{11i} & \Psi_{12i} & 0 \\ * & \Psi_{22i} & 0 \\ * & * & \bar{\Psi}_{33i} \end{bmatrix} < 0, \tag{9.56}$$

$$c_1 < c^* < c_2, \tag{9.57}$$

$$\bar{\sigma}_P c_1 + \delta^2 T \varphi_i < e^{-\varphi_i T} \underline{\sigma}_P c^*, \tag{9.58}$$

$$\bar{\sigma}_P c^* + \delta^2 T \varphi_i < e^{-\varphi_i T} \underline{\sigma}_P c_2, \tag{9.59}$$

where $\bar{\sigma}_P$ and $\underline{\sigma}_P$ are defined in Lemma 9.1, and

$$\bar{\Psi}_{11i} \triangleq \operatorname{diag}\{\Upsilon_i, \Omega_i\}, \quad \bar{\Psi}_{33i} \triangleq \begin{bmatrix} -P_i & P_i M_i \\ * & -\xi_i I \end{bmatrix}.$$

It is easy to see $\Upsilon_i < \hat{\Upsilon}_i$ and $\Omega_i < \hat{\Omega}_i$ under the fact that $\lambda_{ii} = -\sum_{j=1, j\neq i}^{N} \lambda_{ij}$ for $\varepsilon_i > 0$, where

$$\hat{\Upsilon}_i \triangleq P_i A_i + A_i^{\mathrm{T}} P_i + P_i(B_i K_i) + (B_i K_i)^{\mathrm{T}} P_i + \sum_{j=1, j\neq i}^{N} \lambda_{ij}(P_j - P_i) + \gamma_i N_i^{\mathrm{T}} N_i$$

$$+ \zeta_i \epsilon_i I - \varphi_i P_i + \sum_{j=1, j\neq i}^{N} \frac{\varepsilon_j \lambda_{ij}^2}{4} + \varepsilon_i (B_i K_i)^{\mathrm{T}} (B_i K_i), \tag{9.60}$$

$$\hat{\Omega}_i \triangleq P_i A_i + A_i^{\mathrm{T}} P_i + A_i P_i A_i + \sum_{j=1, j \neq i}^{N} \lambda_{ij}(P_j - P_i)$$
$$+ \gamma_i N_i^{\mathrm{T}} N_i + \xi_i N_i^{\mathrm{T}} N_i - \varphi_i P_i + \sum_{j=1, j \neq i}^{N} \frac{\varepsilon_j \lambda_{ij}^2}{4}. \tag{9.61}$$

Let $\bar{\lambda}_{ij} \triangleq \frac{\varepsilon_j}{2} \lambda_{ij}$. (9.60) and (9.61) can be, respectively, rewritten as (9.62) and (9.63).

$$\hat{\Upsilon}_i \triangleq P_i A_i + A_i^{\mathrm{T}} P_i + \gamma_i N_i^{\mathrm{T}} N_i + \zeta_i \epsilon_i I - \varphi_i P_i + (P_i + B_i K_i)^{\mathrm{T}} \varepsilon_i^{-1} I (P_i + B_i K_i)$$
$$- \varepsilon_i^{-1} P_i P_i + \sum_{j=1, j \neq i}^{N} (P_j - P_i + \bar{\lambda}_{ij})^{\mathrm{T}} \varepsilon_j^{-1} I (P_j - P_i + \bar{\lambda}_{ij})$$
$$- \sum_{j=1, j \neq i}^{N} \varepsilon_j^{-1} (P_j - P_i)(P_j - P_i), \tag{9.62}$$

$$\hat{\Omega}_i \triangleq P_i A_i + A_i^{\mathrm{T}} P_i + A_i P_i A_i + \gamma_i N_i^{\mathrm{T}} N_i + \xi_i N_i^{\mathrm{T}} N_i - \varphi_i P_i$$
$$+ \sum_{j=1, j \neq i}^{N} (P_j - P_i + \bar{\lambda}_{ij})^{\mathrm{T}} \varepsilon_j^{-1} I (P_j - P_i + \bar{\lambda}_{ij})$$
$$- \sum_{j=1, j \neq i}^{N} \varepsilon_j^{-1} (P_j - P_i)(P_j - P_i) \tag{9.63}$$

Similar to Theorem 9.1, there exist two adjustable scalars ν_1 and ν_2 such that $-\varepsilon_j^{-1}(P_j - P_i)(P_j - P_i) \leq -2\nu_1(P_j - P_i) + \nu_1^2 \varepsilon_j I$ and $-\varepsilon_i^{-1} P_i P_i \leq -2\nu_2 P_i + \nu_2^2 \varepsilon_i I$. Thus, it can yield that $\hat{\Upsilon}_i < \tilde{\Upsilon}_i$ and $\hat{\Omega}_i < \tilde{\Omega}_i$, where

$$\tilde{\Upsilon}_i \triangleq P_i A_i + A_i^{\mathrm{T}} P_i + \gamma_i N_i^{\mathrm{T}} N_i + \zeta_i \epsilon_i I - \varphi_i P_i + (P_i + B_i K_i)^{\mathrm{T}} \varepsilon_i^{-1} I (P_i + B_i K_i)$$
$$- \varepsilon_i^{-1} P_i P_i + \sum_{j=1, j \neq i}^{N} (P_j - P_i + \bar{\lambda}_{ij})^{\mathrm{T}} \varepsilon_j^{-1} I (P_j - P_i + \bar{\lambda}_{ij})$$
$$- \sum_{j=1, j \neq i}^{N} \varepsilon_j^{-1} (P_j - P_i)(P_j - P_i), \tag{9.64}$$

$$\tilde{\Omega}_i \triangleq P_i A_i + A_i^{\mathrm{T}} P_i + A_i P_i A_i + \gamma_i N_i^{\mathrm{T}} N_i + \xi_i N_i^{\mathrm{T}} N_i - \varphi_i P_i$$
$$+ \sum_{j=1, j \neq i}^{N} (P_j - P_i + \bar{\lambda}_{ij})^{\mathrm{T}} \varepsilon_j^{-1} I (P_j - P_i + \bar{\lambda}_{ij})$$
$$- \sum_{j=1, j \neq i}^{N} \varepsilon_j^{-1} (P_j - P_i)(P_j - P_i). \tag{9.65}$$

Due to $\bar{\Psi}_{11i} \leq \tilde{\Psi}_{11i}$ with $\tilde{\Psi}_{11i} \triangleq \text{diag}\{\tilde{\Upsilon}_i, \tilde{\Omega}_i\}$, (9.56) holds if the following inequality holds:

$$\begin{bmatrix} \tilde{\Psi}_{11i} & \Psi_{12i} & 0 \\ * & \Psi_{22i} & 0 \\ * & * & \bar{\Psi}_{33i} \end{bmatrix} < 0, \tag{9.66}$$

which can be ensured by (9.51). Hence, (9.51) can guarantee (9.56).

On the other hand, by the condition (9.53), it has $\vartheta_i < \underline{\sigma}_P, \bar{\sigma}_P < \varpi_i$. Thus, it is easily shown that the inequalities (9.57)–(9.59) can be ensured by (9.52)–(9.55). The proof is finished. □

Theorem 9.3 proposes a hybrid design strategy by synthesizing the sliding mode controller and the TR matrix simultaneously to ensure the SFTB of the closed-loop MJS (9.33) over the finite-time interval composed of reaching phase and sliding motion phase, which answers **Q3**.

9.3.3 Three-Dimensional Search Algorithm

It is worth mentioning that in the matrix inequality (9.51), the decision variables ν_1 and ν_2 are exploited to reduce conservatism via introducing two extra degrees of freedom. By applying a simple three-dimensional search over $\nu_1 > 0$, $\nu_2 > 0$ and $\varpi_i > 0$, the feasibility of the LMIs (9.51)–(9.55) can be checked.

The above design procedure on hybrid design scheme can be summarized as follows.

Algorithm 9.2 Searching parameters ν_1, ν_2 and ϖ_i simultaneously

---

Step 1. Assign the scalars $\nu_1 > 0$ and $\nu_2 > 0$ (for example $\nu_1 = 0.6$, $\nu_2 = 0.6$). By a linear searching over (0.1, 100), if at least a set of values $\{\varpi_i, i \in \mathbb{S}\}$ is found such that LMIs (9.51)–(9.55) are feasible, go to Step 2; else, modify the scalars ν_1, ν_2 (for example $\nu_1 = \nu_1 + 0.6$, $\nu_2 = \nu_2 + 0.6$), and then, restart the above search for ϖ_i.

Step 2. Obtain the matrices $\bar{K}_i$, scalar ε_i and a set of scalars $\bar{\lambda}_{ij} > 0\,(i, j \in \mathbb{S}, i \neq j)$.

Step 3. The optimized control gain in SMC law (9.25) is $K_i = \varepsilon_i^{-1}\bar{K}_i$ and a design parameter ϱ_i is chosen according to the scalar selection criterion (9.27) with the initial state $x(0)$.

Step 4. Calculate $\bar{\lambda}_{ii} = -\varepsilon_i \sum_{j=1,j\neq i}^{N} \varepsilon_j^{-1}\bar{\lambda}_{ij}$. The designed TR matrix is given as $\Lambda = 2\bar{\Lambda}\Xi^{-1}$, where Ξ is defined in Theorem 9.1.

Step 5. The SMC law (9.25) and TR matrix Λ will be obtained.

9.3.4 Simulation Example 2

This example is given to illustrate that the hybrid design scheme on the sliding mode controller and TR matrix in Theorem 9.3 can guarantee the SFTB of the controlled MJS over the finite-time interval $[0, T]$ composed of reaching phase $[0, T^*]$ and sliding motion phase $[T^*, T]$. The system parameters are given as

$$A_1 = \begin{bmatrix} 0 & 0.02 \\ 0.03 & -0.02 \end{bmatrix},\ A_2 = \begin{bmatrix} -0.01 & 0.02 \\ 0 & 0.01 \end{bmatrix},\ B_1 = \begin{bmatrix} -0.1 \\ 0.1 \end{bmatrix},$$
$$B_2 = \begin{bmatrix} 0.3 \\ -0.1 \end{bmatrix},\ D_1 = \begin{bmatrix} 0.1 \\ 0.2 \end{bmatrix},\ D_2 = \begin{bmatrix} 0.1 \\ 0.3 \end{bmatrix},\ M_1 = \begin{bmatrix} 0.3 \\ 0.4 \end{bmatrix},$$
$$M_2 = \begin{bmatrix} 0.5 \\ 0.2 \end{bmatrix},\ N_1 = \begin{bmatrix} 0.1 & 0.2 \end{bmatrix},\ N_2 = \begin{bmatrix} 0.2 & 0.1 \end{bmatrix},$$

and $F_1(t) = 0.1\sin(t)$, $F_2(t) = 0.2\cos(t)$, $f_1(x(t)) = 0.2\sin(\|x(t)\|)$, and $f_2(x(t)) = 0.3\|x(t)\|$ with $\epsilon_1 = 0.2$, $\epsilon_2 = 0.3$. The peak bounded external disturbance satisfies $w(t) = 0.05e^{-t}$ with $\delta = 0.05$.

Utilizing Algorithm 9.2 with the adjustable scalars $\nu_1 = 0.6$, $\nu_2 = 0.6$ and the scalars $\varpi_i = 6.5$ $(i = 1, 2)$, we can obtain the optimized controller gains as

$$K_1 = \begin{bmatrix} 13.4210 & -39.0787 \end{bmatrix},\ K_2 = \begin{bmatrix} -5.1162 & 10.8243 \end{bmatrix},$$

the Lyapunov matrices

$$P_1 = \begin{bmatrix} 0.0063 & -0.0096 \\ -0.0096 & 0.0242 \end{bmatrix},\ P_2 = \begin{bmatrix} 0.0053 & -0.0091 \\ -0.0091 & 0.0252 \end{bmatrix},$$

and the corresponding TR matrix

$$\Lambda = \begin{bmatrix} -0.2541 & 0.2541 \\ 0.0042 & -0.0042 \end{bmatrix}.$$

For the initial condition $x(0) = [-0.5\ \ 0.2]^{\mathrm{T}}$, the finite time $T = 1$, $c_1 = 0.01$, $c_2 = 4.81$ and

$$R_1 = \begin{bmatrix} 0.0010 & -0.0015 \\ -0.0015 & 0.0037 \end{bmatrix},\ R_2 = \begin{bmatrix} 0.0008 & -0.0014 \\ -0.0014 & 0.0039 \end{bmatrix},$$

we have the state response of the closed-loop MJS in Fig. 9.2, from which it is seen that the hybrid control scheme on the sliding mode controller and TR matrix can effectively ensure the SFTB of the controlled MJS. In addition, the curves of sliding variable and control input are shown in Fig. 9.2c and 9.2d, respectively. From this simulation, we can discover that the asymptotic stability cannot be guaranteed in this example but it is not we care about, which further illustrates that what we pay

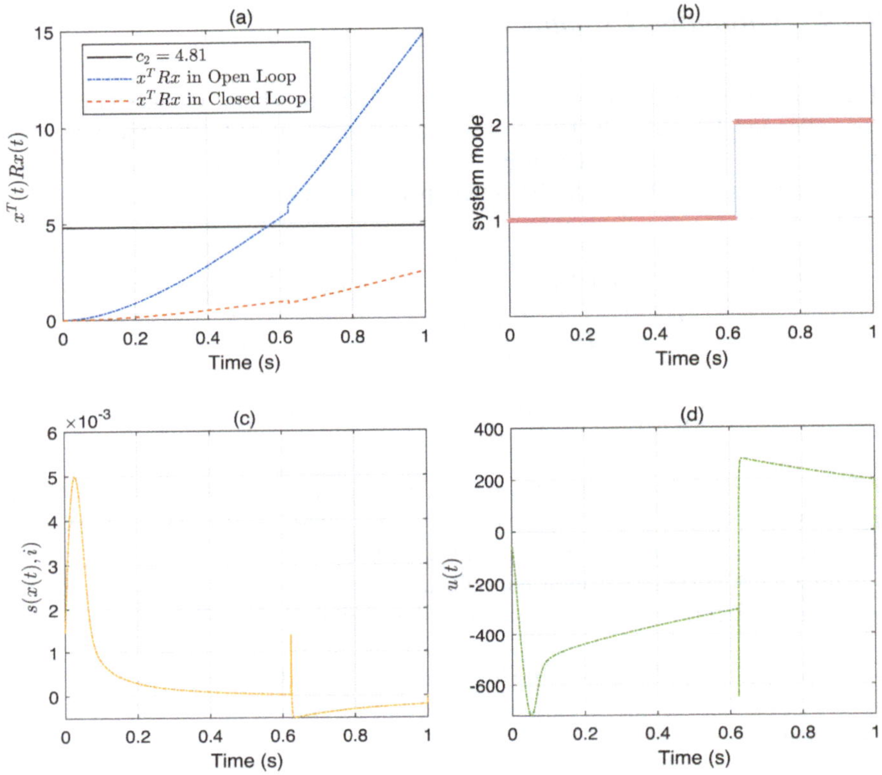

Fig. 9.2 Simulation results in Example 2: **a** evolution of weighted system state variable $x^{\mathrm{T}}(t)Rx(t)$; **b** state responses $x(t)$; **c** sliding variable $s(x(t), i)$; **d** control signal $u(t)$

attention is the transient performance of the controlled system during the given short interval $[0, T]$ by means of SMC technique.

9.4 Hybrid Control Scheme Under Actuator Saturation

9.4.1 Actuator Saturation Issue

Consider a class of MJSs on a complete probability space $(\Omega, \mathscr{F}, \Pr)$ as in system (1.6) $(\Delta A_i(t) \equiv 0)$. The transition probabilities (TPs) are generated by (1.7). Accordingly, for mode $r_t = i \in \mathbb{S}$, the considered MJS is expressed as

$$\dot{x}(t) = A_i x(t) + B_i\big(\Phi(u(t)) + f_i(x(t))\big) + D_i w(t), \tag{9.67}$$

where $x(t) \in \mathbb{R}^n$ is the state; $\Phi(u(t)) \in \mathbb{R}^m$ stands for the nonlinear saturation characteristic with $u(t)$ being the control input; $f_i(x(t))$ represents the unknown nonlinear function satisfying Assumption 1.3; $w(t)$ is the external disturbance satisfying Assumption 1.4.

In this section, define $u(t) \triangleq [u_1(t)\ u_2(t)\ \cdots\ u_m(t)]^{\mathrm{T}}$ and assume the maximum control input as $u_{l\max}$ ($u_{l\max} > 0$) for $u_l(t)$ ($l = 1, 2, \ldots, m$). According to Refs. [9, 10], the nonlinear saturation characteristic can be described as

$$\Phi(u(t)) = \phi(u(t))u(t), \tag{9.68}$$

where $\phi(u(t)) \triangleq \mathrm{diag}\{\phi_1(u_1(t)), \ldots, \phi_m(u_m(t))\}$ with

$$\phi_l(u_l(t)) \triangleq \begin{cases} u_{l\max}/u_l(t), & \text{if } u_l(t) > u_{l\max}, \\ 1, & \text{if } |u_l(t)| \le u_{l\max}, \\ -u_{l\max}/u_l(t), & \text{if } u_l(t) < -u_{l\max}. \end{cases} \tag{9.69}$$

According to the density property of a real number [11], since $0 < \phi_l(u_l(t)) \le 1$, there exists a known constant ξ_l satisfying $0 < \xi_l \le \phi_l(u_l(t)) \le 1$ for $l = 1, 2, \ldots, m$. Thus, we have $0 < \Xi \le \phi(u(t)) \le I$ with $\Xi \triangleq \mathrm{diag}\{\xi_1, \xi_2, \ldots, \xi_m\}$.

Consequently, considering the actuator saturation (9.68), system (9.67) can be rewritten as

$$\dot{x}(t) = A_i x(t) + B_i(\phi(u(t))u(t) + f_i(x(t))) + D_i w(t). \tag{9.70}$$

The control objective is to provide a hybrid design scheme for the TRs and a sliding mode controller such that the SFTB of the MJS (9.70) can be attained.

9.4.2 Saturated Controller Design

The sliding function is constructed as

$$s(x(t), i) = G_i x(t), \quad i \in \mathbb{S}, \tag{9.71}$$

where sliding gain $G_i \triangleq B_i^{\mathrm{T}} P_i$ with $P_i > 0$ to be determined such that $G_i B_i$ is nonsingular.

Now, we construct the finite-time sliding mode controller as

$$u(t) = K_i x(t) - \varrho_i(t)\Xi^{-1}\mathrm{sgn}(s(x(t), i)) \tag{9.72}$$

with

$$\varrho_i(t) \triangleq \hat{\varrho}_i + \|(G_i B_i)^{-1} G_i D_i\|\delta + \big(\epsilon_i + \bar{\varrho}_i(t)$$

$$
\begin{aligned}
&+ \|K_i\| + \|(G_iB_i)^{-1}G_iA_i\|\big)\|x(t)\|, \\
\bar{\varrho}_i(t) &\triangleq \frac{\|x(t)\|}{2\|s(x(t),i)\|}\left\|\sum_{j=1}^{N}\lambda_{ij}G_j^{\mathrm{T}}(G_jB_j)^{-1}G_j\right\|, \\
\hat{\varrho}_i &\geq \frac{\lambda_{\max}((G_iB_i)^{-1})}{T-\chi}\|G_ix(0)\|,
\end{aligned}
$$

where controller gains $K_i \in \mathbb{R}^{m\times n}$ and TRs λ_{ij} $(i, j \in \mathbb{S})$ will be designed, and χ is a given small constant satisfying $0 < \chi < T$.

In the sequel, for a given finite-time interval $[0, T]$, we will prove that the state trajectories can be driven onto the specified sliding surface before a finite instant T^* with $T^* < T$.

Theorem 9.4 *For the MJS* (9.70) *subject to actuator saturation* (9.68)*, the controller* (9.72) *can make sure that the system state is attracted into the specified sliding surface* $s(x(t), i) = 0$ *before* T^* *(i.e.,* $T^* < T$*).*

Proof Choose the mode-dependent Lyapunov function as:

$$
V_1(x(t), i) = s^{\mathrm{T}}(x(t), i)(G_iB_i)^{-1}s(x(t), i). \tag{9.73}
$$

We have from (9.73)

$$
\begin{aligned}
&\mathscr{L}V_1(x(t), i) \\
&= 2s^{\mathrm{T}}(x(t), i)\big((G_iB_i)^{-1}G_iA_ix(t) + f_i(x(t)) + (G_iB_i)^{-1}G_iD_iw(t) \\
&\quad + \phi(u(t))K_ix(t)\big) + \sum_{j=1}^{N}\lambda_{ij}s^{\mathrm{T}}(x(t), j)(G_jB_j)^{-1}s(x(t), j) \\
&\quad - 2s^{\mathrm{T}}(x(t), i)\varrho_i(t)\phi(u(t))\Xi^{-1}\mathrm{sgn}(s(x(t), i)).
\end{aligned} \tag{9.74}
$$

Note that

$$
\begin{aligned}
&s^{\mathrm{T}}(x(t), i)\phi(u(t))\Xi^{-1}\mathrm{sgn}(s(x(t), i)) \\
&= \big[s_1(x(t), i) \ \cdots \ s_m(x(t), i)\big] \\
&\quad \times \mathrm{diag}\{\phi_1(u_1(t))/\xi_1, \ldots, \phi_m(u_1(t))/\xi_m\} \\
&\quad \times \big[\mathrm{sgn}(s_1(x(t), i)) \ \cdots \ \mathrm{sgn}(s_m(x(t), i))\big]^{\mathrm{T}} \\
&= \sum_{l=1}^{m}\phi_l(u_l(t))/\xi_l s_l(x(t), i)\mathrm{sgn}(s_l(x(t), i)) \\
&\geq \sum_{l=1}^{m}s_l(x(t), i)\mathrm{sgn}(s_l(x(t), i)) \\
&\geq |s(x(t), i)| \geq \|s(x(t), i)\|.
\end{aligned}
$$

Thus, we can easily get $\mathscr{L}V_1(x(t), i) \leq -2\hat{\varrho}_i\|s(x(t), i)\|$. In summary, by the SMC law (8.10)–(8.15), the state trajectories of the system will be pushed into the specified sliding surfaces $s(x(t), i) = 0$.

Furthermore, one has from (9.73) that $V_1(x(t), i) \leq \lambda_{\max}((G_i B_i)^{-1})\|s(x(t), i)\|^2$. Then, referring to Chapter 6, the state trajectories will be pushed into the specified sliding surfaces $s(x(t), i) = 0$ within finite time T^* with $T^* < T$.This proof is accomplished. □

9.4.3 SFTB Analysis Integrated by TR Design

In this subsection, we shall consider to stabilize the *forced* MJS (i.e., system (9.70)) with the constructed sliding mode controller (9.72) to be SFTB via hybrid designing the TR matrix Λ and controller gains K_i $(i \in \mathbb{S})$.

By analyzing SFTB problem of the MJS (9.70) against actuator saturation (9.68) during the reaching phase and the sliding motion phase, respectively, the SFTB problem of the MJS (9.70) will be investigated and some sufficient conditions will be given to guarantee the SFTB over the specified finite-time interval $[0, T]$.

(I) **SFTB over the reaching phase [0, T^*].**

Substituting SMC law (9.72) into the system (9.70), we obtain the following closed-loop system:

$$\begin{aligned}\dot{x}(t) =&A_i x(t) + B_i f_i(x(t)) + D_i w(t) + B_i\phi(u(t))K_i x(t)\\ &- B_i\phi(u(t))\Xi^{-1}\varrho_i(t)\text{sgn}(s(x(t), i)).\end{aligned} \tag{9.75}$$

In the following theorem, some sufficient conditions are derived to guarantee SFTB over the reaching phase $[0, T^*]$.

Lemma 9.5 *For the given the parameters (c_1, c_2, T, R, δ), scalars $\varphi_i > 0$, and TR matrix Λ, if there exist constants $\sigma_i > 0$, $c^* > 0$, and matrices $P_i > 0$, $Q_i > 0$ for any $i \in \mathbb{S}$, satisfying the following matrix inequalities:*

$$\begin{bmatrix} \Gamma_{1i} & P_i D_i \\ * & -\varphi_i I \end{bmatrix} < 0, \tag{9.76}$$

$$c_1 < c^* < c_2, \tag{9.77}$$

$$\frac{\bar{\varpi}_P c_1 + \delta^2 T \varphi_i}{\underline{\varpi}_P} < e^{-\varphi_i T} c^*, \tag{9.78}$$

where

$$\bar{\varpi}_P \triangleq \max_{i \in \mathbb{S}} \left\{\lambda_{\max}\left(R^{-\frac{1}{2}} P_i R^{-\frac{1}{2}}\right)\right\},$$

$$\underline{\varpi}_P \triangleq \min_{i\in\mathbb{S}} \left\{\lambda_{\min}\left(R^{-\frac{1}{2}} P_i R^{-\frac{1}{2}}\right)\right\},$$

$$\Gamma_{1i} \triangleq \text{He}\big(P_i A_i + P_i B_i \phi(u(t)) K_i\big) + \sum_{j=1}^{N} \lambda_{ij} P_j + \sigma_i^{-1} P_i B_i B_i^T P_i + \sigma_i b_i^2 I - \varphi_i P_i.$$

then, the closed-loop system (9.75) *is SFTB with respect to* $\aleph_{[0,T^*]}$.

Proof Consider a Lyapunov functional candidate:

$$V_2(x(t), i) = x^{\mathrm{T}}(t) P_i x(t). \tag{9.79}$$

For $G_i = B_i^{\mathrm{T}} P_i$, it is easy to get

$$\begin{aligned}&2x^{\mathrm{T}}(t) P_i B_i \phi(u(t)) \Xi^{-1} \text{sgn}(s(x(t), i))\\ =& s^{\mathrm{T}}(x(t), i) \phi(u(t)) \Xi^{-1} \text{sgn}(s(x(t), i)) \geq \|s(x(t), i)\| > 0.\end{aligned} \tag{9.80}$$

Referring to Lemma 9.3, we can obtain $\mathbb{E}\{x^{\mathrm{T}}(t) R x(t)\} < c^*$ for all $t \in [0, T^*]$ from (9.76)–(9.78). Thus, according to Definition 1.5, the resulting closed-loop system (9.75) is SFTB with respect to $\aleph_{[0,T^*]}$. This proof is fulfilled. □

(II) **SFTB over sliding motion phase [*T**, *T*].**

Now, we will analyze the SFTB of the sliding mode dynamics during the interval $[T^*, T]$. According to SMC theory, when the state trajectories maintain onto the sliding surface $s(x(t), i) = 0$, from (9.70) and (9.71), we have the following equivalent controller:

$$\begin{aligned}u_{eq}(t) =& -\phi^{-1}(u(t))(G_i B_i)^{-1} G_i A_i x(t) - \phi^{-1}(u(t)) f_i(x(t))\\ &- \phi^{-1}(u(t))(G_i B_i)^{-1} G_i D_i w(t).\end{aligned} \tag{9.81}$$

By substituting (9.81) into (9.70), we get the following sliding mode dynamics:

$$\dot{x}(t) = \tilde{G}_i A_i x(t) + \tilde{G}_i D_i w(t), \tag{9.82}$$

where $\tilde{G}_i \triangleq I - B_i (G B_i)^{-1} G_i$.

Lemma 9.6 *For the given parameters* (c_1, c_2, T, R, δ), *scalars* $\varphi_i > 0$, *and TR matrix* Λ, *if there exist positive scalar* c^* *and matrices* $P_i > 0$ *for* $i \in \mathbb{S}$, *satisfying condition* (9.77) *and the following matrix inequalities:*

$$\begin{bmatrix} \Upsilon_{1i} & P_i \tilde{G}_i D_i \\ * & -\varphi_i I \end{bmatrix} < 0, \tag{9.83}$$

$$\frac{\bar{\varpi}_P c^* + \varphi_i \delta^2 T}{\underline{\varpi}_P} < e^{-\varphi_i T} c_2, \tag{9.84}$$

where

$$\Upsilon_{1i} \triangleq \mathrm{He}\big(P_i \tilde{G}_i A_i\big) + \sum_{j=1}^{N} \lambda_{ij} P_j - \varphi_i P_i,$$

then the sliding mode dynamics (9.82) *are SFTB with respect to* $\aleph_{[T^*,T]}$.

Proof This proof is similar to the one in Lemma 9.4, which is omitted. □

(III) **Solving algorithm over the whole phase [0, *T*].**
It is worthy of pointing out that the sliding mode controller gain K_i and TR matrix Λ should be designed to ensure the conditions in Lemmas 9.5 and 9.6 to be feasible simultaneously. However, it is noted that the matrix inequalities in Lemmas 9.5 and 9.6 are nonlinear. In the sequel, we shall give sufficient conditions based on LMIs, which can guarantee that the conditions in Lemmas 9.5 and 9.6 hold simultaneously.

Theorem 9.5 *For the given parameters* (c_1, c_2, T, R, δ), *and positive scalars* φ_i, ρ_i, ν, *if there exist positive constants* τ_i, ζ_i, σ_i, ϑ_i, c^*, $\bar{\lambda}_{ij}$ *and matrices* $P_i > 0$, $\bar{K}_i$ $(i, j \in \mathbb{S})$ *satisfying the following matrix inequalities:*

$$\begin{bmatrix} \bar{\Omega}_{11i} & \tilde{\Omega}_{12i} & \bar{\Omega}_{13i} \\ * & \tilde{\Omega}_{22i} & 0 \\ * & * & \bar{\Omega}_{33} \end{bmatrix} < 0, \tag{9.85}$$

$$\begin{bmatrix} \bar{\Omega}_{11i} & \hat{\Omega}_{12i} & \bar{\Omega}_{13i} \\ * & \hat{\Omega}_{22i} & 0 \\ * & * & \bar{\Omega}_{33} \end{bmatrix} < 0, \tag{9.86}$$

$$c_1 < c^* < c_2, \tag{9.87}$$

$$\vartheta_i R < P_i < \rho_i R, \tag{9.88}$$

$$\begin{bmatrix} -e^{-\varphi_i T} c^* & \sqrt{\rho_i c_1 + \delta^2 \varphi_i T} \\ * & -\vartheta_i \end{bmatrix} < 0, \tag{9.89}$$

$$\rho_i c^* + \delta^2 T \varphi_i - c_2 \vartheta_i e^{-\varphi_i T} < 0, \tag{9.90}$$

where

$$\bar{\Omega}_{11i} \triangleq \nu^2 \sum_{j=1, j\neq i}^{N} \tau_j I - 2\nu \sum_{j=1, j\neq i}^{N} P_j + 2\nu(N-1)P_i + \mathrm{He}(P_i A_i) - \beta_i P_i,$$

$$\bar{\Omega}_{13i} \triangleq \big[P_1 - P_i + \bar{\lambda}_{i1} I \cdots P_{i-1} - P_i + \bar{\lambda}_{i(i-1)I} \\ P_{i+1} - P_i + \bar{\lambda}_{i(i+1)I} \cdots P_N - P_i + \bar{\lambda}_{iN} I\big],$$

$$\bar{\Omega}_{33} \triangleq -\mathrm{diag}\{\tau_1 I, \ldots, \tau_{i-1} I, \tau_{i+1} I, \ldots, \tau_N I\},$$

$$\tilde{\Omega}_{12i} \triangleq \begin{bmatrix} P_i D_i & P_i B_i & \bar{K}_i^{\mathrm{T}} & P_i B_i & \sigma_i \epsilon_i I \end{bmatrix}, \ \tilde{\Omega}_{22i} \triangleq -\mathrm{diag}\{\varphi_i I, \zeta_i I, \zeta_i I, \sigma_i I, \sigma_i I\},$$

$$\hat{\Omega}_{12i} \triangleq \begin{bmatrix} P_i D_i & P_i B_i & A_i^{\mathrm{T}} P_i B_i \end{bmatrix}, \ \hat{\Omega}_{22i} \triangleq \begin{bmatrix} -\varphi_i I & 0 & D_i^{\mathrm{T}} P_i B_i \\ * & -B_i^{\mathrm{T}} P_i B_i & 0 \\ * & * & -B_i^{\mathrm{T}} P_i B_i \end{bmatrix},$$

then the MJS (9.70) *with the controller* (9.72) *is SFTB with respect to* $\aleph_{[0,T]}$ *over the whole finite-time interval* $[0, T]$. *In the meanwhile, TR matrix is given as* $\bar{\Lambda} = [\bar{\lambda}_{ij}]_{N\times N}$ *with* $\bar{\lambda}_{ii} = -\tau_i \sum_{j=1, j\neq i}^{N} (\tau_j^{-1} \bar{\lambda}_{ij})$. *Moreover, the TR matrix can be obtained as* $\Lambda = 2\bar{\Lambda}\Theta^{-1}$ *with* $\Theta \triangleq \mathrm{diag}\{\tau_1, \tau_2, \ldots, \tau_N\}$ *and the controller gains can be obtained as* $K_i = \zeta_i^{-1} \bar{K}_i$ $(i \in \mathbb{S})$.

Proof Consider the conditions (9.76)–(9.78) in Theorem 9.5 and conditions (9.83)–(9.84) in Theorem 9.6 simultaneously. It is shown that

$$\begin{bmatrix} \Gamma_{1i} & P_i D_i \\ * & -\varphi_i I \end{bmatrix} = \begin{bmatrix} \Gamma_{2i} & P_i D_i \\ * & -\varphi_i I \end{bmatrix} + \mathrm{He}\left(\begin{bmatrix} P_i B_i \\ 0 \end{bmatrix} \phi(u(t)) \begin{bmatrix} K_i & 0 \end{bmatrix} \right), \tag{9.91}$$

$$\begin{bmatrix} \Upsilon_{1i} & P_i \tilde{G}_i D_i \\ * & -\varphi_i I \end{bmatrix} = \begin{bmatrix} \Upsilon_{2i} & P_i D_i \\ * & -\varphi_i I \end{bmatrix} - \mathrm{He}\left(\begin{bmatrix} P_i B_i \\ 0 \end{bmatrix} (B_i^{\mathrm{T}} P_i B_i)^{-1} \begin{bmatrix} B_i^{\mathrm{T}} P_i A_i & G_i D_i \end{bmatrix} \right) \tag{9.92}$$

with $\Gamma_{2i} \triangleq \Omega_{11i} + \sigma_i^{-1} P_i B_i B_i^T P_i + \sigma_i b_i^2 I$.

Noting $\phi(u(t)) < I$, we have for $\zeta_i > 0$ and $G_i B_i > 0$

$$\begin{aligned} &\mathrm{He}\left(\begin{bmatrix} P_i B_i \\ 0 \end{bmatrix} \phi(u(t)) \begin{bmatrix} K_i & 0 \end{bmatrix} \right) \\ \le& \zeta_i^{-1} \begin{bmatrix} P_i B_i \\ 0 \end{bmatrix} \begin{bmatrix} B_i^{\mathrm{T}} P_i & 0 \end{bmatrix} + \zeta_i \begin{bmatrix} K_i^{\mathrm{T}} \\ 0 \end{bmatrix} \begin{bmatrix} K_i & 0 \end{bmatrix}, \end{aligned} \tag{9.93}$$

$$\begin{aligned} &- \mathrm{He}\left(\begin{bmatrix} P_i B_i \\ 0 \end{bmatrix} (B_i^{\mathrm{T}} P_i B_i)^{-1} \begin{bmatrix} B_i^{\mathrm{T}} P_i A_i & G_i D_i \end{bmatrix} \right) \\ \le& \begin{bmatrix} P_i B_i \\ 0 \end{bmatrix} (B_i^{\mathrm{T}} P_i B_i)^{-1} \begin{bmatrix} B_i^{\mathrm{T}} P_i & 0 \end{bmatrix} \\ &+ \begin{bmatrix} A_i^{\mathrm{T}} P_i B_i \\ D_i^{\mathrm{T}} P_i B_i \end{bmatrix} (B_i^{\mathrm{T}} P_i B_i)^{-1} \begin{bmatrix} B_i^{\mathrm{T}} P_i A_i & B_i^{\mathrm{T}} P_i D_i \end{bmatrix}. \end{aligned} \tag{9.94}$$

Thus, by using Schur Complement (Lemma 1.3), conditions (9.76) and (9.83) can be ensured by

$$\begin{bmatrix} \Omega_{11i} & \tilde{\Omega}_{12i} \\ * & \tilde{\Omega}_{22i} \end{bmatrix} < 0, \tag{9.95}$$

$$\begin{bmatrix} \Omega_{11i} & \hat{\Omega}_{12i} \\ * & \hat{\Omega}_{22i} \end{bmatrix} < 0 \tag{9.96}$$

with $\bar{K}_i \triangleq \zeta_i K_i$.

Furthermore, by means of $\sum_{j=1}^{N} \lambda_{ij} = 0$ and complete square technique, one has for $\tau_j > 0$ that

$$\begin{aligned}\Omega_{11i} &= \mathrm{He}(P_i A_i) + \sum_{j=1}^{N} \lambda_{ij} P_j - \varphi_i P_i \\ &\leq \mathrm{He}(P_i A_i) - \varphi_i P_i + \sum_{j=1, j\neq i}^{N} \lambda_{ij}(P_j - P_i) + \sum_{j=1, j\neq i}^{N} \frac{\tau_j \lambda_{ij}^2}{4} I.\end{aligned}$$

There exist an adjustable scalars $\nu > 0$ such that

$$\begin{aligned}&((P_j - P_i) - \nu\tau_j)^T \tau_{1j}^{-1} I((P_j - P_i) - \nu\tau_j) \\ =&(P_j - P_i)^T \tau_{2j}^{-1} I(P_j - P_i) + \nu^2 \tau_j I - 2\nu(P_j - P_i) \geq 0,\end{aligned} \tag{9.97}$$

from which we have the following condition

$$-(P_j - P_i)^T \tau_{2j}^{-1} I(P_j - P_i) \leq \nu^2 \tau_j I - 2\nu(P_j - P_i). \tag{9.98}$$

From (9.97) and (9.98), we have $\Omega_{11i} < \breve{\Omega}_{11i}$, where

$$\begin{aligned}\breve{\Omega}_{11i} \triangleq& \mathrm{He}(P_i A_i) - \varphi_i I + \sum_{j=1, j\neq i}^{N} (P_j - P_i + \bar{\lambda}_{ij} I)^{\mathrm{T}} \tau_j^{-1} I(P_j - P_i + \bar{\lambda}_{ij} I) \\ &+ \sum_{j=1, j\neq i}^{N} \nu^2 \tau_j I - 2\nu \sum_{j=1, j\neq i}^{N} (P_j - P_i)\end{aligned} \tag{9.99}$$

with $\bar{\lambda}_{ij} \triangleq \frac{1}{2}\tau_j \lambda_{ij}$. It is easily shown that conditions (9.95) and (9.96) can be ensured by conditions (9.85) and (9.86), respectively. Further, the conditions (9.85) and (9.86) can guaranteed (9.76) and (9.83), respectively.

In addition, we obtain $\underline{\varpi}_P > \vartheta_i$ and $\rho_i < \bar{\varpi}_P$ from (9.88). Moreover, by means of Schur Complement (Lemma 1.3), it is easily shown that the conditions (9.78) and (9.84) can be ensured by (9.88)–(9.90), respectively. Thus, conditions (9.85)–(9.90) can ensure the SFTB of the closed-loop system (9.75) with respect to $\aleph_{[0,T]}$ over the whole finite-time interval $[0, T]$, which completes this proof. □

9.4.4 Simulation Example 3

In this example, we will verify the effectiveness of the proposed hybrid design method for TR matrix and controller gains. To this end, consider the system parameters:

$$A_1 = \begin{bmatrix} 0 & 0.02 \\ 0.06 & -0.04 \end{bmatrix}, \; A_2 = \begin{bmatrix} -0.02 & 0.04 \\ 0 & -0.02 \end{bmatrix},$$
$$B_1 = \begin{bmatrix} -0.1 \\ 0.1 \end{bmatrix}, \; B_2 = \begin{bmatrix} 0.3 \\ -0.1 \end{bmatrix}, \; D_1 = \begin{bmatrix} 0.1 \\ 0.2 \end{bmatrix}, \; D_2 = \begin{bmatrix} 0.1 \\ 0.3 \end{bmatrix},$$

$f_1(x(t)) = 0.2\sin(\|x(t)\|)$, $f_2(x(t)) = 0.3\sin(\|x(t)\|)$, and the external disturbance is $w(t) = 0.05e^{-t}$, and thus, $\delta = 0.05$.

By solving the conditions (9.85)–(9.90) in Theorem 9.5, we can obtain the TR matrix as

$$\Lambda = \begin{bmatrix} -0.5567 & 0.5567 \\ 0.5899 & -0.5899 \end{bmatrix},$$

and the controller gains as

$$K_1 = \begin{bmatrix} -0.2906 & 0.5812 \end{bmatrix}, \; K_2 = \begin{bmatrix} 0.5687 & -0.2844 \end{bmatrix}.$$

For the initial state $x(0) = [-0.5 \quad 0.2]^{\mathrm{T}}$, given the finite time $T = 0.5$, parameters $c_1 = 0.0430$, $c_2 = 5.0430$, weight matrices $R_1 = \begin{bmatrix} 0.1010 & -0.0045 \\ -0.0045 & 0.0999 \end{bmatrix}$, $R_1 = \begin{bmatrix} 0.1107 & -0.0068 \\ -0.0068 & 0.0991 \end{bmatrix}$.

The simulation results are shown in Figs. 9.3 and 9.4, from which it is exhibited that (1) the proposed hybrid design method for TR matrix and controller gains can ensure the SFTB of the considered system; (2) lager the maximum control input is, the better the controlled performance.

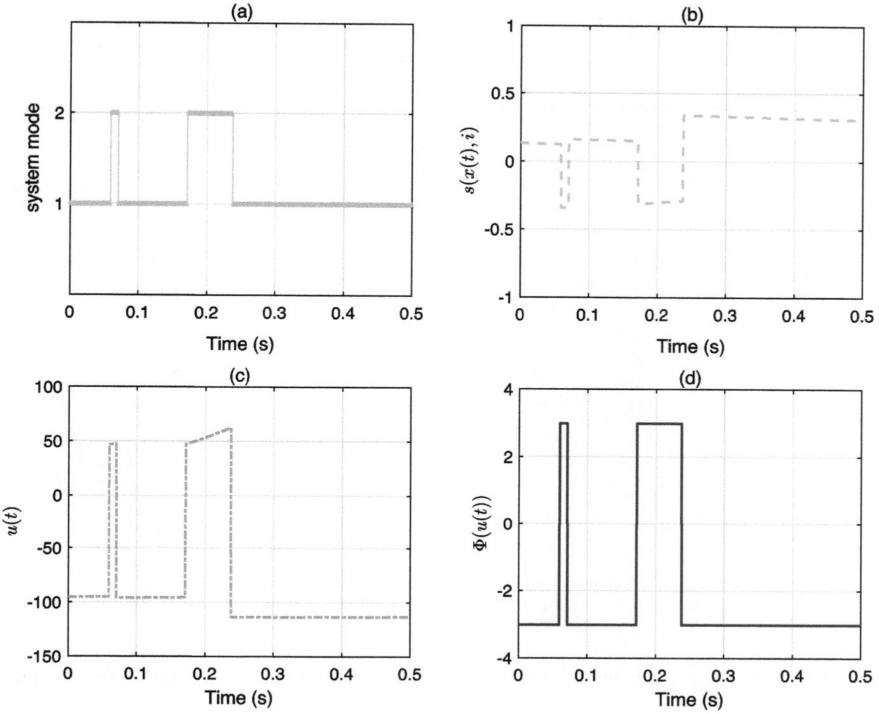

Fig. 9.3 Simulation results on system mode, sliding variable and control input in Example 3: **a** possible mode transition sequence r_t; **b** sliding variable $s(x(t), i)$; **c** original control signal $u(t)$; **d** saturated control signal $\Phi(u(t))$

9.5 Conclusion

In this paper, the finite-time hybrid design scheme on SMC problem of uncertain MJSs has been investigated. By the mode-dependent parametric approach, a sufficient condition for designing the TR matrix has been established, and a sliding mode controller has been designed to ensure the reachability of the specified sliding surface in a given finite-time interval. Moreover, an optimization solving algorithm has been provided on the proposed hybrid design scheme. Further considering the issues of actuator saturation, the hybrid design scheme of the TR matrix and sliding mode controller is established, and the sufficient conditions for both the reaching phase and the sliding motion phase of the closed-loop system to simultaneously satisfy the SFTB are provided.

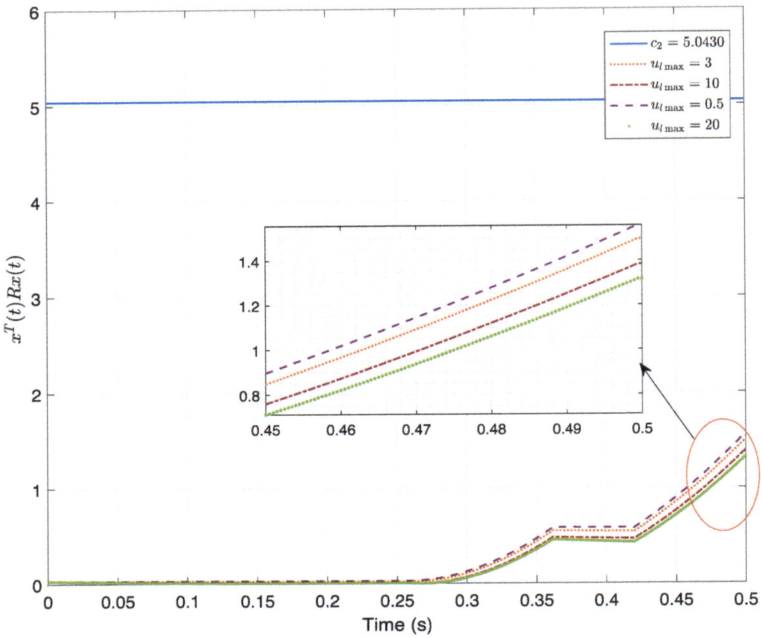

Fig. 9.4 Simulation results on $x^T(t)Rx(t)$ under different maximum control inputs in Example 3

References

1. Feng, J.E., Lam, J., Shu, Z.: Stabilization of Markovian systems via probability rate synthesis and output feedback. IEEE Trans. Autom. Control **55**(3), 773–777 (2010)
2. Ma, S., Xiong, J.: Stabilization of time-delay Markovian jump systems via probability rate synthesis and state feedback. IFAC Proc. Vol. **47**(3), 10415–10420 (2014)
3. Cao, Z., Niu, Y., Lam, J., Song, X.: A hybrid sliding mode control scheme of Markovian jump systems via transition rates optimal design. IEEE Trans. Syst. Man Cybern. Syst. **51**(12), 7752–7763 (2021)
4. He, X., Wang, Z., Gao, C., Zhou, D.H.: Consensus control for multi-agent systems under asymmetric actuator saturations with applications to mobile train lifting jack systems. IEEE Trans. Indust. Inform. **19**(10), 10224–10232 (2023)
5. Geng, H., Wang, Z., Hu, J., Dong, H., Cheng, Y.: Distributed recursive filtering over sensor networks under random access protocol: when state saturation meets censored measurement. IEEE Trans. Cybern. **53**(12), 7760–7772 (2023)
6. Song, J., Wang, Z., Niu, Y., Yi, X., Han, Q.L.: Co-design of dissipative deconvolution filter and Round-Robin protocol for networked 2-D digital systems: optimization and application. IEEE Trans. Syst. Man Cybern. Syst. **53**(10), 6316–6328 (2023)
7. Liu, J., Yin, T., Shen, M., Xie, X., Cao, J.: State estimation for cyber-physical systems with limited communication resources, sensor saturation and denial-of-service attacks. ISA Trans. **104**, 101–114 (2020)
8. Deng, Y., Lu, H., Zhou, W.: Security event-triggered control for Markovian jump neural networks against actuator saturation and hybrid cyber attacks. J. Frankl. Instit. **358**(14), 7096–7118 (2021)

9. Zhu, Z., Xia, Y., Fu, M.: Adaptive sliding mode control for attitude stabilization with actuator saturation. IEEE Trans. Indust. Electron. **58**(10), 4898–4907 (2011)
10. Li, H., Wang, J., Shi, P.: Output-feedback based sliding mode control for fuzzy systems with actuator saturation. IEEE Trans. Fuzzy Syst. **24**(6), 1282–1293 (2016)
11. Hu, Q.: Robust adaptive sliding mode attitude maneuvering and vibration damping of three-axis-stabilized flexible spacecraft with actuator saturation limits. Nonlinear Dyn. **55**(4), 301–321 (2009)

Chapter 10
Conclusion and Future Research Directions

10.1 Conclusion

In conclusion, this book provides a comprehensive exploration of sliding mode control (SMC) techniques for ensuring the finite-time transient performance of Markov jump systems (MJSs) under various constraints and attack scenarios. The research presented in this book is structured around a detailed framework that addresses both theoretical and practical challenges in the control of MJSs, with a focus on finite-time stability and robustness.

Chapter 2 lays the foundational groundwork by proposing two research frameworks for SMC under restricted states and outputs. These frameworks ensure the finite-time boundedness (FTB) of MJSs with fully known transition rates (TRs) and input-output finite-time stability (IO-FTS) for MJSs with partially known TRs. The sliding mode controller designed in this chapter effectively drives state trajectories to the desired sliding surface within a specified finite time, employing a partitioning strategy to analyze both the reaching and sliding motion phases.

Chapters 3–5 delve into finite-time SMC methods for MJSs under different physical constraints. Chapter 3 addresses actuator faults, designing a fault-tolerant sliding mode controller that ensures system stability despite incomplete control signal executions. An incremental technique is introduced to optimize the controller for minimal energy consumption. Chapter 4 tackles actuator nonlinearities, including dead zone and sector nonlinearity, proposing a set of component-based controllers that handle uncertainties in the TR matrix. Chapter 5 considers random parameter uncertainties and controller gain variations, using a non-fragile SMC scheme to maintain stability and reachability of the sliding surface.

Chapters 6–8 focus on finite-time SMC schemes to counter various cyber attacks. Chapter 6 deals with randomly occurring injection attacks in the control/actuation channel, designing a robust controller that mitigates attack impacts and ensures system stability. Chapter 7 addresses channel fading and deception attacks in the sensor/communication channel, proposing a security SMC law that accounts for fading

Z. Cao and Y. Niu, *Finite-Time Sliding Mode Control for Markov Jump Systems*,
Studies in Systems, Decision and Control 603,
https://doi.org/10.1007/978-981-96-7651-4_10

probabilities and attack deception information. Chapter 8 integrates both physical and communication constraints, specifically actuator saturation and injection attacks, employing neural networks to estimate attack bounds and enhance control reliability.

Chapter 9 broadens the SMC technologies for MJSs by proposing TR stabilization methods and TR-fuzed hybrid SMC approaches for the sliding mode finite-time stability (SFTS). This chapter introduces a collaborative control framework that synthesizes the SMC and TR matrix, breaking the assumption of prior knowledge of TRs. It provides conditions for SFTB and a design scheme that integrates TR matrix with the sliding mode controller, ensuring stability over both reaching and sliding phases.

Overall, this book significantly advances the field of SMC for MJSs by addressing a wide range of challenges, from physical constraints and uncertainties to cyber attacks. The proposed methods and frameworks offer robust and efficient solutions for enhancing the stability and performance of MJSs in various applications.

10.2 Future Research Directions

In conclusion, this book has systematically explored the sliding mode control (SMC) problem for Markov jump systems (MJSs), addressing several critical issues and providing innovative solutions. Despite these contributions, there remain numerous valuable areas for further research:

(1) The sliding mode controllers discussed in this book, whether synchronous or asynchronous, are designed based on a given number of controller modes. However, the optimal relationship between controller mode information and system dynamic performance has not been fully explored. Future research could focus on the optimization scheme of control modes, by which the desired or satisfactory system performance can be attained via the controller with fewer modes. This direction could lead to more efficient and cost-effective control strategies.

(2) When employing event-triggered mechanisms to reduce data transmission frequency in MJSs, both system mode and state information cannot be transmitted in real-time, leading to the system mode information unavailable for the controller side. Although this book uses a mode detection mechanism to estimate system modes, this approach will result in asynchronous phenomena and performance degradation. A promising research direction is to consider mode holding methods, where the controller updates the system mode only at triggering moments and maintains the most recently received mode value between triggers. This approach could mitigate the issues associated with asynchronous behavior.

(3) Chattering is an inherent issue in SMC. While this book uses a simple substitution method to mitigate chattering in simulations, more advanced techniques could be explored. Future work could involve designing higher-order sliding mode controllers based on the Super-Twisting algorithm or integrating intelligent algorithms such as fuzzy logic and neural networks to effectively suppress chattering. This would enhance the practical applicability of SMC in various systems.
(4) Although this book comprehensively addresses SMC under false data injection attacks, other attack types prevalent in practical scenarios, such as replay attacks, have not been fully explored. Further research could investigate the negative impacts of replay attacks on MJSs and the design of sliding mode controllers. Developing robust security control schemes to counter such attacks would be a significant contribution to the field.

By addressing these areas, future research can continue to advance the theory and application of SMC for MJSs, enhancing their reliability and performance in complex and dynamic environments.

MIX
Papier aus verantwortungsvollen Quellen
Paper from responsible sources
FSC® C105338

If you have any concerns about our products,
you can contact us on
ProductSafety@springernature.com

In case Publisher is established outside the EU,
the EU authorized representative is:
Springer Nature Customer Service Center GmbH
Europaplatz 3, 69115 Heidelberg, Germany

Printed by Libri Plureos GmbH
in Hamburg, Germany